HEATH EARTH SCIENCE

Authors **Rolland B. Bartholomew**
Associate Professor of Science Education and Geological Sciences
University of Texas at Austin, Texas
Former Science Department Head, Boulder Public Schools
Boulder, Colorado

Bill W. Tillery
Professor of Science Education
Arizona State University, Tempe, Arizona
Former Junior High School Teacher
Jefferson County Public Schools
Colorado

Consulting Author Henry N. Gary
K-9 Science Coordinator, Science Teacher and Chairperson
Summit Public Schools, Summit, New Jersey

Content Consultants John Arthur Day METEOROLOGIST
Emeritus Professor, Linfield College
McMinnville, Oregon

Gerald S. Hawkins ASTRONOMER
Washington, D.C.
Former Chairperson, Department of Astronomy
Boston University, Boston, Massachusetts

Les Picker MARINE BIOLOGIST
Professor, Department of Marine Biology
University of Delaware, Newark, Delaware
Director of Project Coast

Robert W. Ridky GEOLOGIST
Associate Professor of Geology
University of Maryland, College Park, Maryland

D.C. Heath and Company
Lexington, Massachusetts Toronto

Executive Editor: Roger R. Rogalin
Project Editor: Harley R. Mutzfeld
Assistant Editor: Toby Klang

Series Design: William Tenney
Book Designer: Dawn Ostrer
Production Coordinator: Donna Porter

Cover Photo: Sunset Point, Bryce Canyon National Park, Utah; Grant Heilman

Teacher Consultants

Jeffrey H. Alexander
Earth Science Instructor and Planetarium Lecturer
Charlotte-Mecklenburg Schools
Charlotte, North Carolina

Howard Arrowood
Earth Science Teacher
Meadowbrook Junior High School
Orlando, Florida

Richard E. Cooper
Director, K–12 Science and Mathematics
Danvers Public Schools
Danvers, Massachusetts

Diane P. Horn
Earth Science Teacher
Norfolk Public Schools
Norfolk, Virginia

Larry A. Irwin
Principal, Franklin Middle School
Weld County District Six
Greeley, Colorado

Paul Quay
Science Teacher
Ford Junior High School
Brook Park, Ohio

Series Consultants

Napoleon Bryant
Professor of Education
Xavier University
Cincinnati, Ohio

James H. McGill
Science Department Chairperson
Hill Country Middle School
Austin, Texas

Sister Rita Meredith, O.P.
Former Science Teacher
St. Agnes High School
Rockville Center, New York

Richard Merrill
Curriculum Specialist
Mt. Diablo Unified School District
Concord, California

William D. Thomas
Science Supervisor
Escambia County Schools
Pensacola, Florida

Reading Level Consultant

J & F
Milton D. Jacobson, Founder
Charlottesville, Virginia

Field Test Teachers

Grateful acknowledgement is given to the teachers and students who participated in field tests for this program.

ARIZONA
Harold Pruitt, Pima Middle School, Scottsdale

CALIFORNIA
L. Meyer, C. W. Tewinkle Middle School, Costa Mesa
Catherine Baird, Rancho San Joaquin Intermediate School, Irvine
Patrick Croner, Venado Middle School, Irvine
Patrick Gleason, Vista Verde School, Irvine

ILLINOIS
Kristin Ciesemier, Naperville North High School, District 203, Naperville
David Klussendorf, Michael Sommers, Jefferson Junior High School, District 203, Naperville
Jean Kriebs, Butler Junior High School, Oakbrook
Kelly Douglas, Wilson Middle School, Rockford

INDIANA
Nancy Watson, Burris Lab School, Ball State University, Muncie

MASSACHUSETTS
Brian Flaherty, Ronald Gomes, Plymouth Carver Intermediate School, Plymouth
Warren Phillips, Nathaniel Morton Elementary School, Plymouth
Paula Garten, Russell Magee, Middle School East, Salem
Alan Chasse, North Intermediate School, Wilmington
Joseph Gilligan, West Intermediate School, Wilmington
Priscilla Lockwood, Wilmington High School, Wilmington

MISSOURI
Beverly Edwards, Visual and Performing Arts Center, St. Louis
Marie Globig, Investigate Learning Center, St. Louis

NEW JERSEY
Joan Camarigg, Eisenhower School, Freehold
Susanne Flannelly, Joseph Herman, Barkalow School, Freehold

NEW YORK
John Spring, Trinity Lutheran School, Hicksville
Sam Alaimo, Hoover Middle School, Kenmore
Elaine Johnson, Kenmore Middle School, Kenmore
James Morgan, Benjamin Franklin Middle School, Kenmore
Edith Newfield, Hendrick Hudson High School, Montrose
A. Charles Rossi, Blue Mountain School, Peekskill
Alan Goodman, Pamona Junior High School, Suffern

PENNSYLVANIA
Anna McCartney, Tidioute Junior/Senior High School, Tidioute

TENNESSEE
Jo Quarles, Jefferson Middle School, Jefferson City
Ted Frisby, Westwood Junior High School, Manchester
Peggy Mason, Gary Mullican, Central Middle School, Murfreesboro

TEXAS
Mary Jane Vasquez, Lake Travis Junior High School, Austin

Published simultaneously in Canada.
Printed in the United States of America.
International Standard Book Number: 0-669-07541-8

TABLE OF CONTENTS

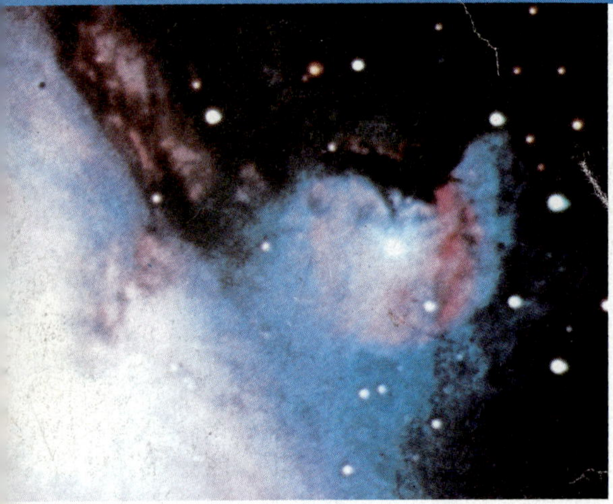

1 Earth Science: A Beginning

Imagine you are a long way from earth. You think you are as far as you can go! Well, then, go much farther and double that distance. You probably aren't far enough yet. Go even farther right out to the "edge" of the universe. Now imagine you are able to climb on a light beam and ride it for about 10 billion years back to earth. What would you see?

You'd see the objects of outer space whiz by. Then at long last you'd begin to see the edge of our galaxy, which contains millions and millions of stars. As you moved through the galaxy, you would finally come to the solar system—the sun and planets. Just as you see the earth, the open end of a big telescope appears to be coming at you. With a *whomp,* the journey is over and you are now an image on a piece of photographic film.

Let's think about that journey and what you saw. You discovered one thing that exists nowhere else. It is a telescope made by humans.

Remember that. Humans appear to be unique. They live, play, work, and die on a tiny speck in space called earth. Earth science is the study of that earth and its surroundings.

Figure 1.1 *opposite:* Earth, a tiny speck floating in a vast and infinite universe

1.1 Why Is Earth Science Important to You?

You will find out

- how and where information is obtained about the earth;
- what some of the instruments are that are used to study the earth;
- why you study the earth.

What are some of your ideas about the earth? Where can you get new information about the earth? Many scientists help you learn about the earth. **Astronomers** [uh-STRAHN-uh-muhrs] use telescopes to see and photograph objects in space. They may also listen to the sky with radio telescopes. Astronomers use what they learn from their studies to infer certain information about the earth as a body in space.

Meteorologists [mee-tee-uh-RAHL-uh-jists] study the atmosphere. Their information about the weather helps people plan trips at the right time and decide how to

Figure 1.2 *left:* An astronomer looking through a large telescope at Mauna Kea Observatory, Hawaii

Figure 1.3 *right:* A meteorologist reviewing weather charts

dress. You may have used information provided by weather forecasters to decide what to wear to school today. Meteorologists save millions of dollars when they correctly predict a big storm.

Oceanographers [oh-shuh-NAHG-ruh-fers] study the oceans. From oceanographers people learn about the nature of the oceans and the ocean floor. For example, the tallest mountains in all the world are found on the ocean floor. Oceanographers can't see them directly because these mountains are covered by water. However, instruments can show you that they are there.

Other scientists, called **geologists** [jee-AHL-uh-jists], study the solid outer and liquid inner parts of the earth. They help reconstruct the history of the earth. From studying rocks they can even learn about animals that lived long ago. Geologists also find oil, which supplies both power and chemicals. They even find the raw materials that make up the metal or plastic in the chair you are sitting in.

Figure 1.4 *left:* An oceanographer diving to examine underwater oceanography equipment

Figure 1.5 *right:* A geologist collecting rock and mineral samples in the field

Astronomers, geologists, meteorologists, and oceanographers are all earth scientists. How do they learn about the earth? One way is to make observations. They observe the earth by using their senses of sight, touch, hearing, taste, and smell. Sometimes they use a combination of these senses. If you think about it, you used these senses earlier today in getting to school.

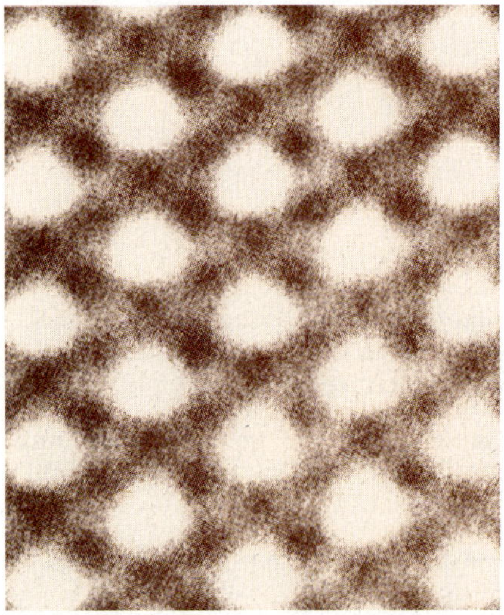

Figure 1.6 *above:* A picture of a rock magnified by an electron microscope showing the rock's internal structure

Figure 1.7 *right:* A thin section of granite magnified 20 times by a microscope shows the crystals in the rock.

Do You Know?
If one of your fingernails was enlarged as much as the photographs above right, it would be 500 m wide.

Instruments can be used to extend your senses. The **optical microscope** [AHP-ti-kuhl MY-kruh-skohp] lets you look at a small piece of a rock. A microscope is an instrument for magnifying objects too small to be seen by the naked eye alone. The **electron microscope** [ee-LEK-trahn MY-kruh-skohp] magnifies objects even more. It uses a stream of electrons to magnify pieces of an object as much as 50,000 times. By using instruments like these, you can sometimes see unexpected things.

Earth scientists observe and explain a changing earth. That's what you'll do as you take this course. You'll use instruments and make measurements. You'll observe patterns and solve problems about the earth.

You study earth science to understand the earth and space in which you live. By knowing how the earth "works," you can enjoy living on it to the fullest.

Figure 1.8 Anyone can make a mistake.

Study Questions for 1.1

1. What information about the earth do you learn from astronomers? Geologists? Meteorologists? Oceanographers?
2. List three instruments that are used in studying earth science.
3. Why is earth science important for you to study?

1.2 Thinking like an Earth Scientist

You will find out
- what kinds of events make scientists curious;
- what the steps are in the scientific process;
- what a scientific model is.

Figure 1.9 *above:* Geologists sometimes find an interesting discovery that leads them to more questions about the earth.

Figure 1.10 *right:* Are you curious about how this boy rides his bicycle in the air?

Imagine that you are this geologist looking at some rocks high up on a mountain. As you pick away gently at the rock layers you suddenly see the clear fossil outline of a fish. But fish can't climb mountains. What would you think? Would you try to find out how the fish fossil got where it was? What questions would you ask?

Earth scientists act like that too. They make observations of the earth and sky. Suppose they find something wrong. Something they see doesn't seem right. What do they do? They ask questions. Often it's a lot of questions. Then they try to find answers to those questions.

Here is an example of how this works. Suppose workers have just finished building a road through a hill. Now imagine you are looking at the cut hillside. You are a geologist trying to understand how the rocks have made the pattern you see. What do you do?

First you back off and observe the cut from a distance to see the broad picture. This overall visual inspection of the broad details of the road cut is called **observation** [ahb-zuhr-VAY-shuhn].

Figure 1.11 What do you see in this hillside? How were these rocks formed?

Figure 1.12 A close-up shows details that can be used to answer your questions.

Next you **analyze** [AN-uh-lyz] what you are seeing by asking yourself questions. When you analyze, you separate things in order to see what is important and what is unimportant. Some things look normal and the questions are easily answered. The answers to other questions may require a closer look. You move up close so you can touch the rocks in the road cut. You look carefully at the rocks. You move from place to place until you begin to see some answers to the other questions you asked.

Then you move back and look at the road cut again. Perhaps you begin to relate the process that deposited the layers to the forces that pushed and bent those layers into the pattern you see. In this process you begin to **synthesize** [SIN-thuh-syz] your questions and answers. When you pull together ideas to make a complete picture, you are synthesizing.

Figure 1.13 Visual models sometimes make explanations easier to understand.

Figure 1.14 An ancient Chinese model portrayed the earth as resting on a frog.

The complete picture or pattern you construct to explain or answer your questions is also called a **scientific model.** A model may be an idea of the way something is, or it may be a physical construction that is similar in form to something. An airplane model is a physical model of the real thing. A picture, a story, or an explanation can be a model that accounts for the way something is. The early Chinese had a model to explain earthquakes. They thought the earth was supported on the back of a frog. When the frog moved, it caused earthquakes. A model is useful because it represents the best explanation based on what is known.

Once you have a model, you begin to test it. For this reason models are changeable things. They change as more is learned about what is being explained. You change your model as you gain new information.

That's the way it is with models. If new information conflicts with the model, then the model is changed to fit what has been observed. Models are tested again and again by scientists. If the model continues to seem correct when tested, people begin to believe in it. If the model fails some test, it is rechecked, changed, or rejected.

Some people describe scientists as seekers. A seeker is a person who looks for something. Scientists search for new facts or for new ideas about old facts. In earth science as in any other science, this is done by observing, analyzing, synthesizing, and testing. First you get the broad picture. Then you subdivide the picture into little pieces that are easier to study. Later you pull back and put your thoughts together. You test and revise as necessary. This procedure is called the scientific process. By applying this process you can learn about the earth and the space around it.

Figure 1.15 Solving a puzzle is like using the scientific process. You study the pieces, analyze how they may fit together, and try possible arrangements until the puzzle is done.

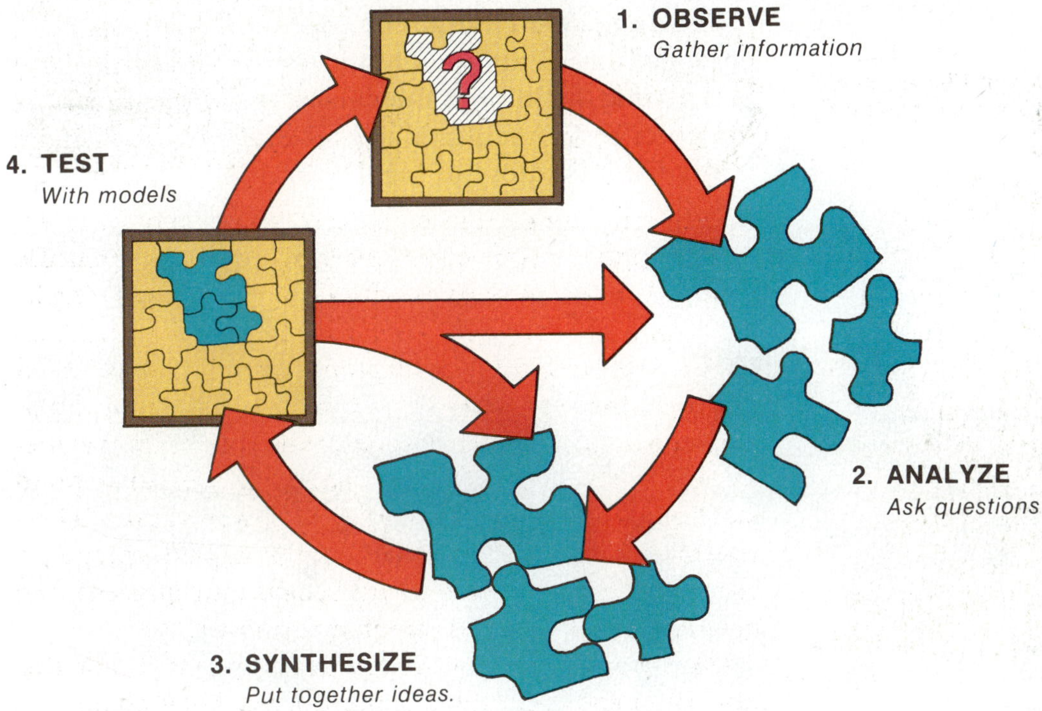

Study Questions for 1.2

1. What kinds of events will likely catch the eye of an earth scientist?
2. List the processes earth scientists use in studying the earth and space.
3. What happens if a model fails its test?

1.3 The Laboratory

You will find out

- what a science laboratory is;
- the role that laboratory work plays in an earth science course;
- some basic laboratory safety procedures.

What do you think when you hear the words scientific **laboratory?** Do you picture some place with huge instruments? Maybe you think about a neat, clean place filled with animals that are fed special foods. Or perhaps you see a person carefully examining some objects that were recently collected. Just what is a laboratory?

The dictionary defines the word *laboratory* as a place equipped for scientific study and testing. Earth scientists accept that definition. But where is the earth science laboratory? All the earth and all that surrounds it are really the laboratory of earth scientists.

Figure 1.16 In the laboratory you can observe, test, and handle materials firsthand. Be sure you follow all directions carefully.

You can use laboratory work as a special way to learn something you don't already know. In the laboratory you can solve problems by searching for answers. Each laboratory activity has a question for you to answer. It is in the laboratory where you apply what you know. You think as you manipulate equipment or measure. You apply the scientific processes of observing, analyzing, synthesizing, and testing. For a moment you are an earth scientist. In the laboratory you experience science directly.

In laboratory activities you sometimes have to handle chemicals or to use things that break easily. For your own safety as well as that of others, it is important for you to follow safe laboratory practices. Here is a list of safety procedures.

Safety Procedures

Tie up long hair and tuck in loose clothing.

Wear laboratory aprons.

Wear eye protection when doing activities that may be hazardous to your eyes.

Watch for any hazards around you and report them to your teacher.

Report all accidents or injuries immediately to your teacher.

Use safety tongs on heat-insulated mittens when handling hot objects.

Do not perform unauthorized experiments.

Do not play around. Unauthorized activity is the greatest hazard in the laboratory.

Clean thoroughly all work surfaces after each use. Do not eat or drink in the laboratory.

You have been reading about the processes that earth scientists use to obtain information. Now you can apply these processes to a laboratory activity.

Here's the setting. Near the rubble of what appears to have been a city, there is a hill. The hill was made by dumping waste in a pile and covering it with dirt. Then more waste was dumped and covered. This process was repeated over and over to make a big hill filled with pockets of waste surrounded by dirt. See Figure 1.17.

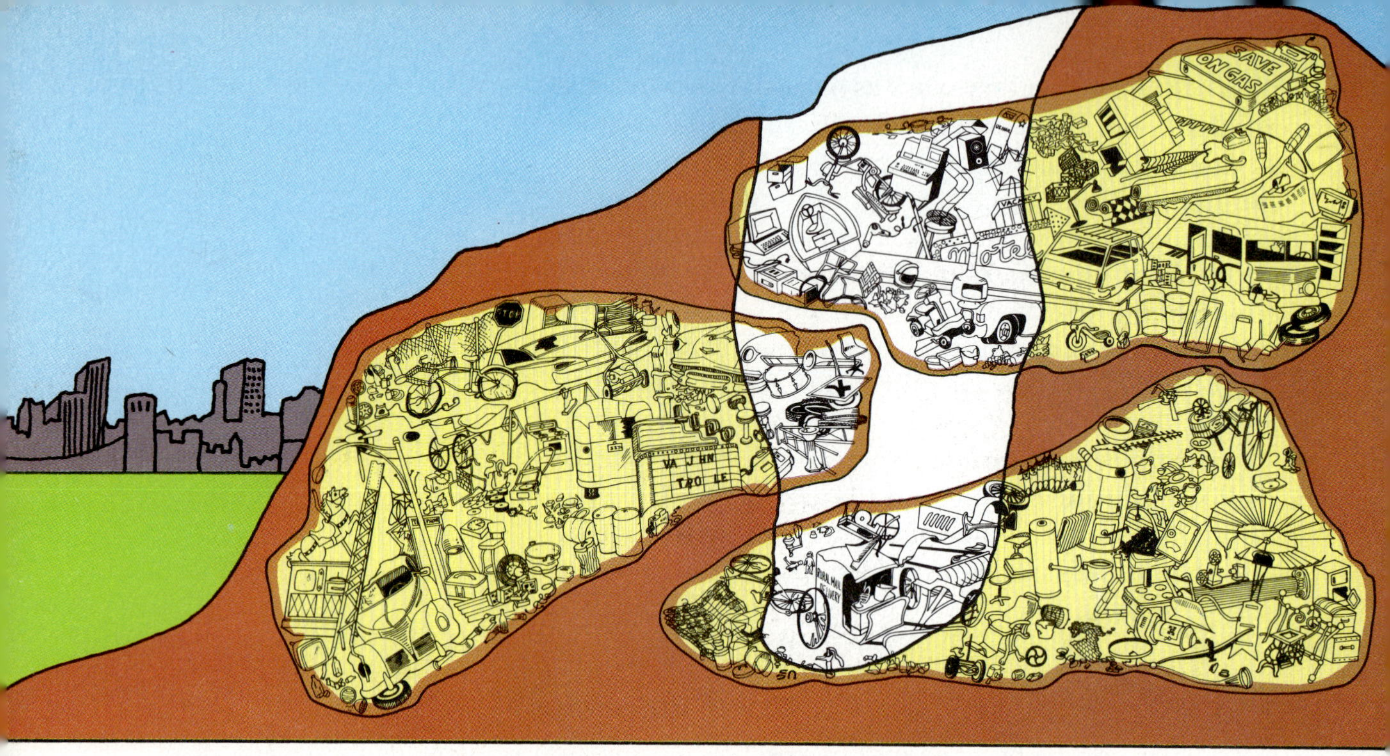

It is now 10,000 years later, and holes have just been dug into the hill. You are to find out when this waste was deposited and what the people did then.

ACTIVITY

When Was This?

Materials

magnifying glass
pencil and paper

Procedure

1. Look at each waste pocket and make a list of the items it contains. Use the magnifying glass if necessary. Try to guess at what time in the past each might have been used.

2. List the waste pockets in order from oldest to youngest.

Questions

1. Which item is the oldest in each waste pocket? How do you know?

2. Which item is the youngest in each waste pocket? How do you know?

3. Explain your reasoning for listing the pockets in the order that you did.

4. Are any of the waste pockets out of order? How do you know? How can you explain any that are?

5. Make up a story that could explain all your findings. This is your model.

Figure 1.17 A 10,000 year old dump

Think about this laboratory activity. You looked at the hill and figured out the relative ages of the waste pockets. Then you made up a story describing how the hill was made. These are exactly the kinds of things earth scientists do in their laboratories. They look at the sky, the rocks, the water, or the slope of the land. They try to describe how things got to be as they are—a model. It's like putting together a giant puzzle.

Study Questions for 1.3

1. Define the word *laboratory* as an earth scientist might.
2. List four things that are done in a science laboratory.
3. List at least five important safety procedures for the laboratory.

1.4 Collecting and Interpreting Data

You will find out
- what things earth scientists measure;
- what the metric system is;
- how to make a representative measurement;
- how to graph data.

Figure 1.18 "It was this big."

Did you ever catch a fish and then try to describe it to someone? How did you do it? You probably mentioned how long it was and how much it weighed. Before you could do this, you had to measure the fish.

Since scientists generally use the metric system for expressing measurements, you'll use that system in this course. The metric system is used by most people in the world.

The most common basic things measured in the metric system are distance, time, and mass. Distance is the interval separating two points in space. Time is the interval between two events. Mass is the amount of matter in an object. Each measurement is expressed by using a number and a unit. For example, your home may be more than 10 kilometers from school. The length of time for this class may be 45 minutes. You may have a mass greater than 40 kilograms. The units in these examples are kilometers, minutes, and kilograms. The other units of the metric system and the relationship between units are found in Figure 1.20.

Figure 1.19 *above:* What does this picture tell you about the mass of these two objects? *right:* Do you believe that each of these containers has the same amount of liquid in it?

Other things frequently measured in earth science are speed, area, volume, density, and temperature. A thermometer is used to measure temperature. In the metric system the units for measuring temperature are Celsius degrees (°C).

Figure 1.20 Units of the metric system

Metric Units of Measurement					
Measurement	**1/1000**	**1/100**	**1/10**	Basic **1** Unit	**1000**
Length	**millimeter** (mm)	**centimeter** (cm)	**decimeter** (dm)	**meter** (m)	**kilometer** (km)
Mass	**milligram** (mg)	**centigram** (cg)	**decigram** (dg)	**gram** (g)	**kilogram** (kg)
Volume	**milliliter** (mL)	**centiliter** (cL)	**deciliter** (dL)	**liter** (L)	**kiloliter** (kL)

Temperature	Degrees Celsius (°C)
Speed	Distance/Time = Meters/Hour (m/h)
Density	Mass/Volume = Grams/Liter (g/L)

Some units of measurement are combinations of the basic units. Here is a closer look at some of these. Area is expressed as square units such as square centimeters. It is found by multiplying units of length by units of width. Volume is an amount of enclosed space. Volume is determined by multiplying units of length by units of width by units of height. Volume is expressed in cubic units such as cubic meters or as a liter for liquid measurement. Speed, for example, can be expressed as meters (distance) per second (time). The density of an object is found by determining its mass and dividing that by its volume.

Sometimes there are problems in making measurements. Suppose you are trying to measure the temperature of the air. Perhaps you have noticed that sometimes it is much hotter in a parking lot than it is in the park. Which place would you use for measuring the temperature of the air?

Figure 1.21 Which thermometer should this girl believe?

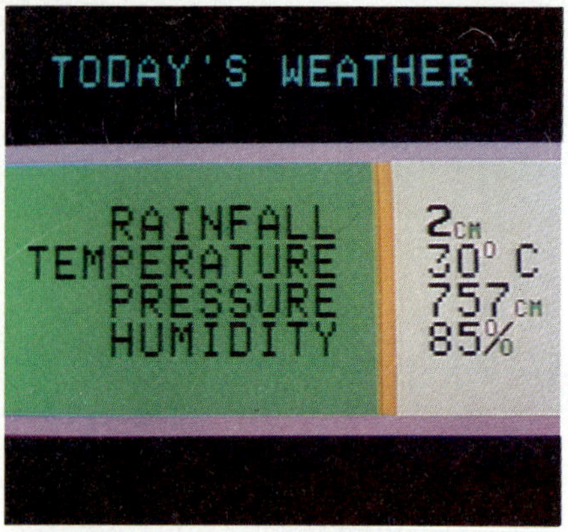

Figure 1.22 The reports from the weather bureau are representative measurements.

You need measurements that are comparable from place to place. To do this requires measurements made in a prescribed manner. Holding a shaded thermometer over grass is an example of a prescribed procedure. Measurements made following a prescribed procedure are called representative measurements. Temperature of air, air pressure, and speed of stream flow are examples of measurements that must be made following prescribed procedures. All the air temperature or pressure readings on a TV weather map are representative measurements. They are all comparable.

ACTIVITY

What Is the Temperature of the Air Today?

Materials

thermometer
pencil and paper

Procedure

1. Select a place outside to measure the temperature. Hold the thermometer in that location for at least one minute. Make a note of where the thermometer is located and record the temperature.

2. Now choose another location that you think is either cooler or hotter than the first one. Place the thermometer in that location for one minute. Again note the location and record the temperature.

3. The third measurement is to be made as follows: Hold the thermometer about 1.5 m above a grassy area and shade the thermometer from the sun. Read and record the temperature after one minute.

Questions

1. What was the highest temperature you recorded? Where was the thermometer located?

2. What was the range in temperature of your readings (lowest to highest)?

3. If you were to go outside again, where would you place the thermometer to record where it was coolest?

4. Which of your three readings do you think best represents the temperature of the air? Explain your choice.

Another class of students did the activity you just completed and recorded the **data** [DAY-tuh] shown in Table 1-1. Data are information you record from your observations.

That class then made a **graph** of its data. A graph is a picture of the data. Graphs are helpful in explaining observations because they organize the data. To make a graph like this, the horizontal scale is used to represent the temperature. Every temperature reading can be located some place on the horizontal scale. The vertical scale was used to show how many students found the same temperature reading. This kind of graph is called a **histogram** [HIS-tuh-gram]. From the histogram you can see that the temperature readings ranged from 15 to 25. The most frequent reading was 20°C.

Table 1-1

Student at 1st Location	Temperature °C
A	20
B	23
C	15
D	25
E	18
F	21
G	20
H	23
I	16
J	17
K	24
L	18
M	21
N	20
O	19
P	20

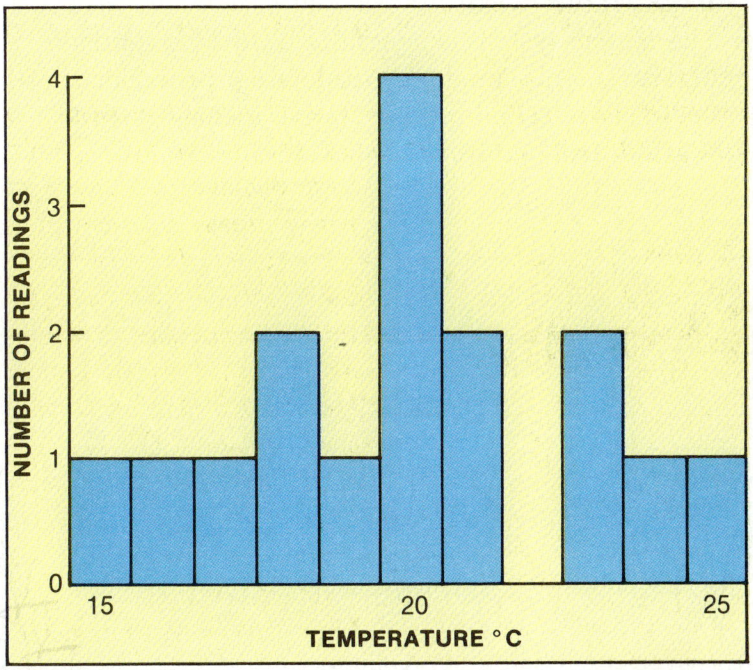

Figure 1.23 An example of a histogram made from the data found in Table 1-1

Study Questions for 1.4

1. Name three basic measurements. What is one unit for each?
2. List five things that earth scientists measure.
3. Define *representative measurement.*
4. What is a histogram?

What Did You Learn?

- Earth scientists are people who study the earth and sky.
- Astronomers, geologists, meteorologists, and oceanographers are all earth scientists.
- Scientific instruments are used to extend the senses.
- You study earth science so that you can understand the earth and space in which you live.
- The scientific processes are observing, analyzing, synthesizing, and testing.
- A model is an idea or a physical device used to explain or depict something else.
- In the laboratory you apply the scientific processes to stated problems.
- You are required to follow approved safety procedures in laboratory work.
- The metric system is generally used by scientists.
- Measurements made by following prescribed procedures are called representative measurements.
- A graph is a picture record of data.

Key Terms

astronomer
geologist
meteorologist
oceanographer
optical microscope
electron microscope
observation
analyze
synthesize
scientific model
laboratory
data
graph
histogram

Biography

William Smith (1769-1839)

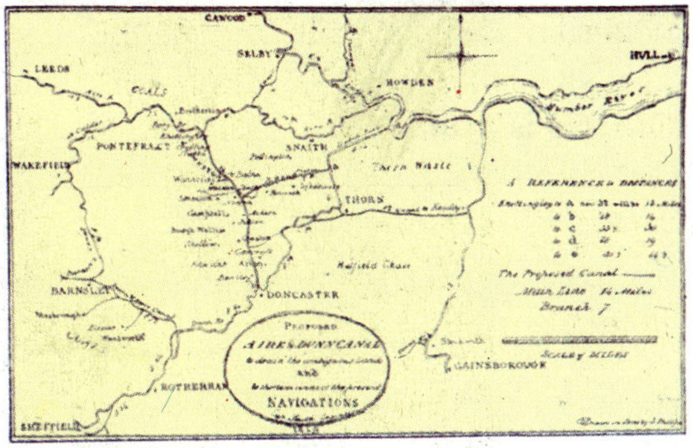

William Smith, one of the earliest great observers, lived in England between 1769 and 1839. At a young age Smith, a surveyor, was asked to plan the route of a new canal across southern England. Smith knew that the type of rock to be encountered would directly affect the cost of the canal. So he conducted some geological research. He traveled 16,000 kilometers a year on horseback and by carriage for his geological study. It was during this time that he developed his remarkable observational skills.

He noticed that various rock strata could be identified by the kind of fossils they contained. He observed that the fossils in each rock layer differed from those above and below it. He made this discovery when he was only 22 years old. Using this information, he could predict the location of buried rocks. His technique of matching the rock and fossils from different areas is still used today. As a result of his work, he was nicknamed "Strata" Smith.

Strata Smith visited the Reverend Benjamin Richardson, a fossil collector. The minister was astonished when Smith told him exactly where and in what formation specimen after specimen had been found. In a distant hill Smith correctly predicted the rock layers and fossils. It was as though he knew every rock in England.

Later, Strata Smith put his observational skills to good use for England. He constructed the first geologic map of England. He was rewarded in 1831 with a gold medal from the Geologic Society of London.

TO THINK ABOUT AND DO

Vocabulary

On another piece of paper unscramble the letters in the words and define each:

1. TILOOEGSG is a/an ______________________.
2. POGCOAREEHNRA is a/an ______________________.
3. GSORETELOMTOI is a/an ______________________.
4. RAESOORTMN is a/an ______________________.

What Do You Remember?

1. Why do scientists use instruments?
2. Give an example of using each sense.
3. Describe an experience when you applied the scientific processes to solve a problem.
4. Why do scientists use models?
5. Define *laboratory*.
6. What do earth scientists mean when they speak of having a direct experience with something?
7. List five safety procedures that should be followed in the laboratory.
8. List five different metric units.
9. Describe what is meant by a rainfall measurement of 3 centimeters.
10. Why do earth scientists use representative measurements?
11. What are metric units for volume, area, and speed?
12. Name four fields of science that are included in earth science. Tell what each field studies.
13. What is a *graph?*
14. Once you have a model, what do you do to it?

Applying What You Have Learned

1. Make a histogram of these data:

weight (kg)	36	38	40	54	40	52	40
student	A	B	C	D	E	F	G

2. Find an object that is close to each of the following sizes:

	Actual Size	What Is It?
30 cm long	________	________
1 m long	________	________
75 g long	________	________
41 km long	________	________
1.75 L volume	________	________
28 mL volume	________	________

Research and Investigation

1. Read about the metric system in an encyclopedia. Report how it was created.
2. Find the meaning of the following prefixes: *tera, giga, mega, hecto, deka, deci, micro, nano, pico, femto,* and *atto.*

Kitt Peak National Observatory near Tucson, Arizona. Have you ever questioned why observatories are placed on mountaintops?

Unit One

The Universe

2

Outer Space

The photograph of M31 shows a fuzzy looking object, but M31 is really a group of stars. In fact it is millions to hundreds of thousands of millions of stars. This group is also called the Andromeda Galaxy. It is one of the closest of such groups to earth.

How close is this? In reality it is a long way away. Think about this. Light travels at a speed of 300,000 km per second. That's fast—seven times around the earth in one second! Your heart beats about 72 times per minute. In one minute light could travel 60 times 300,000 km or 18,000,000 km. Just think, 18,000,000 km in the time your heart makes 72 beats! In the time of about 1,200 heartbeats light can go to the sun and back. But compared to distances in space, that's not very far.

The distance to Andromeda is so far that it is measured in light-years. A light-year is the distance light can travel in one year. So a light-year then is 60 seconds times 60 minutes times 24 hours times 365 days times 300,000 km or about 9,500,000,000,000 km. Or while your heart is beating a mere 38,000,000 times during a year, light travels 9,500,000,000,000 km. Now get ready for this. Andromeda is 2,200,000 light-years away!

Figure 2.1 *opposite:* The Andromeda Galaxy, M31

2.1 Looking into Space

You will find out

- that astronomers have tried to classify the objects in space for many years;
- what the difference is between refractor and reflector telescopes;
- what some of the objects are that telescopes have helped to find.

As you gaze upward at the sky on a clear night, you get the impression that the sky is a great sphere. The center of the sphere seems to be where you are on the earth. The Greeks called this imaginary sphere to which the stars appear to be attached the **celestial sphere** [suh-LES-chuhl SFEER]. The ancients imagined that the celestial sphere slowly rotated, carrying the stars across the sky. They observed that some objects even wandered across the celestial sphere.

The Greeks, as people had before them, divided the sky into the groupings of stars that are called **constellations** [kahn-stuh-LAY-shuhns]. Certain characters in Greek mythology were honored by assigning them a position on the celestial sphere. Sections of the sky are still referred to by their constellation name.

Figure 2.2 Some ancient people believed the earth was inside a large rotating sphere. Do you recognize this constellation of Hercules?

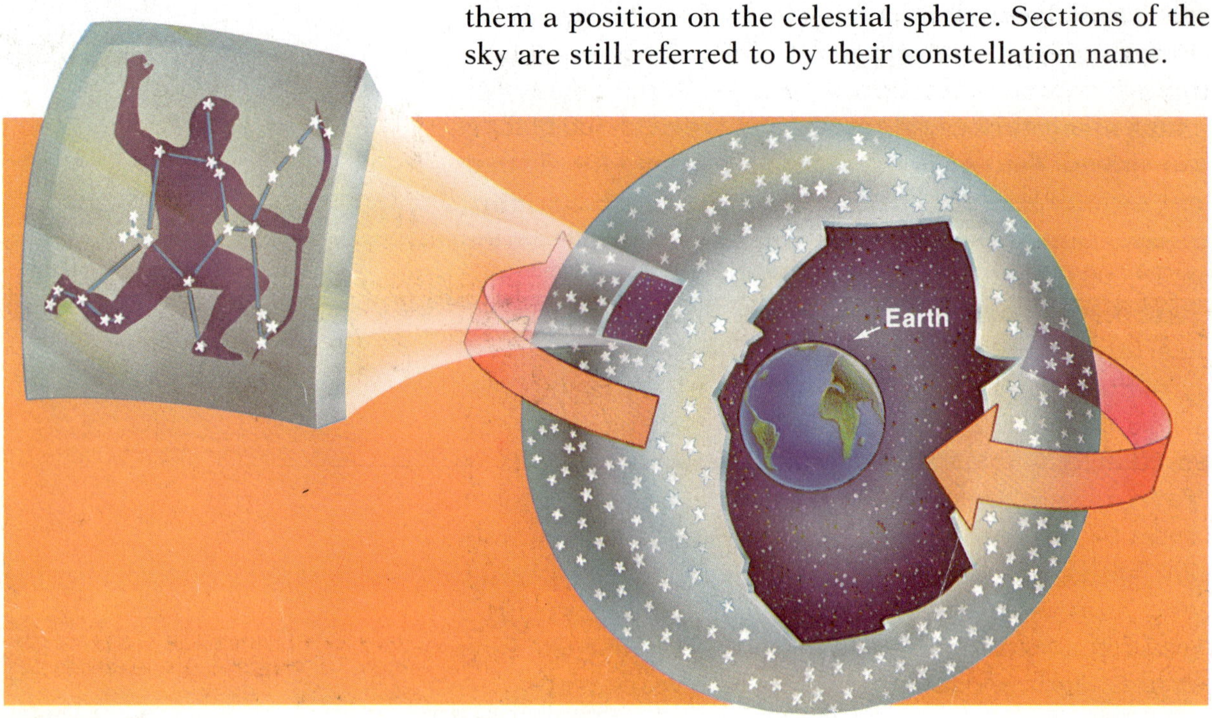

Figure 2.3 *left:* Galileo and his early telescope

Figure 2.4 *right:* Tycho Brahe's early observatory

Through time, people made better and better observations of the sky. In the 1550's Tycho Brahe constructed a superbly equipped observatory near Copenhagen. His observatory was equipped with the most accurate instruments ever designed before the time of the telescope. He planned to observe the sun, moon, planets, and stars regularly.

In 1609 the optic tube was invented by Galileo. This primitive telescope, no more powerful than field glasses of today, revolutionized the study of the sky. Galileo was the first to see Jupiter's moons and craters on the earth's moon. Astronomy has come a long way from Galileo's optic tube to the wonderful telescopes of today.

Some of the modern telescopes work the same way as Galileo's. Light coming from an object in space passes through a lens and forms an image. You look at this image with another lens. Telescopes that work in this way are called **refracting telescopes** [ree-FRAK-ting TEL-uh-skohps]. The largest existing refractor has a 102 cm lens through which the light enters.

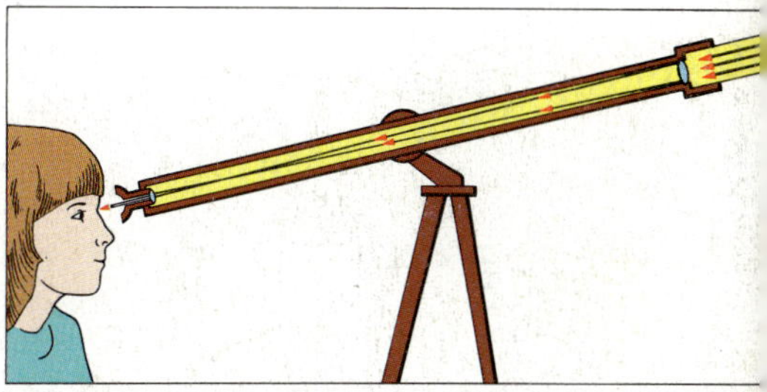

Figure 2.5 *top:* A refracting telescope; *bottom:* The diagram shows the positions of the lenses in a refracting telescope.

Figure 2.6 One-meter refractor telescope at the Yerkes Observatory, Williams Bay, Wisconsin

Figure 2.7 Horsehead Nebulae in the constellation Orion

Refractor telescopes are most useful for looking at objects close to earth. With refractor telescopes you can see the planets, the sun, stars, and moons. You can also see **nebulae** [NEB-yoo-lie]—thin, wispy-colored, glowing-gas clouds. Many **comets** [KAHM-its] have been discovered by observers using refracting telescopes. Comets are small bodies of icy and dusty matter that orbit about the sun.

In the year of Galileo's death a new type of telescope was created. This telescope uses a mirror to reflect light and form an image of an object. The image is seen through a lens. Telescopes that work in this way are called **reflecting telescopes** [ree-FLEK-ting TEL-uh-skops]. This was a wonderful invention because it allowed people to look farther and farther into space. Reflector telescopes have been able to see objects more than 4 billion light-years away.

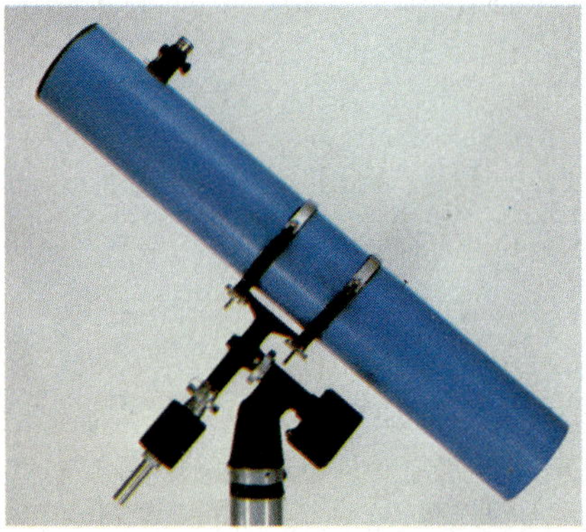

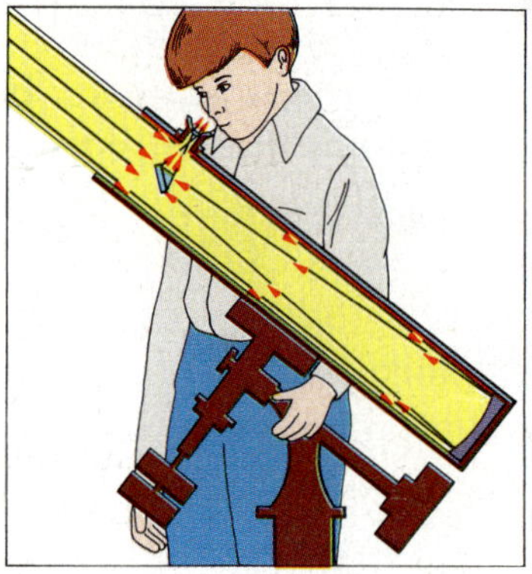

Figure 2.8 *left:* A reflecting telescope; *right:* Special mirrors are used with a lens in a reflecting telescope.

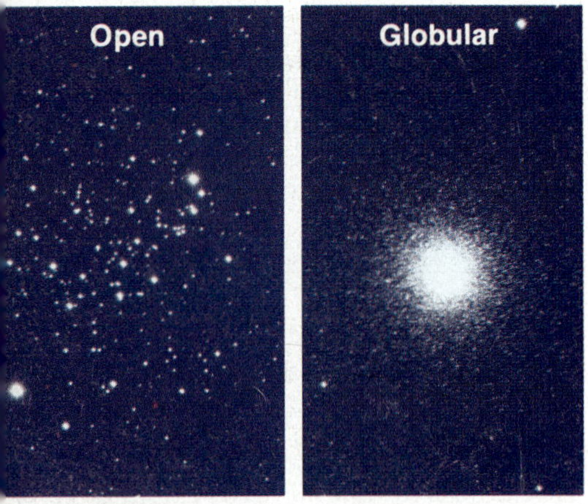

Figure 2.9 Star clusters

What kinds of objects are found in outer space? Glowing nebulae have already been mentioned, but not all nebulae give off light. Some gases absorb light and appear dark as in the constellation Orion. With reflector telescopes groups of stars can also be seen. Loose groupings of stars, containing from a few dozen to a few thousand, are called **open clusters. Globular clusters,** which contain thousands of stars in a close grouping, are also found. The most common objects found in far outer space are called **galaxies** [GAL-uhk-sees]. Galaxies are large assemblies of stars containing millions to billions of stars.

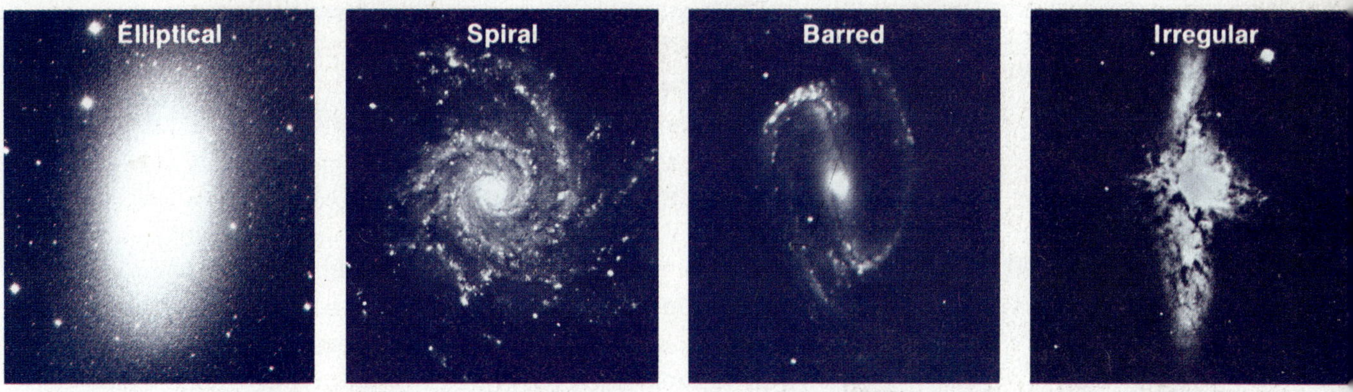

Figure 2.10 Examples of the major types of galaxies

There are four major types of galaxies—ellipticals, spirals, barred spirals, and irregulars. Look carefully for an example of each in these photographs of galaxies.

People using reflector telescopes have discovered that some stars vary in brightness at regular intervals. These stars, called **Cepheid** [SEE-fee-uhd] **variable stars,** get bright, then dim, and then bright in a period of a few days. Some take three days to go through the brightness to dimness cycle. Others take as many as 50 days. Polaris, the North Star, is a familiar example of a Cepheid variable star, but its change is so slight the human eye can't detect it.

Don't you think Galileo would have loved to be living today? He didn't know about galaxies, nebulae, open clusters, globular clusters, or Cepheid variable stars. How he would have loved to know just what you know about the sky! You really do know a lot about the sky. But you have begun to see that the more you know, the more you realize how little you really know. Every new discovery raises more questions than it answers. There are a lot of questions that haven't yet been answered by using the big telescopes.

Figure 2.11 Four-meter Mayall reflector telescope at Kitt Peak National Observatory near Tucson, Arizona

Study Questions for 2.1

1. Describe how early people classified the stars.
2. Explain how refractor and reflector telescopes work.
3. Make a list of the objects that are found in outer space.

2.2 How Can You Tell What's Out There?

You will find out

- what the electromagnetic spectrum is;
- how spectroscopes and radio telescopes are used to tell us about distant objects;
- what objects are identified by radio waves.

Figure 2.12 A spectrum from sunlight

The information astronomers receive from outer space tells them what is going on out there. It's like playing hide-and-seek. You are "it" and looking for your friend. You hear a noise, you see a shadow, or you see something move. You know what's going on by the information you receive. That's the way it is when astronomers use telescopes to look for objects in space. If an object sends out some information, you know it's there.

The information astronomers receive from objects in outer space is energy. The energy comes in the form of waves from space and is called **electromagnetic radiation** [ee-lek-troh-mag-NET-ik ray-dee-AY-shuhn]. Visible light is the most familiar form of this radiation. Other types of radiant energy are radio waves, infrared waves, ultraviolet waves, X rays, and gamma rays. Figure 2.13 shows the complete spectrum of electromagnetic radiation. All types of electromagnetic radiation travel at the speed of light, or 300,000 km per second.

It takes different kinds of instruments to detect each type of radiation. Radio receivers detect radio waves.

Do You Know?
On a clear night the naked eye can see a candle at about 8 km.

Figure 2.13 The electromagnetic radiation spectrum. Wavelengths are longest for radio waves and shortest for cosmic rays.

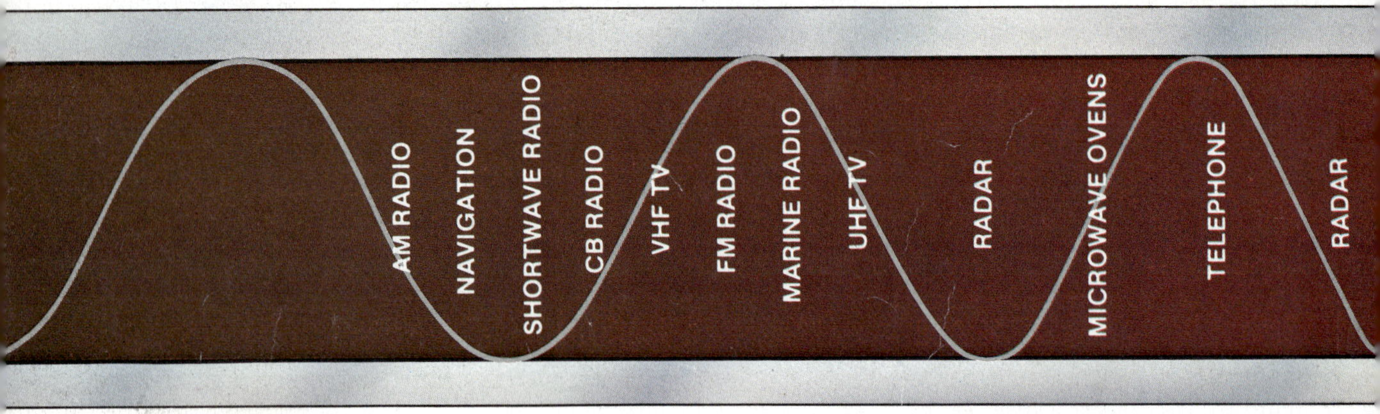

Infrared, light, and ultraviolet radiation are detected by optical telescopes using film. X rays and gamma rays are detected by X-ray and gamma-ray telescopes.

Much of what is known about outer space has been built on what has been learned by studying light energy. The light energy from each star carries its own message about that star.

That message can be studied by examining the light from that star. One way of examining light is to pass it through a **prism.** If light is passed through a prism, it is separated into different colors. The prism bends the light waves passing through it. Different colors are bent in different amounts. The pattern the light makes after passing through a prism is a **spectrum** [SPEK-truhm].

You can get a spectrum of the light from each star by using a **spectroscope** [SPEK-troh-skohp]. A spectroscope is an instrument which can be attached to telescopes and is used to examine light. It uses lenses and a prism or **diffraction grating** [di-FRACK-shuhn GRAY-ting] to form a spectrum. A diffraction grating is a piece of glass or plastic that has hundreds of narrow slits per centimeter. As light passes through it, the waves interfere with each other in such a way as to make a spectrum. By studying the spectrum of each star, you can read the message that the light carries. Large telescopes use the spectroscope more than half of their observation time.

Figure 2.14 A spectrogram of a star cluster

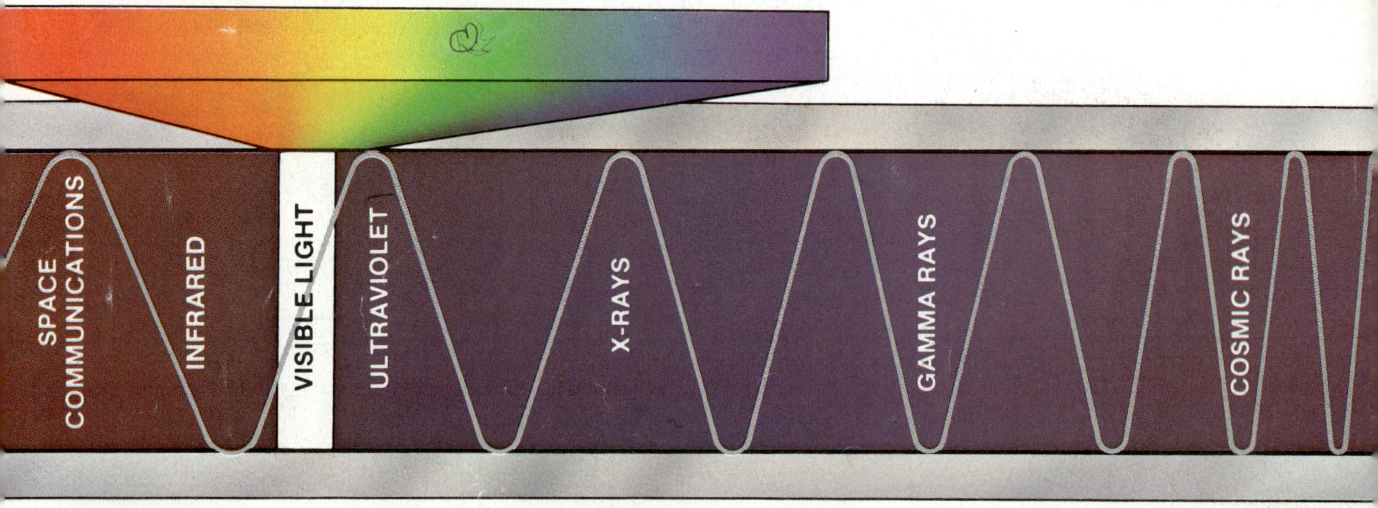

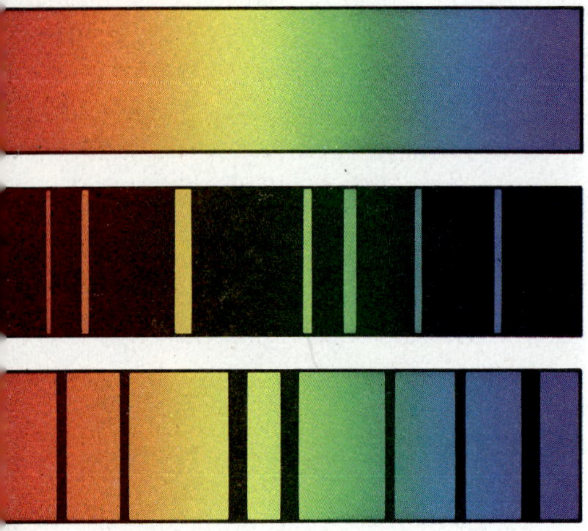

Figure 2.15 *top:* Continuous spectrum; *middle:* Bright line spectrum; *bottom:* Dark line spectrum

There are three kinds of spectra. A **continuous spectrum** is like a rainbow. A **bright line spectrum** has narrow bright lines at certain positions. A **dark line spectrum** has narrow dark lines at certain positions. A bright spectrum comes from a hot, shining object. When light is cut off or absorbed by passing through a cold gas, a dark spectrum is recorded. Much of what is known about the stars in the visible wavelengths has been learned by reading these spectra.

Each substance in a star or galaxy can give off energy. That energy is like a fingerprint, as each substance has its own color pattern. The astronomer focuses the telescope on the distant object. The light from that object passes through a spectroscope and is separated into its spectrum fingerprint. By matching the fingerprint with that produced by substances on earth, you can tell what an astronomical object is made of.

ACTIVITY

What Do the Light's Fingerprints Tell?

Materials

ruler or straight edge
pencil
paper

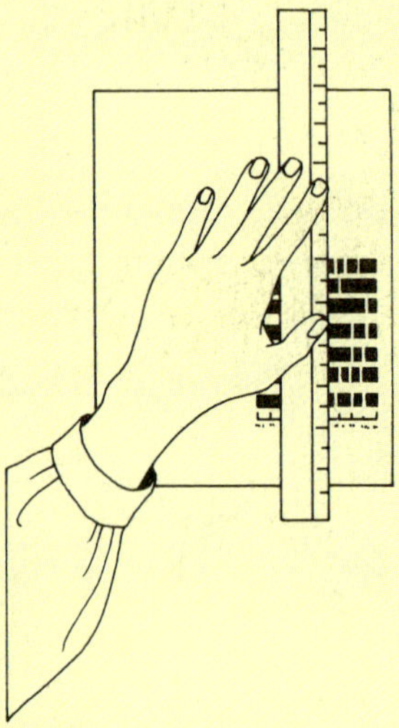

Procedure

1. Use your ruler to see if any lines in spectrum 1 match any lines in the known spectra.

2. List each substance(s) contained in unknown spectrum 1 that matches a known spectrum.

3. Repeat this process for the two other unknown spectra.

Questions

1. What substances does unknown spectrum 1 represent? Spectrum 2? Spectrum 3?

2. Explain how you might find the composition of a newly discovered galaxy.

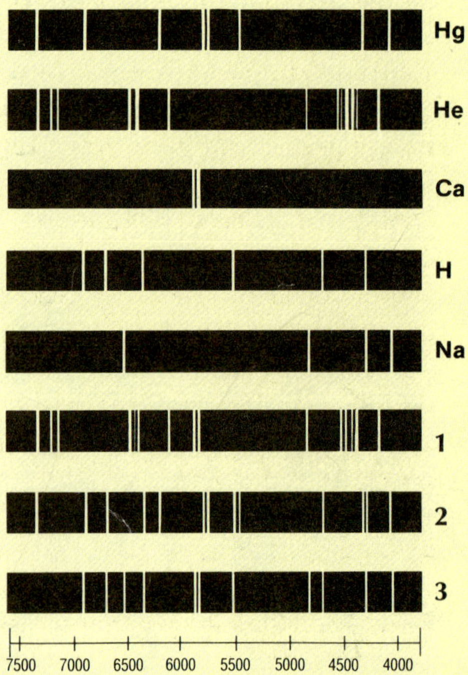

Figure 2.16 Ears on the universe. *left:* Very Large Array, radio telescopes near Socorro, New Mexico; *below:* Radio telescopes receive radio waves from stars and feed the waves to computers for reading.

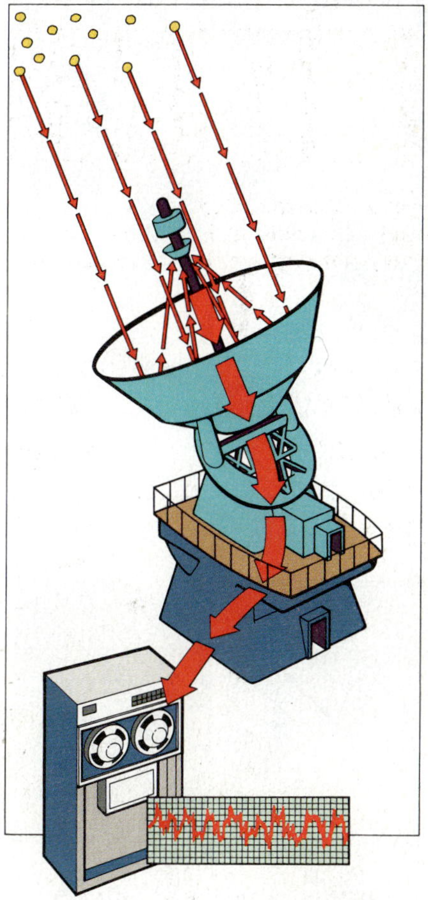

Only within the last 30 years have instruments other than optical telescopes been used to study outer space. Radio telescopes have been developed to read the radio waves received from space. Astronomers use huge radio telescopes to listen to the sky. The weak radio energy from space is collected by one or more reflecting dishes. Then the energy is carried to a receiver. The receiver converts the radio energy into an electrical signal which can be recorded on a graph. In this way the radio telescope receives and records the radio signals from space. Radio maps have been drawn of the sky. Studies of these radio maps have revealed two types of radio energy sources never known to exist before radio telescopes were invented. These sources, **quasars** [KWAY-sahrs] and **pulsars** [PUHL-sahrs], were discovered in the 1960's. Quasars, radio energy stars, are objects in space that emit strong radio signals. Pulsars are radio energy stars that emit energy in sharp, intense, and regular pulses. They were discovered in 1967. In 1960 there were two known radio energy stars. By 1979 more than 20,000 quasars had been identified, and more than 600 pulsars were known.

Study Questions for 2.2

1. Describe the electromagnetic spectrum.
2. Explain how you can determine the composition of a star.
3. List the objects that are studied with radio telescopes.

2.3 Mysteries of the Universe

You will find out

- that many observations in space can't be explained;
- how the Doppler effect is used to explain the shift of spectral lines;
- why the red shift supports the big bang theory.

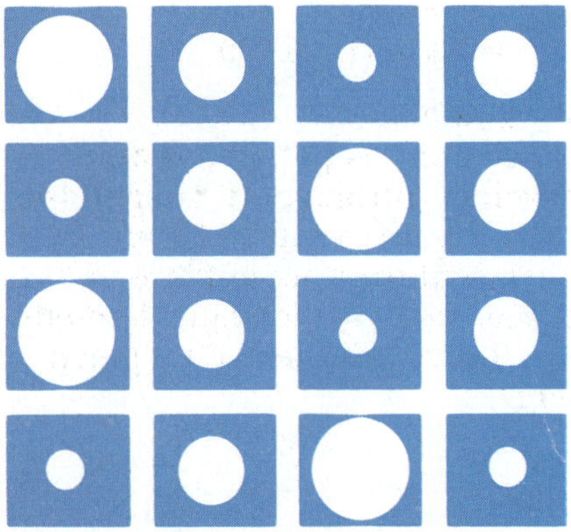

Figure 2.17 Over a period of time this could be the pulsation pattern of a Cepheid variable star.

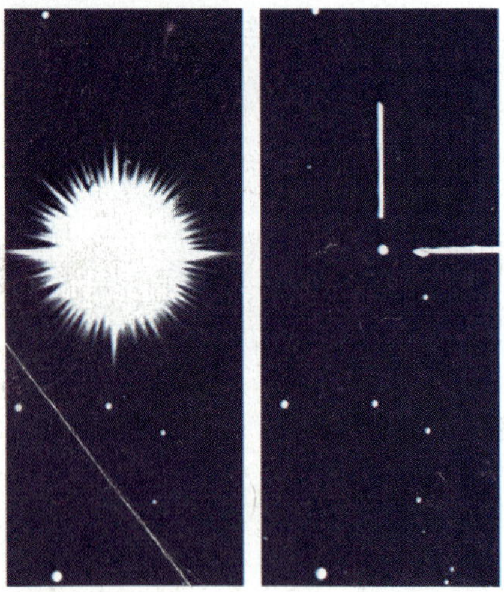

Figure 2.18 *left:* Nova Hercules at the time of its explosion, March 10, 1935; *right:* It then rapidly faded to a low magnitude as shown in the photo taken May 6, 1935.

Have you ever been curious? Consider this example. Your teacher has given you two eggs to study. One egg is hard-boiled; the other is uncooked. You are to find out which is hard-boiled. You can't break or crack the egg. What would you do? You have a mystery to solve. That's what happens to astronomers when they find things they can't explain by what they know. These things are mysteries.

What are some of the mysteries for astronomers? One such mystery is the more than 700 Cepheid variable stars. Imagine the sun swelling in size and then shrinking back to normal every few days. That would be very strange, yet some stars do just that. How can a star do that? Present-day physics can't adequately explain how this is possible. A star's size is dependent upon balanced forces. Energy-making processes in the star's interior tend to make the star increase in size. Gravity tries to hold the material together. Yet in some stars there is a sequence of swell–shrink–swell that is repeated time and again. The star pulsates at regular intervals. Cepheids are thought to cease pulsating as they grow older. How long will a Cepheid pulsate? What might cause one to stop pulsating?

Another mystery object is a star that blows off its outer skin and just keeps getting bigger. It's an exploding star called a nova. These firecrackers in space don't happen very often, but when they do, it's a big explosion. A super giant star can also explode in a glorious burst called supernova. Which stars are next going to become a nova or supernova?

Pulsars are objects that give off lots of radio energy in brief but intense pulses. The amount of radio energy emitted during one pulse is from one million to ten billion times the energy the sun emits from a similar

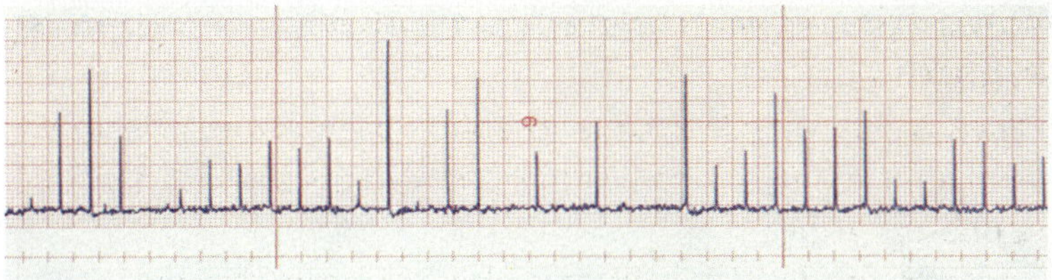

Figure 2.19 Each peak on this graph is a record of the intense pulses of radio energy received from a pulsar.

amount of surface area. These small stars may have about the same mass as the sun, but they are only 10 to 30 km in diameter. One cubic centimeter of a pulsar has a mass of about 10^{11} kg. A cubic centimeter is only about the size of a sugar cube, but there isn't a truck in your town that could haul that much of a pulsar. Pulsars rotate rapidly, and the pulses of radio energy are caused by this rotation. There is a lot yet to be learned about what makes a star so dense and spin so fast.

There are other mysteries of the heavens besides objects that give off lots of energy. For example, some of the lines in the spectra from distant objects appear to be shifted when compared to laboratory spectra.

There may be an explanation for this. You can use this example. Imagine you are watching an auto race. As a car approaches, the sound of the car gets louder and has a certain pitch. Notice how the pitch changes as the car passes by you. The pitch of the car is higher coming toward you than it was going away. Actually the sound didn't change pitch, but you hear a pitch change because the car was moving. The change in pitch is called the **Doppler** [DAHP-luhr] **effect.**

Do You Know?
10^{11} = 100,000,000,000

Do You Know?
When the first pulsar was discovered, astronomers thought the signals were coded and might have been sent from some place in space. They half jokingly dubbed the source LGM for "little green men."

Figure 2.20 The sound pitch of the race car is higher as it comes toward you and lower as it moves away from you.

Figure 2.21 In this picture, the color of the rocket ship coming toward you is blue-shifted, and the color of the rocket ship going away is red-shifted.

It is believed that the shift in the spectral lines in the spectrum of a distant object in space is also due to motion. The faster the object moves, the greater the shift. Lines shifted toward the red, longer wavelength, are caused by objects moving away. Lines shifted toward the blue, shorter wavelength, indicate that the object is moving toward us. This is called the **Doppler shift.** The far away objects of outer space all show such a red shift. Quasars are the objects showing the greatest shift and the fastest motion.

Astronomers now believe that the motion is related to how the universe began. They propose that the universe started with a big bang. From that moment on, the universe expanded. All the objects in space appear to be moving apart from each other. It's like putting spots on a balloon. As you blow it up, the spots move farther apart. All far away objects seem to be fleeing away and getting farther apart. Is this the way the universe began? Or is there some other explanation for the red shift?

The greatest mysteries of all are **black holes.** A black hole is thought to be an object in space from which no light can escape. Without any light you, of course, can't see it. If you can't see it, how do you know it even exists?

Black holes are revealed by X rays. The X rays come from the areas where black holes are suspected to be. It is believed that matter from nearby space is pulled into the black hole. As the matter approaches the black hole, it speeds up and it heats up to give off X rays. There's more to learn about black holes. One theory for black holes is that they start with the collapse of a massive star. Near the end of its life cycle, the star runs out of fuel and collapses. Black holes may grow by swallowing matter from other stars or from swallowing whole stars in a gulp. There may be huge black holes having masses equal to millions of stars. There may be black holes the size of a star. There also may be mini black holes the size of a planet or smaller.

Because black holes are so dense, they have tremendous gravitational fields. Imagine the earth shrunk to the size of a golf ball. That ball would then be about as dense as a black hole. Anything that dense has a strong enough gravitational field to prevent light from escaping. The light would try to get away, but gravity would pull it back.

Do black holes exist? Some astronomers believe they do; others do not. How were black holes formed? How do they work? It is not known for sure. Black holes are one of the greatest mysteries in space.

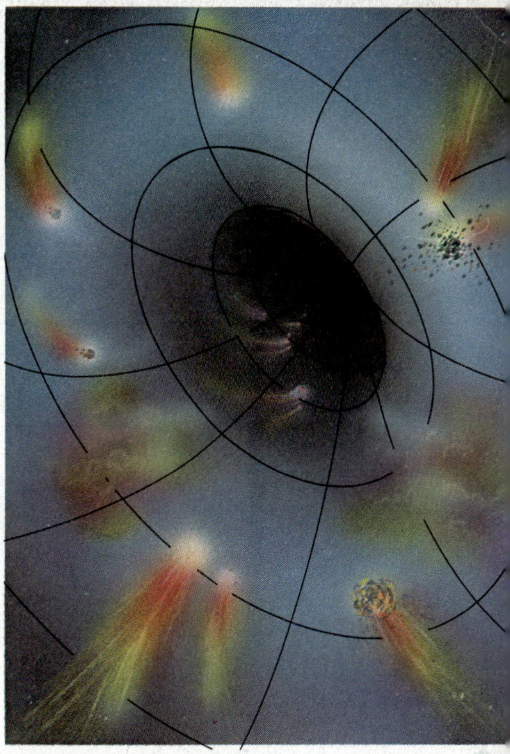

Figure 2.22 This is an artist's conception of a black hole. Why aren't there any photographs of black holes?

Study Questions for 2.3

1. Select one mystery of our universe and explain why it is a mystery.
2. What does the Doppler effect explain?
3. What does the red shift imply about how the universe was started?
4. Use the scientific processes discussed in Chapter 1 to support or reject the big bang theory of an expanding universe.

What Did You Learn?

- A light-year is the distance light can travel in a year.
- The Greeks divided the sky into the groupings of stars called the constellations.
- Refracting telescopes use lenses to form an image of an astronomical body.
- Reflecting telescopes use mirrors and lenses to form an image of an astronomical body.
- Objects found in outer space include the following: open clusters, globular clusters, galaxies, Cepheid variable stars, quasars, and pulsars.
- The energy received from space is called electromagnetic radiation.
- It takes different kinds of instruments to detect each type of electromagnetic radiation.
- A spectroscope is used to separate light into colors.
- Astronomers use three kinds of spectra.
- By studying the spectra you can tell the composition of objects in space.
- Radio telescopes are used to detect quasars and pulsars.
- Not all that has been observed in space can be explained with current knowledge.

Key Terms

celestial sphere
constellation
refracting telescope
nebula
comet
reflecting telescope
open cluster
globular cluster
galaxy
Cepheid variable star
electromagnetic radiation
prism
spectrum
spectroscope
diffraction grating
continuous spectrum
bright line spectrum
dark line spectrum
quasar
pulsar
Doppler effect
Doppler shift
black hole

Biography

Edwin Powell Hubble (1889-1953)

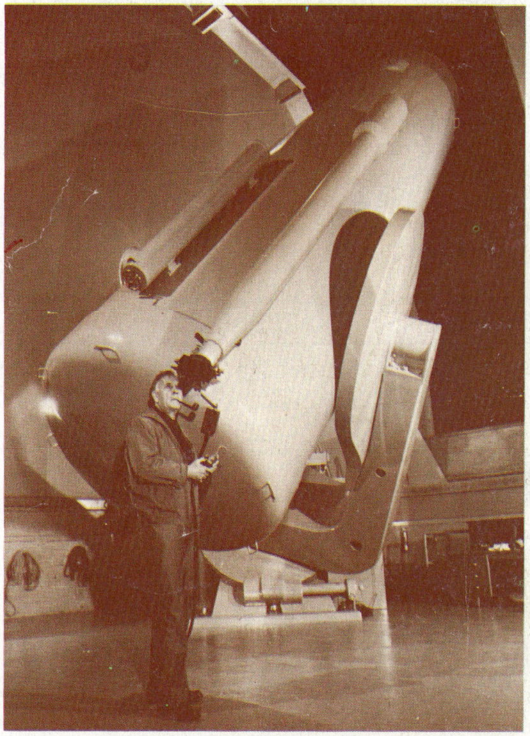

Edwin Hubble was an active athlete. He took part in track, boxing, and rowing. As a boxer he was successful enough to be considered as a contender for the heavyweight championship.

Hubble graduated from the University of Chicago in 1910 with a B.S. degree in astronomy. He studied law at Oxford University, England, and practiced law for a year in Kentucky. Unsatisfied with law, Hubble entered graduate school at the University of Chicago. He graduated in 1917 and joined the army. He was wounded and left the army as a major. His career in astronomy began after World War I.

Hubble's discovery of Cepheids and other variable stars in nebulae served as the basis for investigations into the evolution of the universe. His discovery of the relationship between the shift in spectral lines and the speed of galaxies is known as Hubble's law. Hubble found that for every million light-years a galaxy is away from us, its speed increased 30 km per second. This relationship is used in determining the size of the universe. Results of Hubble's work rank as the most significant contributions to the study of the universe since the time of Copernicus. Hubble was a giant in the field of astronomy.

TO THINK ABOUT AND DO

Vocabulary

1. Unscramble the words on the left.
2. Place the designated letter from the unscrambled word in the column marked "letter" on your paper.
3. What word is spelled out?

Scrambled Words	Unscrambled Words	Letter	Designated Letter from Unscrambled Word
eptoeslc			use 6th letter
tnoloncsaetli			use 2nd letter
ustemprc			use 1st letter
eoavn			use 5th letter
arst			use 2nd letter

What Do You Remember?

1. List the contributions of the Greeks to early astronomy.
2. Who was Tycho Brahe? What did he do?
3. What is an optic tube? How powerful is it?
4. List some of the objects that can be seen with refracting telescopes.
5. Which type of telescope is used to study objects in far outer space?
6. What is a globular cluster?
7. What is the name of stars that shrink and swell?
8. Name the kinds of energy found in the electromagnetic spectrum.
9. How can you learn about far away objects in space?
10. What do a prism and diffraction grating have in common?
11. Name three kinds of spectra astronomers use.
12. Describe what a radio telescope does.
13. Name two radio energy sources in the sky.

14. List four of the mysteries of the universe.
15. Why is a pulsar a mystery?
16. Can you see a black hole? If yes, describe one. If no, why not?

Applying What You Have Learned

1. Explain why the Greeks thought of the heavens as a celestial sphere.
2. How can you tell a refracting telescope from a reflecting telescope?
3. Explain the statement, "Every new discovery raises more questions than it answers."
4. Explain the statement, "You can tell the composition of an object in space without touching it."
5. Describe the balanced forces that keep a star the same size.
6. You are in a moving car, and the horn is blowing. Why didn't you hear a change in pitch?

Research and Investigation

1. Select one of the astronomical objects mentioned in this chapter and write a paper describing that object.
2. Prepare a poster for display on one of the astronomical objects mentioned in this chapter.
3. Write a description of the work of one of these famous astronomers: Henrietta Leavitt, Tycho Brahe, Galileo Galilei, Sir Isaac Newton.
4. Find or write a poem about the heavens.
5. Make a tape recording of the Doppler effect. These are suggestions for a sound source: a car horn, a trumpet player on a bicycle, a swinging tuning fork safely attached to a cord, or anything that makes a continuous sound while moving past you.

A comet

3

The Milky Way Galaxy

As your journey from outer space continues, you begin to find that things are becoming more crowded. What you saw earlier looked like many other galaxies, but now it seems to be millions of stars all around you.

You wonder why it is that you see so many more stars in some directions than others. You also seem to pass through areas where there are more stars and then less stars. This begins to tell you something about the shape of this special galaxy that you are now going to study.

You can't help doing what scientists have done for a long time. You speculate about whether there is life somewhere on a planet circling one of those stars. In this chapter you will learn about some of the new tools astronomers use. What mysteries will they solve? But more exciting, what new mysteries will they turn up?

Figure 3.1 *opposite:* The Milky Way Galaxy as viewed from Earth

3.1 Your Galaxy

You will find out

- how your galaxy is structured;
- how your galaxy is changing;
- the galaxy you live in is called the Milky Way Galaxy.

Do you ever get the chance to view the sky away from city lights? You would notice a sky full of stars. On a dark, clear night you might notice a white milky band of light running the entire length of the sky. This band contains millions and millions of stars. The reason it looks this way is because you are really looking out through our galaxy toward its edge. Because this galaxy looks this way to you it is called the Milky Way Galaxy.

Figure 3.2 *bottom:* This drawing is an artist's idea of the Milky Way Galaxy as it would appear from space. *top:* The photograph shows another galaxy that is believed to be similar to the Milky Way.

If anyone outside the Milky Way Galaxy were to look at it, they would see an ordinary spiral galaxy somewhat like M-81. What is known about the Milky Way Galaxy? A lot is already known. It is lens-shaped with two or more spiral arms. It's big—about 100,000 light-years in diameter. It is about 10,000 light-years thick and is rotating. The center of the Milky Way Galaxy lies in the direction of the constellation Sagittarius. The sun and its family of planets are about 30,000 light-years from the center in one of the spiral arms. It will take the sun in its position about 225 million years to make one complete turn around the galactic center.

Do You Know?
If the sun were the size of a basketball, the earth would be the size of a BB pellet located 30 meters from the basketball and the galactic disk would be 200 million km across.

The center of the galaxy shows up dark in optical telescopes because of clouds of gas and dust. Recently, radio telescopes have been used to study the region of space near the center of the galaxy. Radio energy penetrates through the light, gas, and dust. Consider, for example, a foggy or smoggy day. You may not be able to see very well, but your radio and TV set still work fine. Astronomers use radio telescopes to study the material between the stars near the center of the galaxy. The radio telescope helps them study areas where they are unable to see with optical telescopes.

From radio astronomy, astronomers learn that the galaxy has spiral arms. The spiral arms contain vast clouds of gas and dust between the stars. These clouds are not very thick, as they have only 25 to 50 very small particles per cubic km. Astronomers believe that these clouds are the birthplace of new stars. The material will someday collect together, and new stars will be born.

Measurements of different stars show that they move with respect to each other. They do not all move together. You can notice this when you compare very old star maps with new ones. The ancients saw the stars in a slightly different pattern than you see today. The next time you look at the stars, find the Big Dipper. Fifty thousands years from now the position of these stars will have changed some. They may no longer look so much like a dipper.

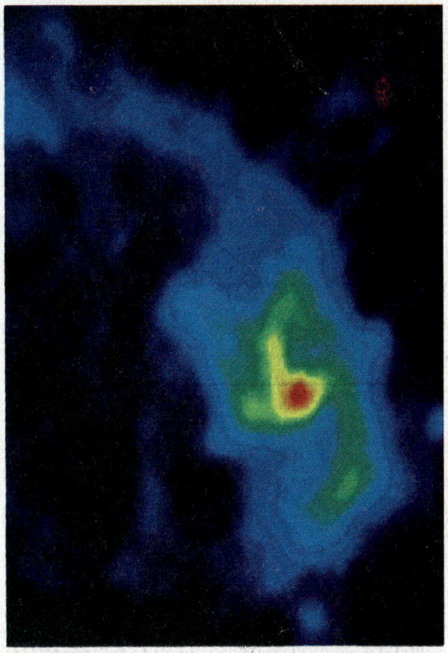

Figure 3.3 A picture of radio energy received from the nucleus of the Milky Way Galaxy

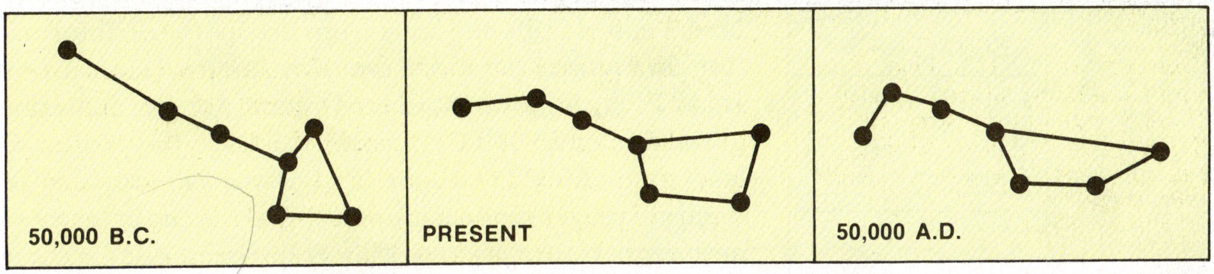

Figure 3.4 Changes in the Big Dipper over time

Study Questions for 3.1

1. Draw a top view of your galaxy. Label its dimensions. Locate the sun.
2. Describe two ways your galaxy is changing.

3.2 Study the Stars in Your Galaxy

You will find out
- the four different types of stars;
- the life history of a star;
- the relationship between temperature and brightness of stars;
- how to make a star count.

Do You Know?
The largest telescopes can see stars that are a million times fainter than those you can see with your naked eye.

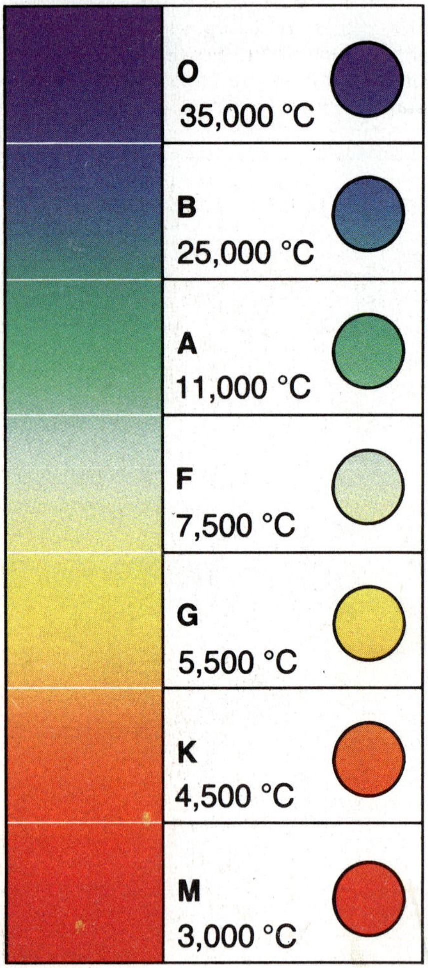

Figure 3.5 Seven spectral types of stars according to their temperature

The most prominent members of the Milky Way Galaxy are the stars you see. Stars are objects in space that make their own light. The best known and nearest star to you is the sun. Even as near as the sun appears to be, it takes light from the sun about eight minutes to reach you. If the sun were to suddenly stop giving off light, how long would it be before total darkness would come? The sun, even though smaller than many other stars, appears bright to you because it is close. There are many brighter stars, but they are much farther away.

One way of judging a star is by how much light it gives off. The brightness of a star can be measured by using telescopes. In most cases the brightest stars are the hottest stars. Just as astronomers analyze the light from galaxies by using spectroscopes, they can do the same thing with starlight. By examining a spectrum of a star, they know much about that star's composition. Each star has its own fingerprint, or spectrum. Astronomers can classify the stars from the spectrum information. In Figure 3.5 you see that the classes of stars are O, B, A, F, G, K, and M. Class O stars are the hottest—greater than 25,000°C. Class M stars are the coolest at less than 3,000°C. About 1910 two astronomers in separate places made a graph comparing the brightness and temperature of stars. This graph is named for those two astronomers. It's called the **Hertzsprung-Russell** (H-R) **diagram** [HUHRTZ-spruhng—rus-sehl]. Look at this graph closely because most stars can be shown in the H-R diagram. On the H-R diagram the sun is a class G star. It isn't very hot and it isn't very bright. The sun is a rather ordinary star.

Astronomers were surprised to learn how the stars showed up on the H-R diagram. They found that there were four types of stars. Most stars are included in the first type, and they lie on the diagonal band from upper left to lower right. Stars that fall close to this band are called **main sequence stars.** You will note that the sun is a main sequence star. The second type of star lies above and to the right of the main sequence band. These brighter stars are called **giants.** Some because of their color are called red giants. The third type includes stars that are even brighter than normal giants. These are called **supergiants,** and you find them in the upper right of the H-R diagram. Although supergiants are cooler, they give off more energy because of their enormous size. The fourth type of star, the **white dwarf,** is found to the left and below the main sequence. They are fainter and smaller than main sequence stars. Sometimes they are called faint hot objects.

Figure 3.6 A Hertzsprung-Russell (H-R) diagram is a plot of temperature versus brightness. The hottest stars are on the left side of the graph. The brightest stars are on the top of the graph. When plotted on an H-R diagram, the stars lie mainly on a diagonal band from upper left to lower right. Thus the hotter stars are brighter than the cooler stars. Stars on the main sequence are called dwarf stars. The sun is a type G dwarf.

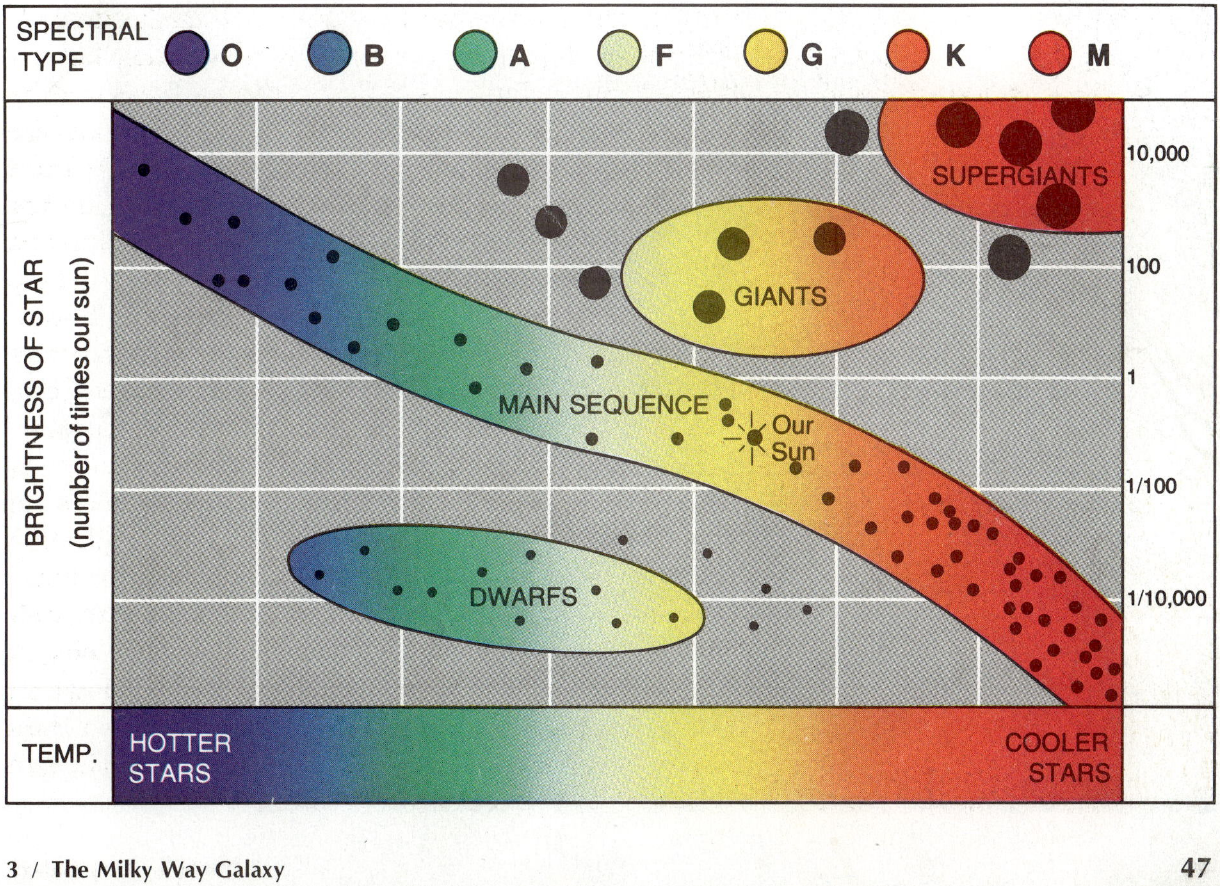

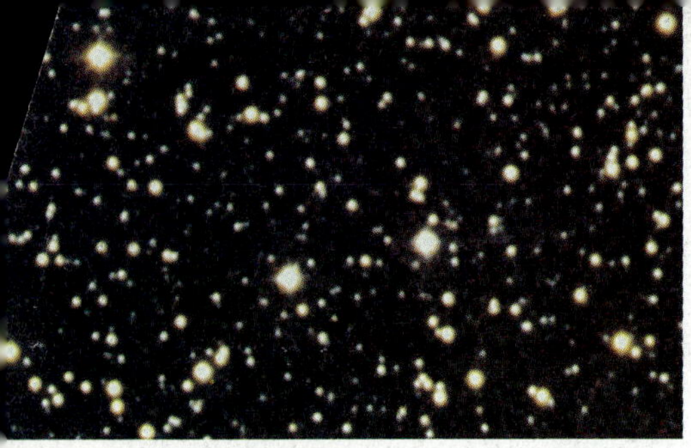

Figure 3.7 A small section of the Milky Way Galaxy is the Sagittarius region.

Astronomers also make surveys of the number of stars in the Milky Way Galaxy. They find that the number of stars is not the same when they look in different directions. Earlier you saw the Milky Way Galaxy as a band of light across the sky. This is because there are so many stars in that direction. How can you count so many stars? One way is to count the stars in one small piece of the sky. Then you can determine the number in the whole sky.

ACTIVITY

How Many Stars Can You See?

Materials

pencil
tape
typing paper
paper for sky charts—three sheets, 1 m × 1.6 cm

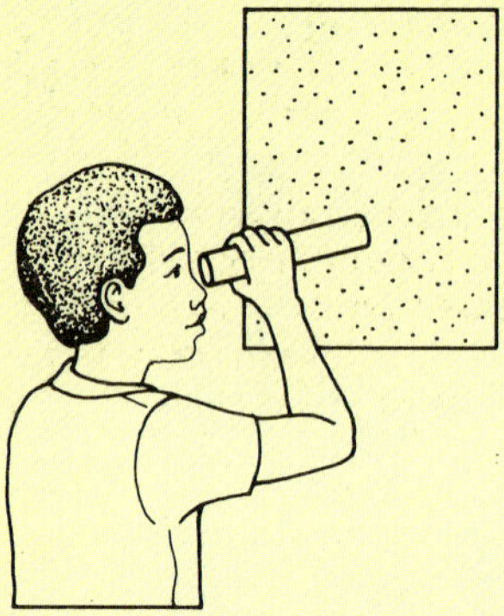

Procedure

Make three sky charts as follows: Place 20 to 30 stars randomly scattered on chart 1, 10 to 25 stars on chart 2, and 30 to 50 stars on chart 3. The charts are then posted on 3 different walls of the classroom.

1. Roll a sheet of typing paper into a tube 28 cm long and 4 cm in diameter. Fasten with tape.

2. Sight through the paper tube toward sky chart 1.

3. Count the number of stars you can see. Record this number.

4. Repeat steps 2 and 3 for sky charts 2 and 3.

5. Add together the number of stars you saw in each count. Divide by three. This is your average number of stars (*As*) seen for each view.

6. Calculate the number of stars by answering these questions. Use the Table. What is the surface area of a sphere with a radius of 28 cm?

_____ sq. cm (*Sa*)

What is the area of view for tube 4 cm diameter?

_____ sq. cm (*A*)

$$\text{Number of stars} = \frac{Sa}{A} \times As$$

$$= \_\_\_\_\_ \text{ stars}$$

Table Information for Calculations

Surface area of sphere with a radius of 28 cm (*Sa*)	Area of view for tube with 4 cm diameter (*A*)
9,847 sq. cm	12.6 sq. cm

Questions

1. What is the area of the end of your tube?

2. What was the average number of stars counted?

3. How many stars would you see in the whole sky?

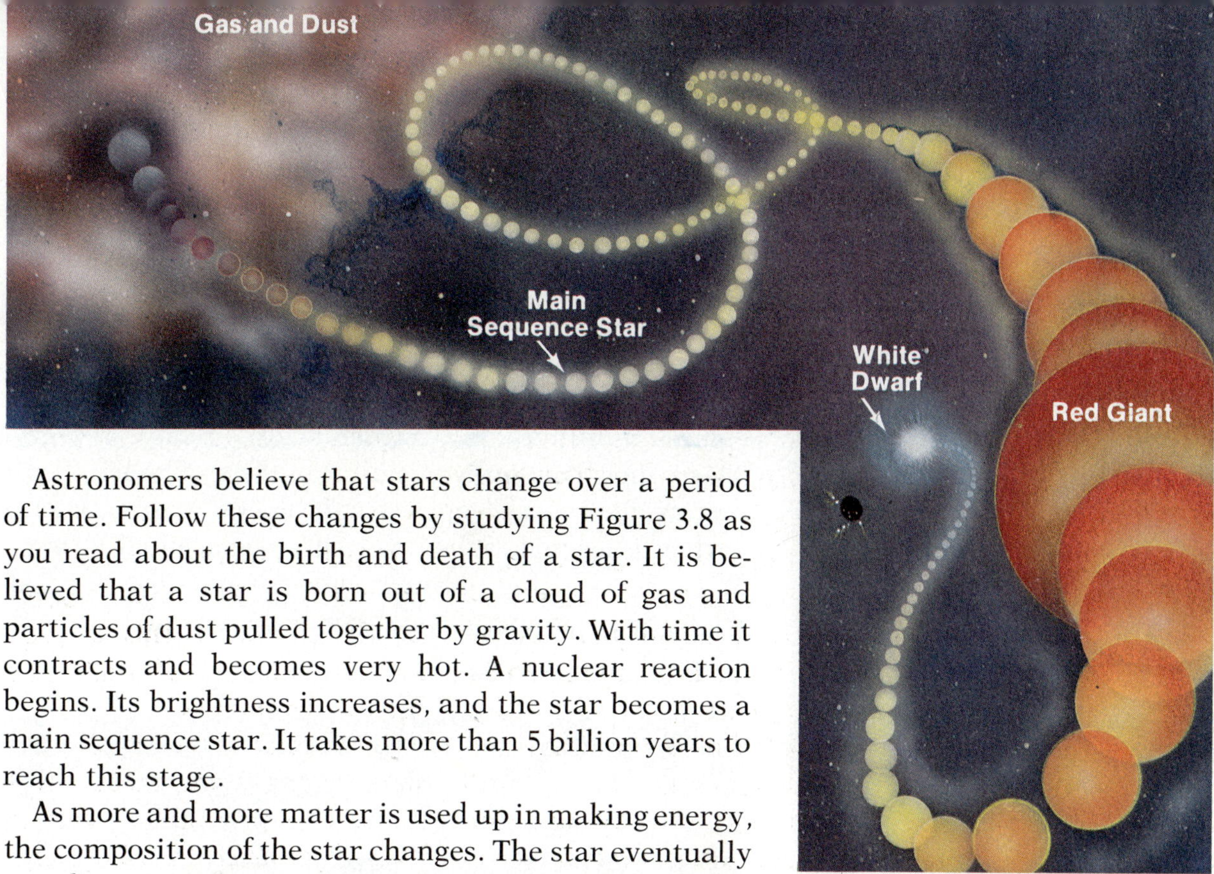

Figure 3.8 An artist's picture of the life of a star from its birth until it ends up billions of years later as a white dwarf. Some stars may also become nova stars or even black holes.

Astronomers believe that stars change over a period of time. Follow these changes by studying Figure 3.8 as you read about the birth and death of a star. It is believed that a star is born out of a cloud of gas and particles of dust pulled together by gravity. With time it contracts and becomes very hot. A nuclear reaction begins. Its brightness increases, and the star becomes a main sequence star. It takes more than 5 billion years to reach this stage.

As more and more matter is used up in making energy, the composition of the star changes. The star eventually gets larger, cooler on the outside, and brighter. It becomes a red giant star. In time the star changes further. It stays about the same brightness but becomes hotter and smaller. The star continues to get hotter and smaller finally ending up as a white dwarf. Some dwarfs become nova stars, which is the death of the star. It takes a long time for this to happen. For most stars this process probably takes more than 10 billion years.

Will the sun follow these steps? No one can be sure, but observations of other stars clearly imply a sequence of stages in the life of a star. Astronomers' observations seem to confirm that the Milky Way Galaxy and the stars within it change with time. You live in a dynamic universe.

Study Questions for 3.2

1. List the four types of stars.
2. Describe the possible life cycle of a star.
3. What is the relationship between temperature and brightness of stars?

3.3 The Search in Space

You will find out

- why space observatories are so desirable;
- what the major parts of a space observatory are;
- how scientists have searched for life in space.

Why would you want to put a telescope into space? One of the best reasons is to take it out of the atmosphere. Think about the atmosphere outside your classroom right this minute. Is it clear? Foggy? Smoggy? Hazy? The atmosphere prevents a clear view of space. It absorbs some energy from space. It blocks out some energy and at times makes the images from space look strange. It is like looking through a dirty window. The atmosphere is something astronomers would like to do without.

Figure 3.9 *above:* The hazy atmosphere over this observatory outside Paris, France, limits ability to peer into space.

Figure 3.10 *above right:* An artist's picture of the NASA space telescope

Space observatories eliminate these problems. In a space observatory the telescopes are outside the earth's atmosphere. Will viewing be better? If the 5-m telescope on Palomar Mountain in Southern California were above the atmosphere, it could see 50 times better.

Another reason to take telescopes outside the atmosphere is to escape light. The atmosphere gives off scattered light from stars and from cities. The amount of darkness sets the limit on the faintest objects that can be seen. It is becoming difficult to find a location on the earth that is as dark as astronomers would like. But above the atmosphere it is completely dark.

Telescopes have been lifted above some of the atmosphere by balloons and jet aircraft. But the most successful way to eliminate the atmospheric problems of telescopes has been to put them in a spacecraft. Spacecraft can place telescopes in orbit about the earth completely above the atmosphere.

In Figure 3.11 you see an artist's concept of a space observatory. The space observatory has one or more telescopes. The large panels convert sunlight into electrical energy that supplies power to operate the observatory. This observatory has a guidance system so that the telescopes can be pointed at any part of the sky. There is also a system to maneuver the observatory and maintain it in a specific position in space. A radio transmitter and receiver will be used to communicate with the observatory.

Do You Know?
Prior to the first orbiting solar observatory, the technique of studying the sun hadn't changed much since Galileo made his first observations some 350 years ago in 1610.

Figure 3.11 A future space observatory may look like this picture. Notice the solar cells for power and the style of construction in the weightlessness of space.

The observatories orbiting in space make the same kinds of studies of the sky that are made with telescopes on the earth. These studies include visual observations and searching for X-ray and gamma-ray sources. Some observatories are designed to study the infrared and ultraviolet part of the electromagnetic spectrum coming from the sky.

More and more is being learned about your galaxy all the time. The main reason for this is that now all parts of the electromagnetic spectrum are studied. Galileo

Figure 3.12 Vehicles such as these carry instruments aloft to study radiation without the interference of Earth's atmosphere.

and all astronomers after him for the next 350 years studied only the visible part of the spectrum. Now with space observatories the entire electromagnetic spectrum is being studied from above any interfering atmosphere. There are many wonders yet to be learned about the Milky Way Galaxy.

One of the wonders that has persisted from ancient times is the question of life. In all of the Milky Way Galaxy which contains billions of stars is there only one sun with an inhabited planet orbiting around it? In every way it appears that humans on the earth are alone in this universe. Even so, the thought of other living beings persists.

One reason for thinking there might be other planets that could support life is the vast number of stars. But of all these stars which ones should you even consider? The very bright, hot stars are unlikely stars to have planets with human life. Other stars are too cool to keep a planet warm. The most likely stars are similar to our sun. But for now there is no information on how many, if any, of those stars also have planets.

Some astronomers say the highest number of such possible places for life in the Milky Way Galaxy is as high as 10 million. They also say the smallest possible number of places may be as few as 20. The only number you can be sure of is one.

Do You Know?
If the thickness of one playing card represents the distance to the moon, you would need 30 km of cards to show the distance to the nearest star.
J. Allen Hynek, Astronomer

Figure 3.13 *left:* Do you think the moon is a good place to locate an observatory?

Figure 3.14 *below:* The Pleiades. These are new stars. Do they harbor new worlds?

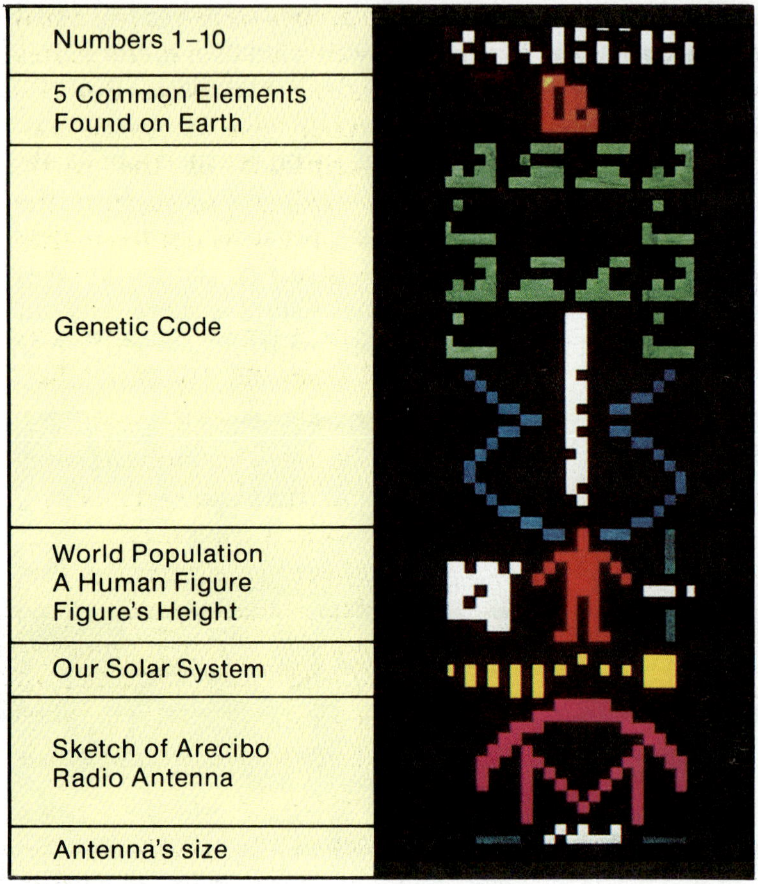

Figure 3.15 *right:* This astrogram was first radioed into space from the Arecibo National Radio Astronomy and Ionosphere Center in Puerto Rico in 1974. Will it be answered?

Figure 3.16 *below:* Radio telescope at Arecibo, Puerto Rico

Space travel has been eliminated in any search for other living beings because the distance to any other star is too great. Even traveling at the speed of light, your trip would take many years. If you can't search by a space voyage, what can you do? One way is to send a radio message into space and hope that another living being receives. Another way is to make radio surveys of the sky in the hope of picking up a message from somewhere else.

This is just what has been done with the radio telescope at Arecibo, Puerto Rico. It has been sending a message, called an **astrogram** [AS-troh-gram]. This message contains information about life on the earth, descriptions of the solar system, and a sketch of the radio telescope. Suppose you received a message from space like Figure 3.15. Could you interpret it?

Several of the earliest space probes were programmed to leave the solar system. They carried thin metal plates with line sketches of people and other messages describing our world. Other space probes have carried phonograph recordings with descriptions of the earth. Although there is no known evidence to support the expectation of a message from space, it is not likely that people will ever stop thinking about it.

Do You Know?
In 1899, an eccentric electrical pioneer Nikola Tesla undertook to transmit a powerful electrical signal into space from his Colorado laboratory and to detect any replies.

ACTIVITY

What Does the Astrogram Say?

Materials
index card
pencil

Procedure

1. Decide what kind of animal you would like to be—a monkey, a dinosaur, a spider, etc.
2. Make up a message that describes this animal. Write your message on the card. Do not write the name of the animal on the card.
3. Trade your card with someone else.

Questions

1. What was your animal?
2. What was the animal on the card you got in trade?
3. What are good clues to show an animal in a message?

Study Questions for 3.3

1. Name three advantages of a space observatory for studying the stars.
2. List the major parts of a space observatory.
3. What kinds of observations outside the visible part of the spectrum are now made by observatories?
4. Describe the information that has been attached to space probes leaving the solar system.

What Did You Learn?

- The Milky Way Galaxy is a lens-shaped, rotating galaxy having at least two arms.
- The solar system is about 30,000 light-years from the center of the galaxy.
- When you look toward the edge of the galaxy you see the Milky Way.
- Radio telescopes are used to study the region near the center of the galaxy.
- New stars develop in the gas and dust found in the spiral arms.
- There are seven classes and four types of stars.
- The Hertzsprung-Russell diagram shows the relationship between brightness and temperature of stars.
- Stars appear to have a life cycle.
- Telescopes above the atmosphere have a clearer view into space.
- Space observatories are controlled from the earth.
- Space observatories use both optical and nonoptical telescopes.
- No evidence has been found of life outside the planet Earth.
- In the search for life, scientists listen for and send messages.

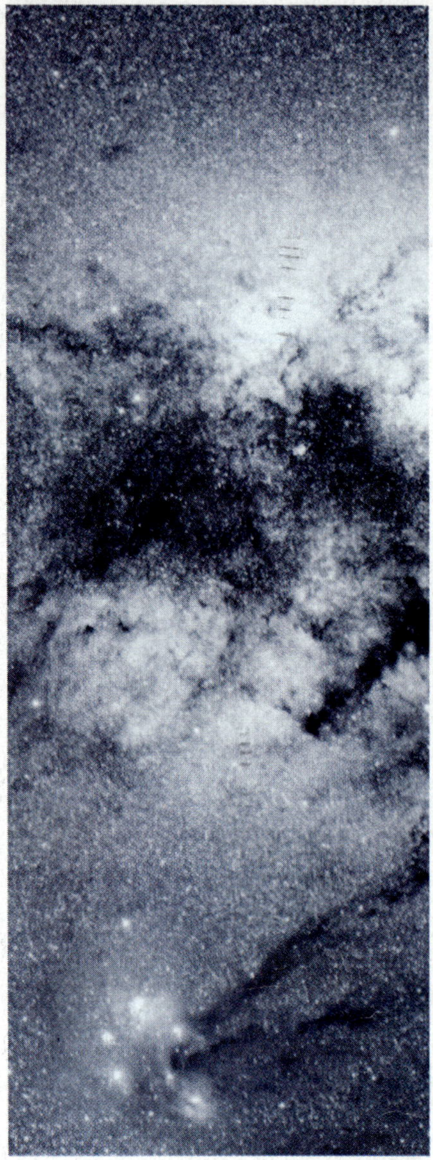

Key Terms

Hertzsprung-Russell (H-R) diagram
main sequence star
giant
supergiant
white dwarf
astrogram

Biography

Maria Mitchell (1818-1889)

Maria Mitchell became interested in astronomy while observing an eclipse with her father. She was 12 years old at the time. Maria Mitchell and her father lived on Nantucket Island, Massachusetts, where people were in the habit of watching the heavens. Everyone there was aware of the changes of the moon and stars.

Later, Mitchell taught school by day, and at night she and her father watched the stars. On October 1, 1847, her father wrote in his notebook, "This evening at half past ten Maria discovered a telescope comet five degrees above Polaris." This was her initial contribution to astronomy. For this discovery she was awarded a gold medal by Frederic VI, King of Denmark. This was the first such award to an American and the first to a woman anywhere.

Mitchell was also the first woman elected to the American Academy for the Advancement of Science. Later she accepted a teaching position at Vassar, a new college for women. This was a time when some people were saying, "Open the doors of your colleges to women and you will accomplish the ruin of the commonwealth."

Mitchell became a superb teacher. She tried to give her students the habit of mind or way of thinking that would serve in whatever they did.

TO THINK ABOUT AND DO

Vocabulary

Define the following terms.

main sequence star
Hertzsprung-Russell diagram
astrogram
supergiant
white dwarf

What Do You Remember?

1. What is the Milky Way Galaxy?
2. Describe the shape of the Milky Way Galaxy.
3. What kind of telescope would you use to study the galactic center? Why?
4. Do we see stars in the same patterns as ancient people? Explain.
5. What is a star?
6. How many types of stars are found on the H-R diagram? List the types of stars.
7. Is the earth the only location known to have life? Explain your answer.
8. What is an astrogram?
9. How does the atmosphere hinder viewing with a telescope?
10. What is the source of power in some space observatories?
11. How does a space observatory communicate with the earth?
12. Why is the exploration for other life in space limited to use of radio telescopes and radio signals?

Applying What You Have Learned

1. Explain why it takes eight minutes for light from the sun to reach the earth.
2. Arrange the following classes of stars in order of increasing temperature. Classes K, B, F, M.
3. Why are observations outside the visible light spectrum important to astronomers?

Table

star	brightness	temperature	type
1	100	10,000	
2	1,000	4,500	
3	1/10	4,500	
4	1/10	15,000	
5	1/100	3,000	
6	100,000	5,000	

4. Use the Hertzsprung-Russell diagram to complete the table above.
5. What would you infer if you had received the following message from space over and over?

```
. . . - . - . . . . - . - . . . . . . - . . . - . - . . . . - . - . . . . . .
  3    1    4      1     6       3    1    4     1     6
```

Research and Investigation

1. Check both the school and public libraries for NASA publications on space observatories.
2. Write a paper on one way you can find the distance to a star.
3. Make a collection of cartoons dealing with astronomy.
4. Write a play about a group of astronomers who have just heard a message from outer space.

4

The Solar System

You are living in an age when science fiction seems to come true. Since 1958 more than 50 space vehicles have flown near to or landed on the moon. Twelve American astronauts in the Apollo program landed on the moon. What discoveries did they make? They found the moon's outer layer had been completely shattered by space objects striking its surface. Many of the elements found on the moon are the same as those found on Earth but in different proportions. Rock samples from the moon have been dated to be 3.9 billion years old. The astronauts also found that the moon's basins, lunar seas, are covered with volcanic rocks and that there are moonquakes.

Even more surprising information has been supplied by space satellites about some of the planets. Who would have believed the surface of Venus could be so hot? No one expected to see ancient riverbeds and great canyons on Mars. And what about those hundreds, perhaps thousands, of rings around Saturn?

Scientists have only begun to learn some of the secrets of the solar system. With each new discovery from spacecraft, new telescopes, and other new technologies, scientists further improve their model of the solar system.

Figure 4.1 *opposite:* An artist's idea of what *Voyager 1* may look like near Saturn

4.1 The Sun

You will find out
- what the parts of the sun are;
- how the sun produces energy;
- why the planets travel around the sun.

Do You Know?
One solar flare may release as much energy as the whole world used in 100,000 years.

Imagine that you could hover just above the visible surface of the sun. As you look down, you see a boiling surface of bright regions surrounded by dark edges. As you leave this hovering place, an enormous mass of flaming gas arches out and then disappears back into the sun. These blasts of flaming gases are called **prominences** [PRAHM-ih-nehn-ses]. As you move farther from the sun you notice dark spots in the brightness. These are called **sunspots.** Sunspots are regions on the sun that are cooler than surrounding areas.

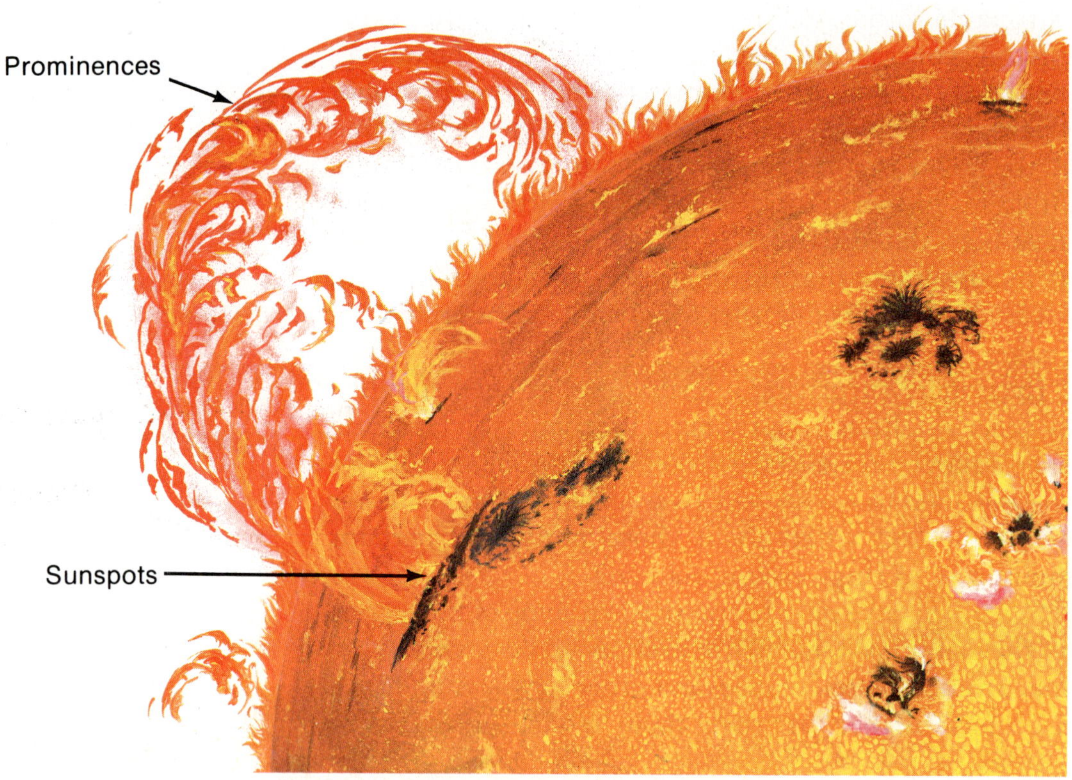

Figure 4.2 The exterior appearance of the sun

Compared to the size of the earth the sun is huge. More than one million earths could fit inside the sun. The sun is divided into three major regions. The first of these regions is the visible surface of the sun. It is called the **photosphere** [FOH-tuh-sfeer]. It has a granularlike texture which is caused by the movement of gases on the surface. The convection movement of the gases is the way energy is carried from the hot interior to the photosphere. The second region is the **chromosphere** [KROH-muh-sfeer]. It extends 2,000 to 3,000 kms above the photosphere. In its upper regions the chromosphere breaks into spikes of flame that rise and fall. The third region lies beyond the chromosphere. It is called the **corona** [Kuh-ROH-nuh]. This outermost part of the solar atmosphere extends millions of kilometers above the chromosphere in long streamers and thin plumes. It takes a tremendous amount of energy to produce all the flames and activity seen on the sun.

Do You Know?
Regions near the equator of the sun rotate in 25 days while regions near the poles take as long as 33 days.

Figure 4.3 The interior appearance of the sun

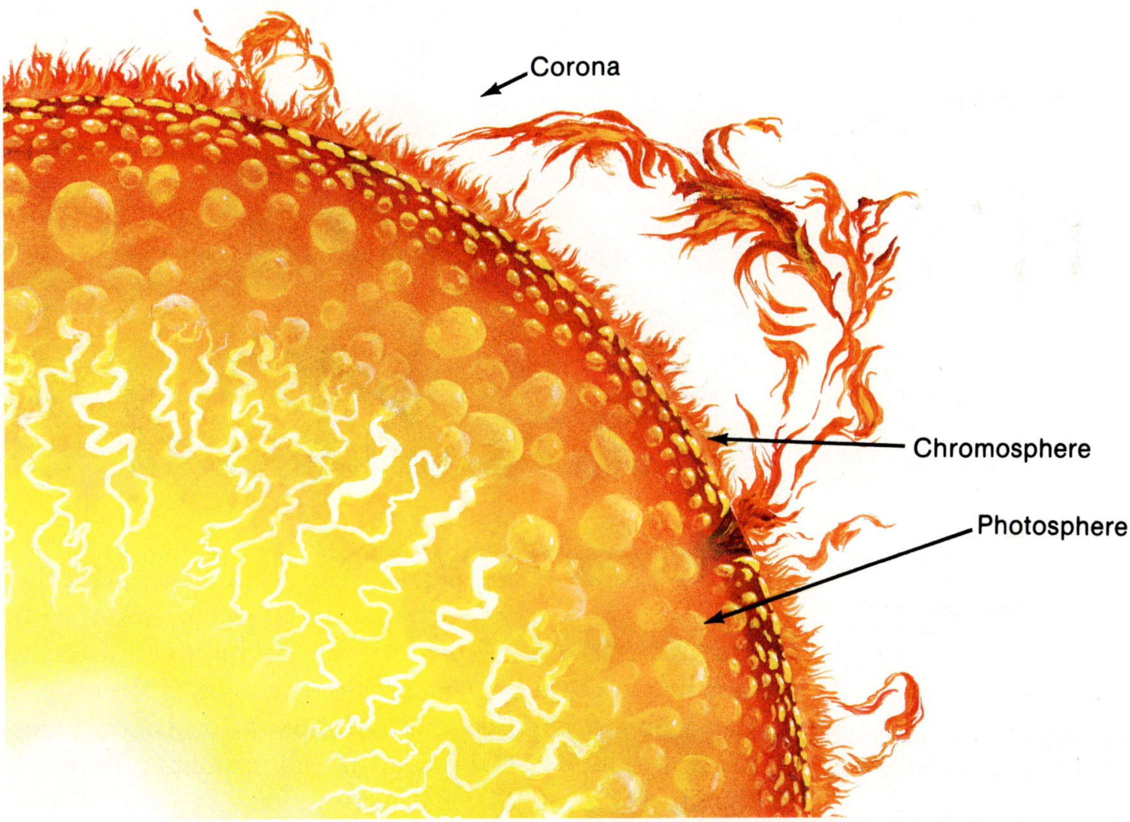

Figure 4.4 Hydrogen particles collide and fuse, producing helium particles and enormous amounts of energy.

Where does the sun's enormous energy come from? The sun produces energy by converting hydrogen into helium. Every second, the sun converts 600 million metric tons of hydrogen into 596 million tons of helium. Four million metric tons of matter are converted into energy. This is a lot of hydrogen to use up in one second. Will the sun burn out soon? Not really, since it is expected to last at least another five billion years. Fortunately the sun has lots of hydrogen.

This enormous source of energy is very important to you as nearly all energy on earth comes ultimately from the sun. This includes coal, oil, gas, and even the winds and water power. Plants as well as animals use solar energy. Changes in solar activity may affect weather and climate and even radio communication.

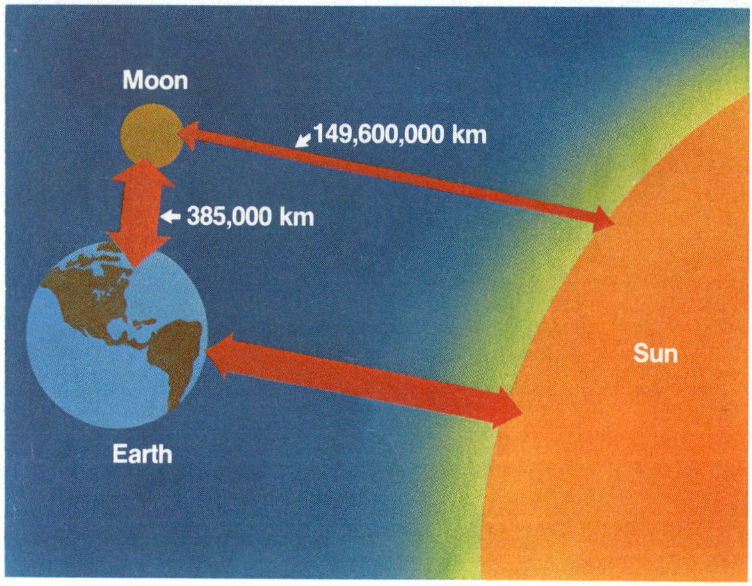

Figure 4.5 Distance also affects gravitational force. Earth, closer to the moon, exerts greater pull on the moon than the distant, but more massive, Sun.

In addition to its energy the sun has another powerful effect. As you learned earlier, every object in the universe is attracted to every other object. This force of attraction is called **gravitation.** The greater the mass of the object, the stronger the force of gravitation. With its enormous mass the sun exerts a powerful gravitational pull on all the objects around it. If it weren't for this gravitational

Do You Know?
The sun contains 99.6 percent of the solar system's mass.

attraction, each of the planets would go zipping away into space rather than staying in orbit around the sun. There are nine planets with their related moons, thousands of chunks of material called asteroids, and the interesting objects called comets all orbiting around the sun. All this is called the **solar system.**

Figure 4.6 The order of the planets as they orbit the sun is shown below. Normally they do not align so perfectly, and some are spaced farther apart than can be shown. In 1979, Pluto crossed inside Neptune's orbit and will travel there until 1999.

Key: 1. Mercury	4. Mars	7. Uranus
2. Venus	5. Jupiter	8. Neptune
3. Earth	6. Saturn	9. Pluto

Study Questions for 4.1

1. Describe how the sun produces its energy.
2. Describe the three regions of the sun.
3. Explain why the planets orbit around the sun.

4.2 The Earth and the Moon

You will find out

- why eclipses occur;
- what the phases of the moon are;
- how the sun and the moon affect the earth;
- why the moon's surface looks the way it does.

Figure 4.7 *above:* Earth is about six times larger than its moon.

Figure 4.8 *right:* Because of the size of the moon, it casts a shadow only on a small part of the earth's surface during a solar eclipse. Ancient people feared eclipses because they believed the sun was being eaten by evil creatures. When the earth comes between the sun and the moon, causing a lunar eclipse, the entire moon is covered. Have you ever seen a solar or a lunar eclipse?

Like the earth traveling around the sun, the moon travels around the earth. Of course that means the moon also travels around the sun as it circles the earth. This makes the moon's path quite different from the earth's path as you can see in Figure 4.8. At times the moon is between the earth and the sun. At other times it is the earth that lies between the moon and the sun. At times the moon passes directly between the sun and the earth blocking the sun's light. When this happens, there is a **solar eclipse** [ee-KLIPS]. Sometimes the earth is between the sun and the moon, and there is a **lunar eclipse.**

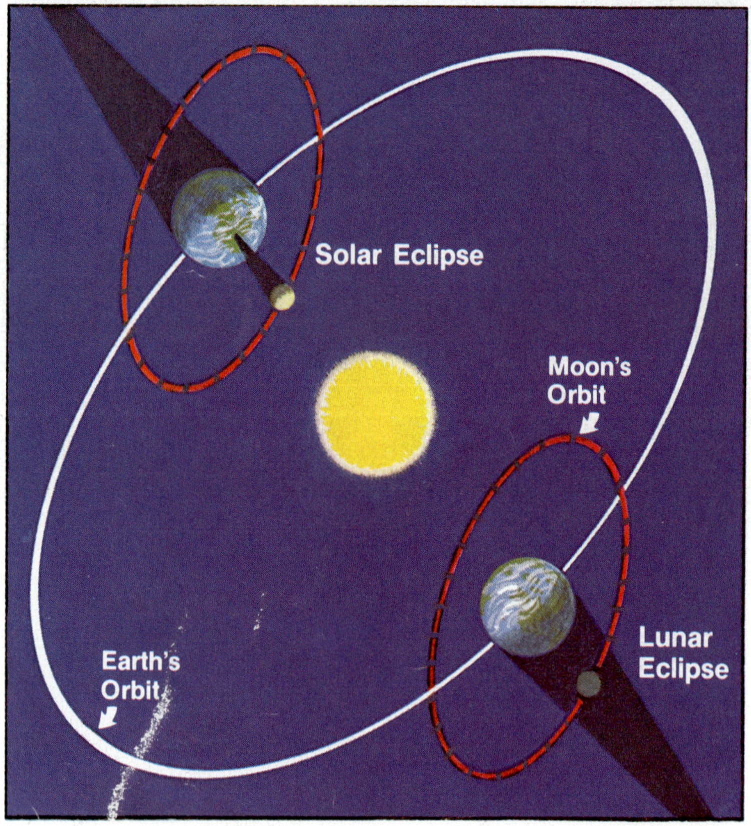

Do You Know?
Laser measurement of the moon's distance from the earth shows that the moon is slipping away from the earth at about four centimeters a year.

You might think that there would be an eclipse of the sun and moon every time the moon passes between the earth and the sun. This doesn't happen because the moon, the earth, and the sun don't always line up in a straight line. Most often the moon's shadow is above or below the earth, or the earth's shadow is above or below the moon.

The orbiting of the moon around the earth causes something else to happen. The moon doesn't always look the same. The changes in the appearance of the moon during a month are called the **phases of the moon.** The phases are the shapes of the unlit areas as seen from the earth.

How does the moon's orbit cause the phases of the moon? Look at Figure 4.10. When the moon is in the direction of the sun, the side toward you is dark. This is called a **new moon.** When the side of the moon toward you is in sunlight, you see a **full moon.** When you see half of the moon's lighted side, it is a called a **quarter phase.** The moon goes around the earth every 29.5 days, and these phases occur over and over.

Figure 4.9 During a total eclipse of the sun, it is possible to view the sun's inner corona for a brief period.

Figure 4.10 Every 29.5 days the moon goes through these phases.

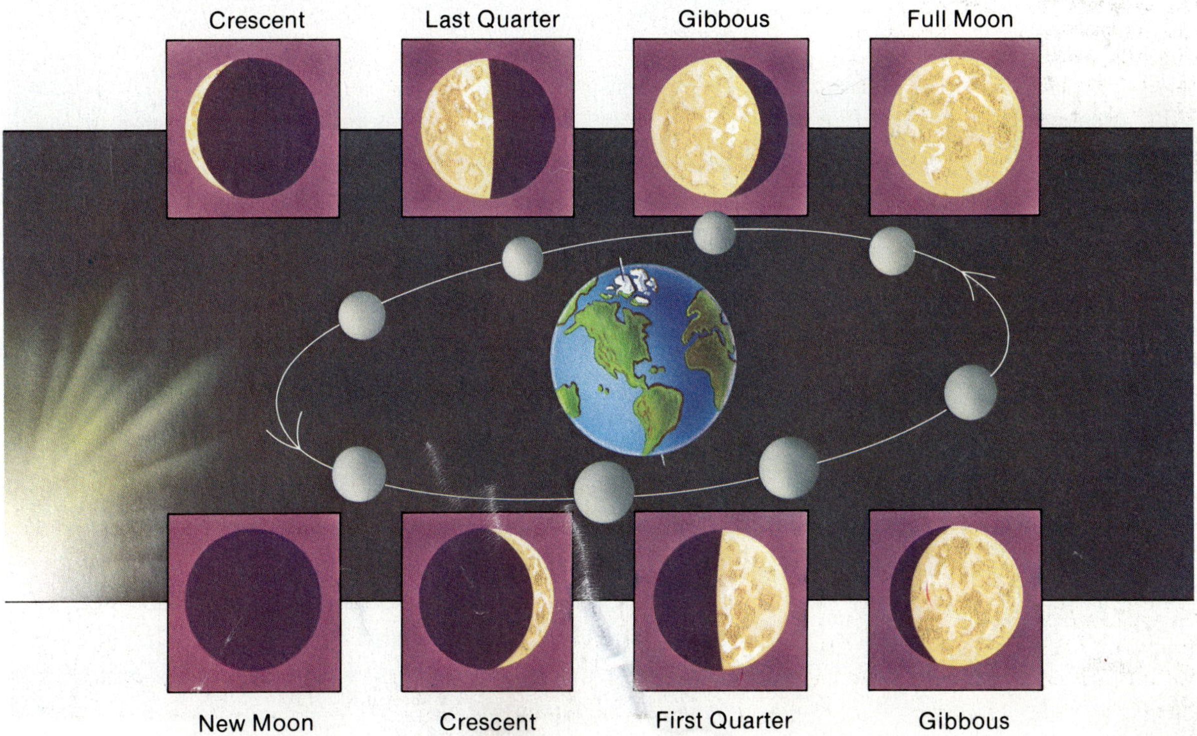

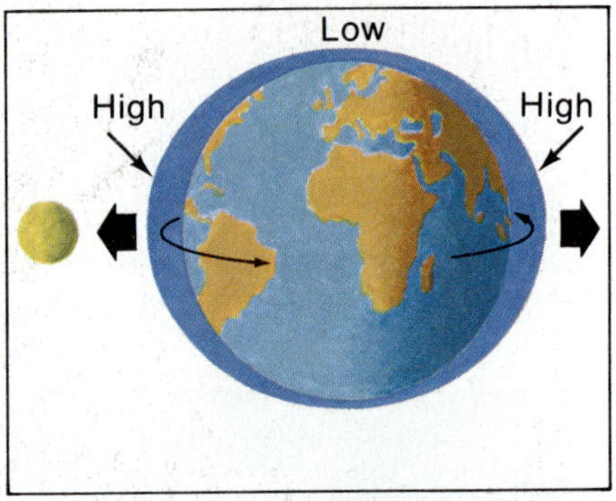

Time Lapse: 6 hours, 12 minutes

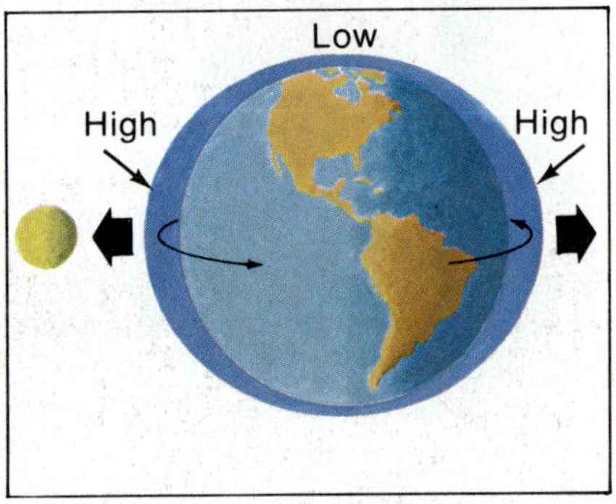

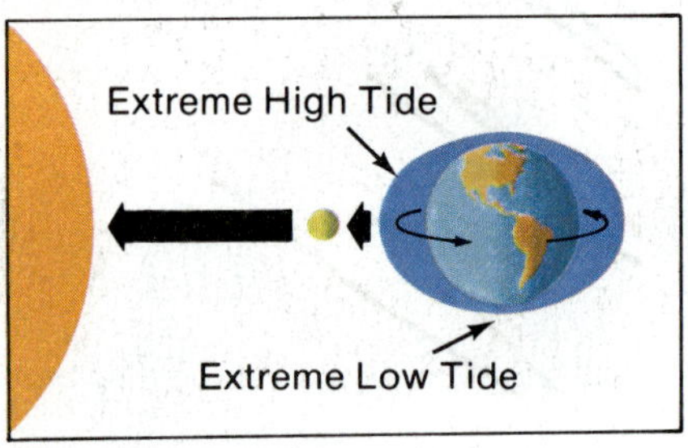

Figure 4.11 *top:* High tide on the eastern coast of North America; *middle:* As the earth rotates, North America is at low tide. *bottom:* The sun's influence can increase tidal forces.

Figure 4.12 Compare high and low tides in the pictures of the Bay of Fundy in New Brunswick, Canada.

Earlier you learned of ways the sun affected the earth. Does the moon affect the earth? Does the earth affect the moon? The answer to both questions is yes. The gravitational pull on each other changes the shape of the moon and the earth ever so little. This pull also causes the ocean water to actually move toward the area below and opposite the moon. As a result the level of the water rises and falls. This is called a **tide.** In Figure 4.11 you can see that the sun sometimes helps the moon to make larger tides than normal. At other times the sun's gravitational pull tends to offset the moon's effect.

With the ability to travel into space, more has been learned about the moon during your lifetime than in all of time before. On July 20, 1969, Neil Armstrong became the first human to step onto another body in space. In all there were six *Apollo* missions carrying people to the moon.

A great variety of experiments were planned for the *Apollo* missions. They tested the lunar soil and took photographs of it. Experiments were set up to measure particles from solar wind and with a seismometer to measure moonquakes. Rocks and soil were collected from each landing site and returned to the earth for detailed analysis. Perhaps you have seen exhibits of some of the lunar rocks.

From the careful analysis of the lunar rocks it appears that the composition of the earth and the moon is more alike than had been thought. One fairly definite finding is that none of the lunar rocks contain any trace of water.

It has been possible from the rock samples to make some estimates about the age of the moon. It is believed that the moon was formed over four billion years ago. In its first billion years the moon was quite active with many volcanoes. Lava flows from the volcanoes filled large basins which, from a distance, look flat and dark like seas. These areas are called **marias** [MAHR-ee-ahs]. But most of the craters on the moon are the result of other causes. The *Apollo* mission discoveries indicate that the craters were the result of meteors falling on the moon. For the last three billion years the moon has been geologically quiet except for some moonquakes. Instruments left by the *Apollo* missions continue to monitor the several thousand moonquakes that occur each year.

The information brought back from the moon has raised many new questions about the moon and the solar system. That always seems to happen in science. One answered question leads to still other unanswered questions. There is no doubt the *Apollo* missions were outstanding achievements.

Figure 4.13 These craters on the moon's surface are believed to be the result of meteors falling on the moon.

Figure 4.14 During the *Apollo 17* mission Astronauts Schmitt and Cernan used a lunar roving vehicle to extend their exploration of the lunar surface.

Study Questions for 4.2

1. Draw the conditions of a solar eclipse.
2. Describe the phases of the moon.
3. At what phase(s) of the moon do we have the greatest tides? Explain why.
4. Name two things that changed the surface of the moon.

4.3 Planet Movement and Features

You will find out
- how Kepler's laws relate to planetary movement;
- some features of the planets;
- how the solar system may have been formed.

Do You Know?
The planets have been assigned symbols. The symbols are the following:

Symbol	Planet
☿	Mercury
♀	Venus
⊕	Earth
♂	Mars
♃	Jupiter
♄	Saturn
♅	Uranus
♆	Neptune
♇	Pluto

The movement of the planets in the sky has been studied since ancient times. Many theories were proposed about their movement. In time astronomers developed better tools and made more accurate observations. In the 1600's Johannes Kepler explained the movement of the planets around the sun by using three laws. These are the following:

- Each planet moves around the sun in a closed path shaped like an **ellipse** [ee-LIPS]—an oval shape.
- A line joining the planet and the sun would sweep out equal areas in space in equal time. Each of the shaded portions in Figure 4.15 represents swept out areas made over equal time. The areas are equal.
- The farther the planet is from the sun, the longer it takes the planet to go around the sun. The closer the planet, the shorter is its time to go around the sun. The length of time to go around the sun is called the **period of revolution.** The earth's period of revolution is 365 1/4 days, which is also called a year.

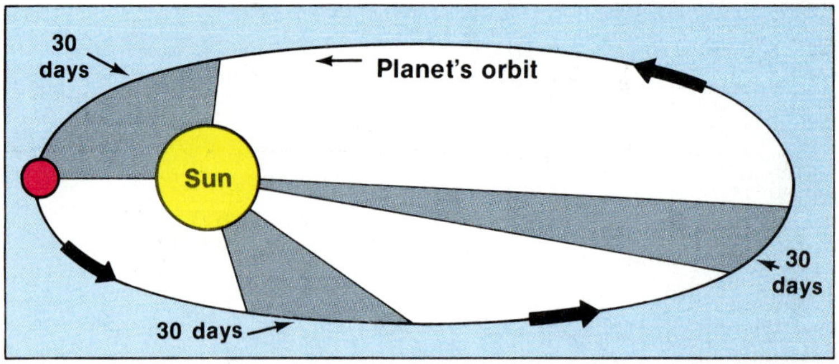

Figure 4.15 Kepler's second law of planetary motion

There is a great variation in the distance between objects in space. Do the planets vary as much in diameter? Figure 4.16 shows a comparison of the relative sizes of the planets. That important planet Earth is one of the smaller planets.

Figure 4.16 Relative size of the planets

ACTIVITY

How Far Apart Are the Planets?

Materials

ruler or meterstick
pencil
paper strip at least
4 m long

Procedure

1. Near the end of the paper strip place a dot for the sun.
2. Plot the planets on your paper strip. The table gives the distances. One astronomical unit (AU) is the distance from the earth to the sun. Use 0.1 m as equal to 1 AU.

Table

Planets	Distance in AU*
Mercury	.39
Venus	.72
Earth	1.0
Mars	1.5
Jupiter	5.2
Saturn	9.2
Uranus	19.2
Neptune	30.0
Pluto	39.4

*One astronomical unit (AU) is the distance from the earth to the sun.

Questions

1. On this scale how many centimeters is it from the sun to Mars?
2. On this scale how far is it from Earth to Jupiter? To Saturn?
3. It takes eight minutes for light to travel from the sun to Earth. How long does it take for light to go from the sun to Uranus?

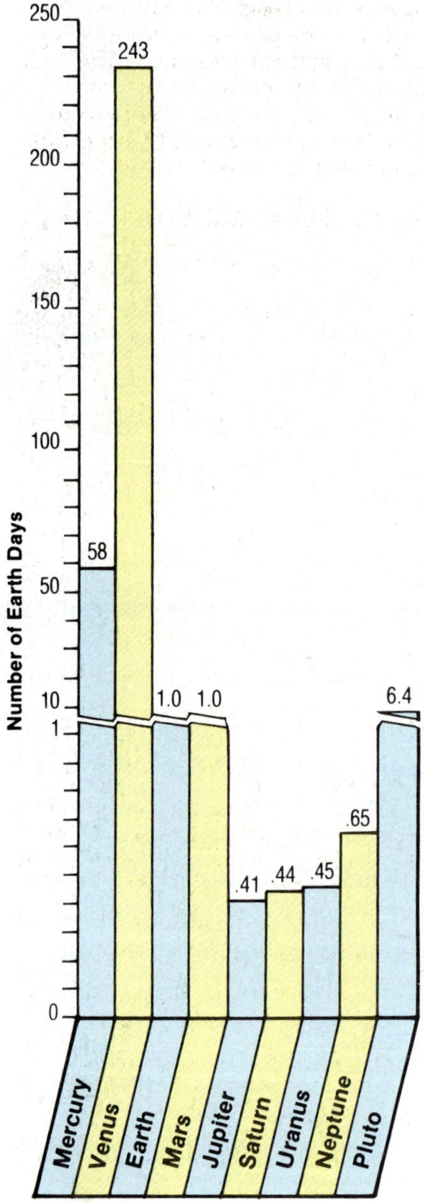

Figure 4.17 A histogram comparing rotation rates of the planets. Can you identify the slowest planet?

Table 4-1				
Planet	**Mass**	**Volume**	**Revolution Period**	**Discovered Satellites**
Mercury	0.05	0.06	0.24	0
Venus	0.81	0.88	0.62	0
Earth*	1.0	1.0	1.0	1
Mars	0.1	0.15	1.9	2
Jupiter	318	1,318	11.9	15
Saturn	95	769	29.5	20
Uranus	1.5	50	94	5
Neptune	17	42	165	2
Pluto	0.002	0.38	243	0

*All values are compared to the earth.

One of the most surprising comparisons of the planets is their different rates of rotation. Generally the largest planets rotate the fastest. A graph of the rotation rates is found in Figure 4.17. Notice the length of the day on Mercury. One "day" on Mercury lasts about 58 earth-days. A "day" on Jupiter lasts less than 10 earth hours.

There are other comparisons that can be made between the planets. Table 4-1 shows the major physical characteristics of each planet compared to Earth. As you look at Table 4-1, put your finger on one of the physical characteristics, for example, mass. Move your finger vertically from planet to planet. Notice which planets are more massive than Earth. Which are less massive?

How did the solar system begin? No one really knows. There are many models regarding its origin, and the popularity of different models varies from time to time. What are some of the things any theory must be able to explain? All of the planets revolve around the sun in nearly the same plane. The planets move in nearly circular orbits about the sun. All of the planets rotate and revolve in the same direction except Uranus and Venus.

Descartes in 1644 proposed that the planets formed from whirlpool-like motions in a rotating cloud of gas. Emmanuel Kant in 1755 visualized a rotating cloud in the beginning that condensed into the sun and planets. Later, Laplace suggested that the rotating cloud threw off rings that condensed to become planets.

The most widely accepted explanation of the solar system's origin is the **protoplanet** [PROH-toh-plan-iht]-**nebular model.** A protoplanet is the earliest form of a planet.

Stage A—At the most distant time you can conceive, there was a huge, slowly rotating mass of gas and dust. This blob of gas and dust, called the solar nebula, extended far past the current orbit of Pluto.

Stage B—With time the solar nebula condensed, rotated more rapidly, and flattened. The nebula had a concentrated mass of material near its center that was to become the sun. Visualize a relatively cool, huge, rotating, flat disk of gas containing whirlpool-like eddies.

Stage C—The gases continue to condense, and larger solid particles form in the eddies. Solid particles in the whirlpool-like eddies collide with each other to grow larger, making protoplanets. Gravity pulls pieces of matter together. The protoplanets in this stage are made of gas and solid particles. Particles rushing together as the sun becomes denser start the sun to warm. The sun begins to produce energy.

Stage D—As the sun grows hotter, the nearby planets lose their atmosphere. The outer planets retain some of the original gases. The sun becomes a G type star, and the solar system stabilizes. Later, Earth, Venus, and Mars develop an atmosphere by holding gases that come from their interiors.

At the moment, the protoplanet-nebular model accounts for many observations better than any other. Future work perhaps will give an even better explanation.

Figure 4.18 An artist's picture of the origins of the planets. From some gas and dust that collected and began to rotate, whirlpool-like eddies, or protoplanets, formed. These protoplanets became denser and formed the planets in orbit about the sun.

Study Questions for 4.3

1. What do Kepler's laws explain?
2. Select one planet and compare its mass, diameter, period of revolution, and time of rotation with Earth.
3. Select one model of how the earth formed. Diagram your model and write an explanation of it.

4.4 Exploring the Planets

You will find out

- what the major surface features of the planet Venus are;
- what the surface features of the planet Mars are;
- what has been learned about Jupiter and its moons;
- why Saturn is called the jewel of the sky.

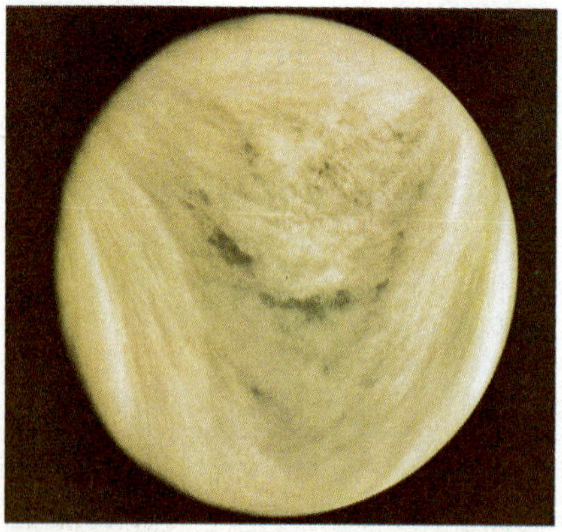

Figure 4.19 *top:* The surface of Venus is covered with dense clouds.

Figure 4.20 *lower:* An artist's drawing of canyons on Venus based on radar surveys of the surface

The brightest object in the sky besides the sun and moon is planet Venus. Venus is about the same size as the earth. Some people think of it as a sister planet, but the relationship is not very close. Imagine trying to live on a planet where the atmospheric pressure is 90 times Earth's atmospheric pressure. There is no water and very little sunlight, and its atmosphere is 96 percent carbon dioxide. The upper atmosphere contains droplets of sulfuric acid. The surface temperature is between 448°C and 459°C, and it's dry and dusty.

Using radar, the *Pioneer Venus Orbiter* has mapped more than 93 percent of Venus' surface. Sixty percent of the surface is a relatively flat, rolling plain. Rising above the plain are three continent-sized plateaus. At one end of the highest landmass, Ishtar Terra, is Maxwell Montes. This is the highest point on Venus—10.8 km above the plains. This is higher than Mount Everest.

Lowlands occupy about ten percent of the surface. The lowest point on Venus lies in a great valley. This is one of the largest valleys in the solar system. It is 1,400 km long, 280 km wide, and 5 km deep. There are many large old craters in the lowland plains. When a Russian spacecraft landed on Venus, it found a very rocky surface.

Figure 4.21 *right:* A photograph of the surface of Venus taken by the Russian *Venera 13* satellite

People have dreamed for centuries about Martians and the possibility of life on Mars. These dreams ended with the *Viking Spacecraft* that landed on Mars. It found no evidence of life.

Figure 4.22 A photograph of the surface of Mars taken by *Viking 1*

Mars has a thin atmosphere with a normal pressure about 1/100th of the earth's. Mars has dust storms, high level carbon dioxide clouds, and lower-level clouds that are believed to be ice crystals. The thin atmosphere is 95 percent carbon dioxide. The temperature at the equator ranges from −75°C to +30°C. Mars has two polar caps containing frozen carbon dioxide and water. These caps grow and shrink with the seasons.

Volcanic activity has taken place on a large scale. There are at least 12 volcanoes. The most spectacular is Olympus Mons. This mountain is 25 km high and 600 km wide. This is five times larger than the greatest volcano on Earth—Mauna Loa, in Hawaii— even including all that is below the surface of the ocean. Another major surface feature is a huge, narrow canyon 5,000 km long, 75 km wide (on the average), and 6 km deep. Since this canyon appears to have been shaped by running water, Mars must have had running water at one time. Ancient islands and dry streambeds are also found. Compared to Venus, Mars would be a better place to visit, but don't expect to run into any Martians.

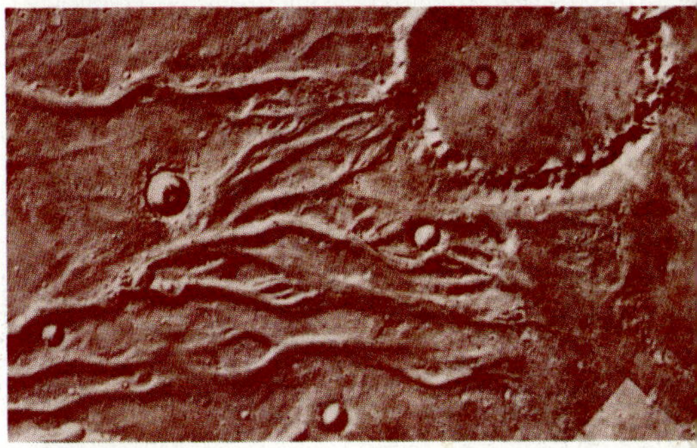

Figure 4.23 This *Viking 1* photograph of Mars' surface indicates there must have been streams and running water at one time on Mars.

Jupiter is the largest planet in the solar system. Two spacecraft, *Voyager I* and *II,* have taken spectacular photographs of this gigantic planet. No surface features are visible on this planet because of a thick cloud cover. Jupiter's atmosphere is primarily hydrogen and helium which show complex patterns of movement. High-speed winds of 120 km per second have been detected. This is more than three times hurricane wind speeds here on Earth.

One of the most prominent features, the Great Red Spot, is a tremendous atmospheric storm. Wind speed at its top, toward the equator, is in one direction. At its bottom, toward the polar area, winds are reversed. Above and below the Great Red Spot the clouds show as alternate light and dark bands.

Jupiter has at least 15 moons. The outer ones revolve around Jupiter backward with respect to most motions in the solar system. The size of the moons varies from Almathea, the size of California, to Ganymede, about as large as Mercury. Jupiter's moons are some of the most interesting objects in the solar system. Seven volcanoes were observed on the moon Io and more have been predicted. A surprising discovery was Jupiter's ring.

Figure 4.24 *top:* Close-ups of Jupiter show bands of colored clouds whirling across the surface. *bottom:* The Great Red Spot on Jupiter is a storm large enough to swallow several earths.

The jewel of the sky is Saturn. Showing up brilliantly through even the lowest-power telescopes, Saturn with its rings is something to see. Now imagine seeing it even closer as from *Voyager I* and *II.* From every view the dominant visual objects are the rings—not one but thousands of rings. The size of particles making up the rings ranges from dust flecks to boulders each in its own orbit around the planet.

Saturn has a banded, layered atmosphere that flows in the same direction around the planet. Wind velocities greater than 500 km per hour have been detected. This is four times the speed of Jupiter's winds. The planet rotates in 10 hours and 39.4 minutes. The atmosphere is about 90 percent hydrogen and helium and 10 percent ammonia, methane, and water vapor.

Do You Know?
The first recorded observation of one of Jupiter's moons (probably Ganymede) was in China in 364 B.C. almost 2,000 years prior to Galileo's discovery of Jupiter's moons.

Four big mysteries surround the exploration of Saturn. Why does Saturn have rings? How is it possible for the rings to remain there? Where did its more than 20 moons come from? What causes the dark radial spokes in one ring? As one astronomer said, "Saturn has become more mysterious than ever before."

Figure 4.25 *left:* The spectacular rings of Saturn have been color enhanced in this photograph taken by *Voyager 2. below:* Close-ups of Saturn's rings have raised many new questions. One mystery that puzzles scientists is the appearance of spokelike patterns in the rings.

Beyond Saturn are two other giant planets. These large planets are Uranus and Neptune and both are apparently cloud covered. *Voyager I* and *II* were programmed to continue on to Uranus after passing Saturn.

The usual outermost planet is Pluto. Pluto has an unusual orbit that at times takes it inside the orbit of Neptune. Since January 22, 1979, Pluto has been closer to the sun than Neptune, and it will remain that way until March, 1999. Pluto's orbit is also inclined with respect to the others. Another very strange feature of this outermost planet is its small size with a mass only 1/400th or so of the mass of the earth.

Do You Know?
It required 1.5 hours for a radio signal to reach Earth from *Voyager II* as it passed Saturn.

Study Questions for 4.4

1. What are some large surface features of Venus?
2. Describe Olympus Mons.
3. What causes the Great Red Spot?
4. Describe Saturn's rings.

What Did You Learn?

- The sun's surface contains bright regions surrounded by dark edges, sunspots, and prominences.
- The sun converts hydrogen into helium. In this conversion, energy is released.
- The sun is divided into three regions: photosphere, chromosphere, and corona.
- Eclipses are caused by either the moon or the earth cutting off sunlight.
- The moon changes appearance as it revolves around the earth.
- Tides are caused by the gravitational pull of the moon and the sun.
- Volcanoes and meteors caused the marias and craters on the moon's surface.
- Kepler's laws explain the movement of the planets.
- The most accepted explanation of the origin of the solar system is the protoplanet-nebula model.
- Planet Venus would not be a hospitable place to live.
- Planet Mars shows evidence of water erosion and volcanic activity.
- Planet Jupiter rotates rapidly and has a turbulent atmosphere of hydrogen and helium.
- Planet Saturn shows a ring structure far more complicated than we can explain.

Key Terms

prominence
sunspot
photosphere
chromosphere
corona
gravitation
solar system
ellipse
period of revolution
solar eclipse
lunar eclipse
phases of the moon
new moon
full moon
quarter phase
tide
maria
protoplanet-nebular model

Biography

Johannes Kepler (1571-1630)

Johannes Kepler, born to a poor family in 1571, had exceptional talents that were recognized by others at an early age. With the help of grants, Kepler was able to enroll in Tubingen University, Germany. He became a Protestant theologian. At that time the study of theology required extensive study in the liberal arts, which included mathematics and astronomy. His astronomical thesis concerned the moon and has to this day never been discovered.

Kepler taught mathematics for a time but continued his study of astronomy by trying to combine geometry and astronomy.

Kepler was a Protestant. When the Catholics came to power in his region, he was forced from his teaching position. Fortunately Tycho Brahe had a secure position and was able to invite Kepler to the observatory in Prague. Kepler's work began with an analysis of data on Mars. It was during this analysis, which involved hundreds of tedious calculations, that Kepler recognized that the planets do not move with constant velocity in orbit. The orbits of the planets were ellipses. This revolutionary idea lead him to further discovery, and he stated the first two of his three laws in 1609. Kepler's third law was published in 1618.

It is remarkable that this man was able to accomplish anything in astronomy. Wars, religious conflict, living with a family that didn't understand his work, seeing the death of loved ones, hounded by requests for unimportant data and charts of the heavens, and lack of acceptance by others were only a few of the obstacles he overcame to publish his theories of the solar system. His friend Pierre Gassendi in Paris wrote "Men like Kepler should never really die." That wish has been fulfilled.

TO THINK ABOUT AND DO

Vocabulary

Unscramble these five words:	What is it that's
cool, but on the sun	P N U S T O S
longer for some than others	D I E P O R
related to the moon	E S P A H S
Kepler's idea	E P L E S L I
related to darkening	P C E L S E I

What Do You Remember?

1. What are the three parts of the sun?
2. How does the sun produce energy?
3. List Kepler's three explanations of planetary motions.
4. Why do astronomers like a total eclipse of the sun?
5. Draw a sketch of a full moon, new moon, quarter moon.
6. What is a tide?
7. List the planets in order from the sun.
8. Graph one of the physical characteristics of the planets shown in Table 4.1.
9. Describe the atmosphere of Venus.
10. Describe the surface of Mars.
11. What is the Great Red Spot?
12. Why would astronomers think of Saturn as the jewel of the sky?

Applying What You Have Learned

1. The sun's surface shows dark spots on its surface. What is the probable cause of these dark spots?
2. Why don't we have eclipses of the sun once every revolution of the moon?
3. Do we first see a full moon at sunrise or at sunset? Explain your answer.
4. Discuss why you think a manned spaceship is better than an unmanned spaceship.

Research and Investigation

Compile a classroom collection of articles, news reports, photographs, etc., about each planet.

Report on the astronauts who went on each of the *Apollo* missions.

What new mystery of the solar system will be discovered next? Will you discover it?

Unit Two

The Earth's Gaseous Envelope

B·11

5 The Atmosphere and Its Movement

As you continue on your journey from outer space, you head for planet Earth. This planet is surrounded by an ocean of air. The surface of the earth lies at the bottom of the ocean of air.

Without air, life could not exist. Air is everywhere, always moving, and it is invisible. Though you cannot see the air, you know it exists because you can feel it. It fills the sails of a boat, and its force moves the boat across the water.

To understand this ocean of air, you must ask some questions. What is air made of? How deep is it? How does it move? And how does it affect you? Each answer to one of these questions is like finding a piece to a puzzle. The more you understand, the more pieces of the puzzle you can fit together.

In this chapter you will find the answers to some of these questions. You will also see how the air and the sun work together to make the weather.

There are still many pieces to the puzzle that are missing. Maybe one day you will find one of them.

Figure 5.1 Windsurfers ride the wind to ride the waves.

5.1 The Air Around You

You will find out

- what four gases make up the the air around you;
- how air exerts pressure;
- how pressure exerted by the air is measured.

Figure 5.2 Air turns a weathervane, indicating the direction of wind.

What is air? You know you can feel air if something moves. You feel it if you wave your arm back and forth. You feel it when you are in a moving car. Air moves through an open window of the car and moves your hair. Moving air can also turn the blades of a windmill. Since you can feel air and it can move things, air must be "stuff," or material. This is one piece of your puzzle.

The material called air is all around the earth. You live in an invisible sea of air. This sea of air is called the **atmosphere** [air sphere]. The atmosphere is a mixture of gases. The gases are invisible, have no color, and have no odor. Nitrogen (78 percent), and oxygen (20.9 percent) make up most of the mixture. Argon (0.9 percent), carbon dioxide (0.03 percent), and traces of many other gases complete the mixture. Near the earth's surface these gases are mixed with water vapor.

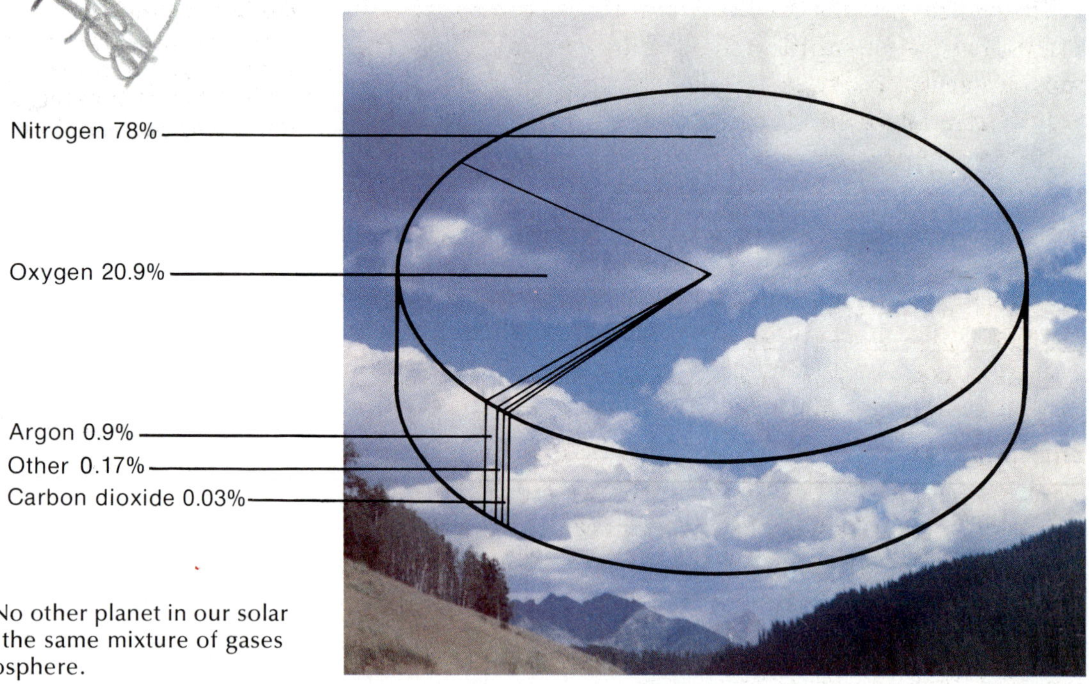

Figure 5.3 No other planet in our solar system has the same mixture of gases as our atmosphere.

Figure 5.4 Tiny gas molecules reveal their rapid motion by the fluttering of flags in the wind.

How can you feel something that is invisible? The atmospheric gases are thought to be made of tiny particles called **molecules** [MAHL-uh-kyools]. Molecules are so very tiny that you cannot see them. Yet, even though you cannot see them, the molecules are materials. It is these tiny particles of materials that you feel when you move your arm through the air.

The gas molecules that make up air move about freely. They also move rapidly. Would you believe that they may move as much as 800 m a second or more? There are billions of gas molecules moving rapidly about and banging into one another. This is taking place all around you this very moment. This movement is also another piece of your puzzle.

Do You Know?
If people were the size of molecules, it would take everyone in the United States standing in line to make a line 2.5 cm long.

There are two results of gas molecules moving rapidly about. One result is that the atmosphere is spread out. If the molecules were not moving, they would all be on the earth's surface. They would be like grains of sand on a beach. You would not be able to breathe! But the molecules are moving so they are found higher than just on the surface. They spread outward and upward. Because of the pull of gravity, however, more are found near the surface than higher up. The atmosphere is not a layer such as the skin on a peach. The atmosphere thins out as you go up from the surface. It gradually merges with the near-emptiness of space.

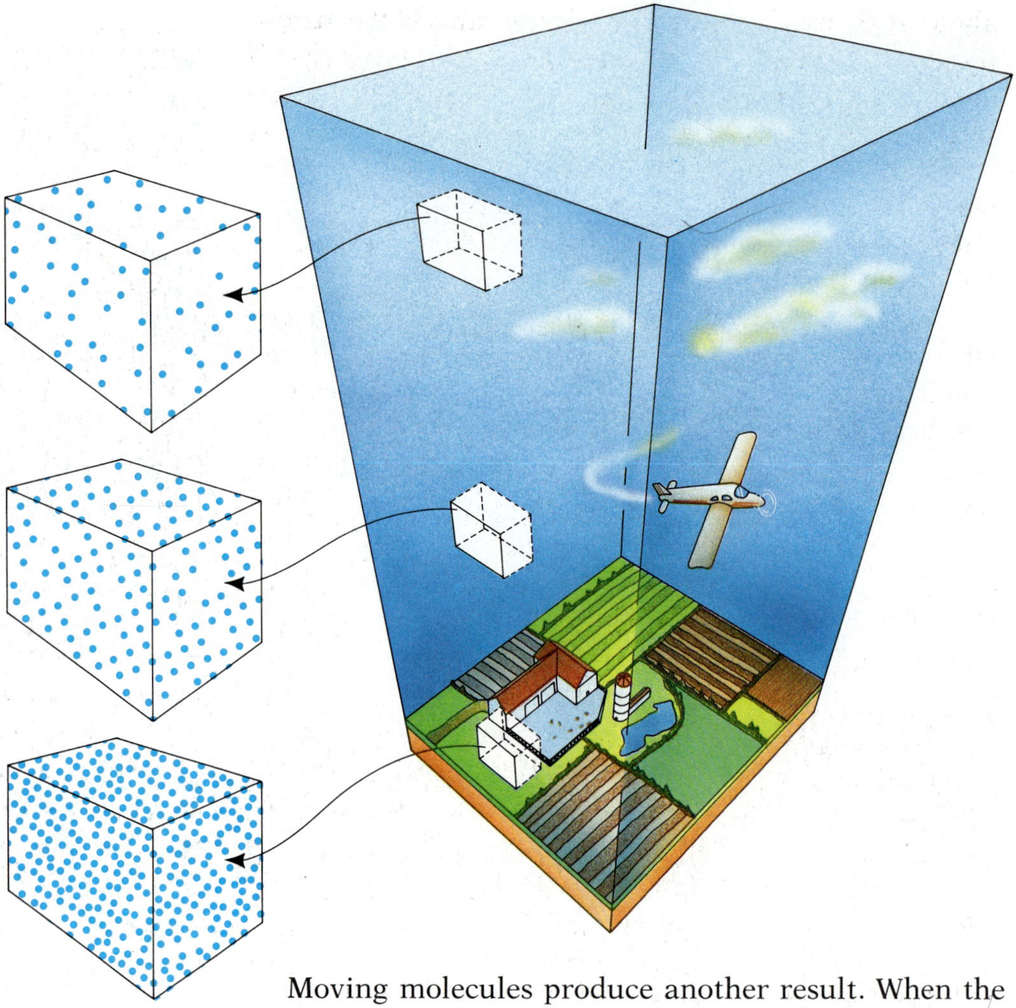

Figure 5.5 At greater heights, the same volume of air contains fewer gas molecules.

Moving molecules produce another result. When the molecules hit a surface they exert a pressure on it. **Atmospheric pressure** is produced by the molecules hitting a surface. Since more molecules are found near the surface, the pressure is greatest there. Higher up, there are fewer and fewer molecules so the pressure becomes less and less. The pressure decreases as you go away from the earth's surface. Atmospheric pressure is a big piece in your puzzle.

Atmospheric pressure on the surface can be measured. One way of measuring it is by using a 10.5-m length of tube that is closed at one end. You fill it with water and place the open end in a pan of water. When raised to an upright position, some of the water will come out of the tube, but not all of it. Atmospheric pressure will cause

Do You Know?
Have your ears ever popped when riding in an elevator or in a car in the mountains? The pop is caused by a tube behind your eardrum adjusting to changes in the atmospheric pressure.

about 10.33 m of water to stay in the tube. If the atmospheric pressure becomes less, however, more water will come out of the tube. Suppose the atmospheric pressure increases. What will happen to the water in the tube? It will be pushed higher. By measuring the height of the water, you can measure the atmospheric pressure. Such a device that measures atmospheric pressure is called a **barometer** [buh-RAHM-uh-tuhr].

Imagine carrying around a 10-m tall water barometer. Of course it would not be useful. So mercury, which is much heavier, is used in place of water. Mercury is 13.6 times as dense as water. The height of mercury in a mercury barometer will be only 1/13.6 as tall as the height of water in a water barometer, or about 76 cm. A weather report stating that, "the pressure is 76 cm and is rising," is talking about the height of mercury in a mercury barometer. Now several of the pieces of your puzzle are beginning to fit together.

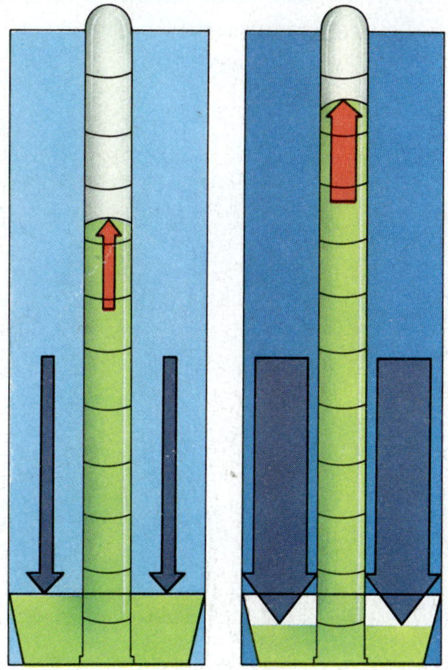

Figure 5.6 *above:* Arrows show the effect of atmospheric pressure on the barometer. Compare the effect of low atmospheric pressure (*left*) with that of high atmospheric pressure (*right*).

Figure 5.7 *left:* An aneroid barometer; *right:* A mercurial barometer.

Study Questions for 5.1

1. List four gases that make up our atmosphere and give their approximate percentage of the total.
2. What is the atmospheric pressure?
3. Explain how a mercury barometer measures atmospheric pressure.
4. Suppose you hear a weather report that the barometer is falling. What does this mean?

5.2 Warming the Atmosphere

You will find out

- what temperature is;
- how the atmosphere is warmed.

Have you ever been tricked by snow? Have you ever had your hands in snow for some time and then washed your hands in cold water? The cold water feels warm. Since the snow made your hands colder than the water, you thought that the cold water was warm. You need a thermometer, since your body can be tricked. A thermometer measures the **temperature** of a material. Temperature is a measure of the average energy of motion of molecules. The higher the temperature, the faster the molecules move about. Warm air has rapidly moving gas molecules, and cooler air has slower moving gas molecules.

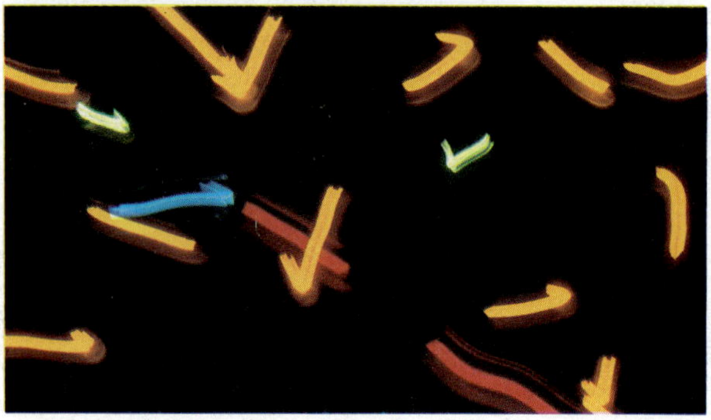

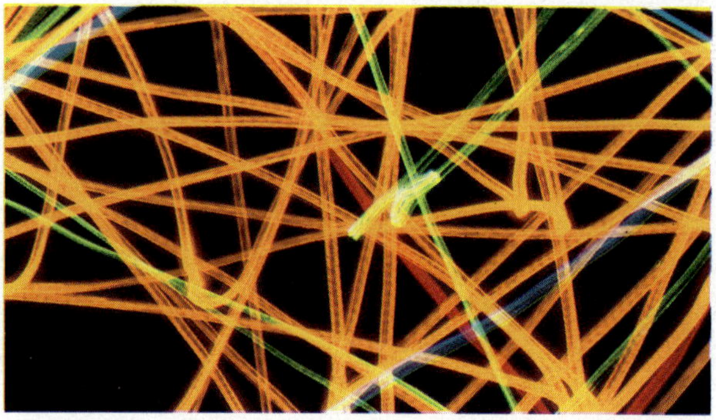

Figure 5.8 Your hands may not be an accurate way to measure temperature.

Figure 5.9 *left:* An illustration of the slow movement of cold gas molecules; *right:* Warm gas molecules move more quickly.

The atmosphere is warmed when the gas molecules gain energy, causing them to move about more rapidly. But where does the energy come from to make them move faster? It comes from two sources—the sun and the earth. Sunlight is energy that has traveled to you, from the sun to the earth. Such energy that travels through space is called **radiant energy.** When you are outside on a bright, hot summer day, you feel the sun's energy. The atmosphere absorbs some of the sun's energy, but much more is absorbed by the earth. The earth then radiates the energy back to the atmosphere. That is the kind of energy you feel coming from warm sand or hot pavement.

You learned in Chapter 2 (pages 30-31) that visible light, which you see is really only a small part of radiant energy. Infrared and ultraviolet are also forms of radiant energy, but you can't see these forms.

For example, consider the heating element of an electric range. When you turn the electric range on to a low setting, the element becomes warmer. You can feel radiant energy from the warm element if you hold your hand close to it. But you cannot see anything from the element, even if the room is dark. The element gives off infrared radiant energy. You cannot see infrared radiation, but you know it is there because you can feel it.

At higher temperatures not only do you feel radiant energy from the element but also you can see some of it. At higher temperatures the element begins to give off shorter wavelengths of radiant energy. These shorter wavelengths are the red light that you can see. Look at Figure 5.10. What color light would the range element give off if it became even hotter?

If the range element became as hot as the sun, you would see the short wavelengths as bright sunlight. Does the earth radiate long or short wavelengths of energy? From what you have learned, you should see that the earth gives off radiant energy at invisible longer wavelengths. The heat you feel from the warm sand at the beach or the hot pavement is infrared radiation.

Do You Know?
Animals can see parts of the radiant energy spectrum that humans cannot. Some snakes (sidewinders) can see infrared. Certain insects (bees, moths) can see ultraviolet.

Figure 5.10 *top:* No light is visible on a low heat setting. *bottom:* Radiant energy is visible at a high setting.

Figure 5.11 An infrared photo taken by satellite of San Francisco Bay; red colors represent the areas radiating the greatest heat.

Here is an atmospheric puzzle piece you may think does not fit. You have probably seen pictures of mountains with snow on the tops. Did you ever wonder why the snow is on top of the mountain and not the bottom? After all, isn't the top of the mountain closer to the sun? One reason for this is the fact that sunlight passing through the atmosphere does not warm it very much.

Figure 5.12 Mountaintops are cooler than the ground below them.

Figure 5.13 Sunlight, shown in yellow, is absorbed by soil and plants. They radiate the energy as infrared rays, shown in orange.

If this is so, how does the atmosphere get warmed? Think about it this way. The warming of the atmosphere is similar to what happens in a greenhouse. Sunlight shines through the glass into a greenhouse. This energy is absorbed by the soil and plants. The soil and plants radiate energy, infrared radiation. Infrared radiation cannot escape through the glass and is absorbed by water and carbon dioxide molecules in the air. The temperature of the air is thus increased by the radiation of longer wavelengths. This is called the **greenhouse effect.**

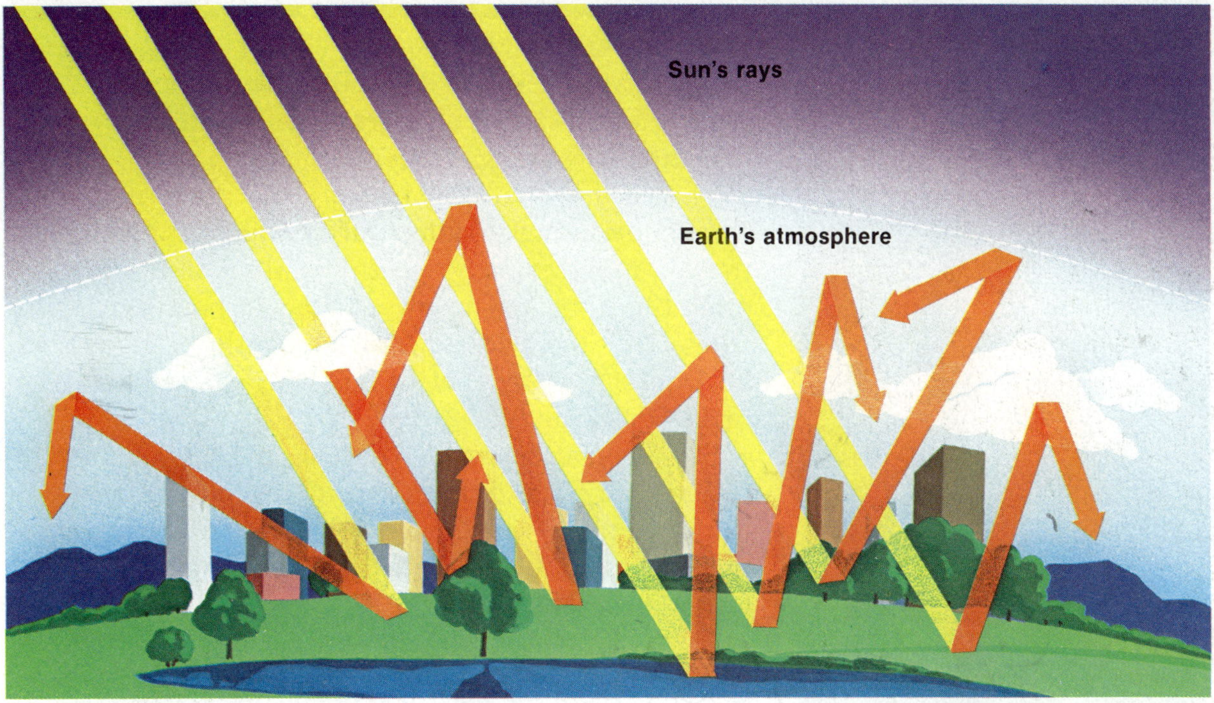

Figure 5.14 The greenhouse effect of the atmosphere; sunlight is absorbed by objects on the earth's surface and then released as infrared radiation.

The earth's atmosphere is warmed in much the same way as the inside of a greenhouse. The radiant energy from the sun that reaches the earth's surface is mostly absorbed. Buildings, plants, water, and the ground become warmer as the energy is absorbed. All these materials radiate this absorbed energy as infrared energy. On their outward journey the longer wavelengths are absorbed by water and carbon dioxide molecules in the atmosphere. The main heating source of the atmosphere is the radiation of longer wavelengths that are then absorbed by the gases in the lower parts of the atmosphere.

Do You Know?
If the year-round average radiant energy received by the earth's surface were not reradiated, it would become 1.5°C warmer each day. How much warmer would it be after one month?

Study Questions for 5.2

1. What happens to the air temperature as the gas molecules that make up air move about more rapidly?
2. What is radiant energy?
3. Explain the greenhouse effect and how it warms the atmosphere.

5.3 Moving Air

Figure 5.15 These socks will dry more quickly because warm air rises from the fire.

You will find out

- that warm air is less dense than cold air;
- how changes in air temperature make the air move;
- what makes the wind blow.

The air around you is almost always moving. Sometimes the air moves slowly, gently rustling tree leaves. At other times the movement is faster, perhaps moving the whole tree top back and forth. Gentle or strong, the horizontal movement of air is usually called **wind.** What makes the wind blow?

Have you ever stood barefoot in front of an open refrigerator? You could feel cold air pouring onto your feet. Where do you hold your hands when warming them at a fire or heater? Warm air can be felt rising above the heat source.

There is a relationship between temperature and the movement of air. Cold air moves downward. Warm air moves upward. This happens because air expands as it is warmed, becoming less dense. Cooler air is more dense than warmer air. The cooler air flows beneath the warmer air and forces it upward. Now you have another piece for your atmospheric puzzle.

ACTIVITY

How Does Temperature Affect Air Density?

Materials

two large jars
candle
refrigerator

Procedure

1. Place a large jar in the freezer for 15 minutes. Leave a second jar at room temperature.
2. Light a small candle on a table.
3. Pour the air from the cold jar onto the candle flame.
4. Repeat with the warm jar.

Questions

1. What do you see happening when you pour air from the jars?
2. How does cooling change air density? How do you know?

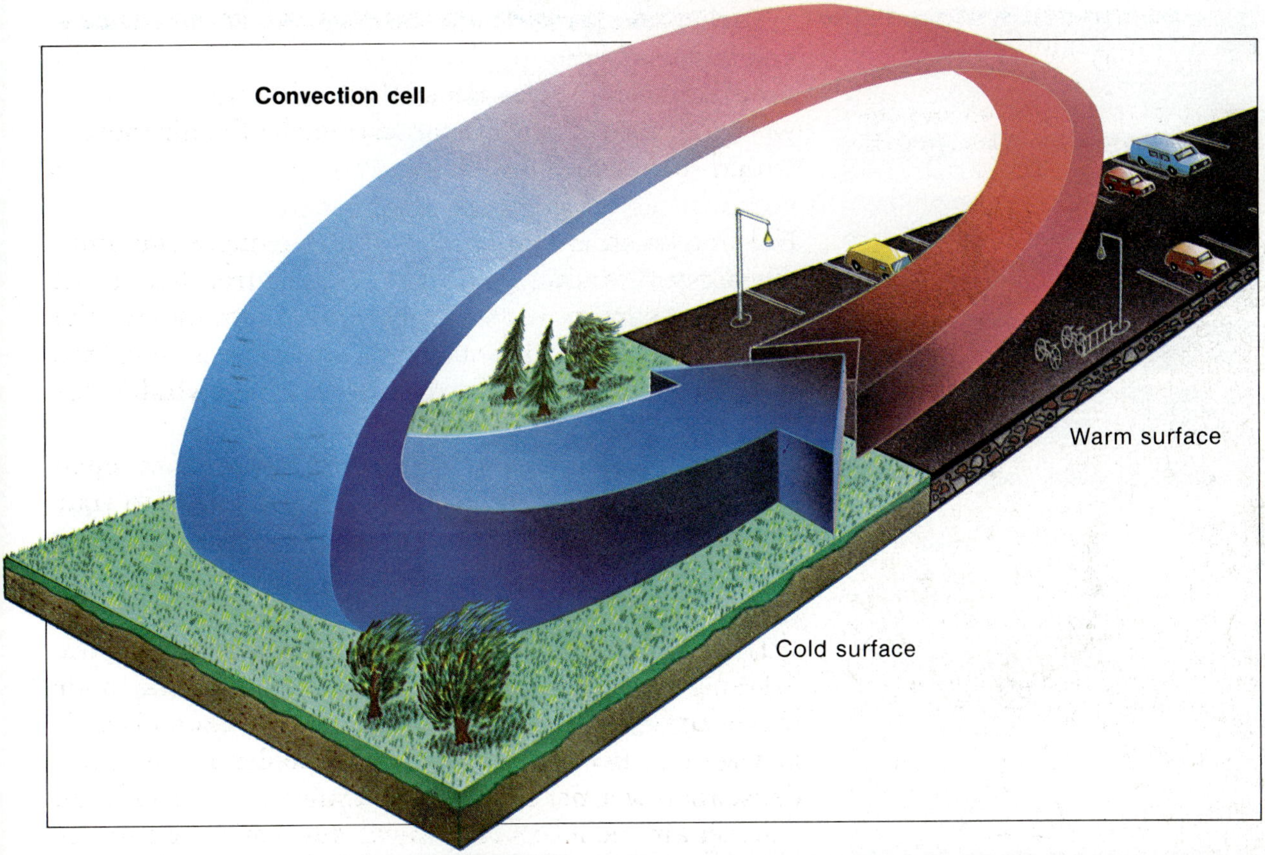

Earlier you noted that buildings, plants, and the ground become warmer when they absorb sunlight. However, not all materials are warmed the same by the sun. You can notice this difference if you walk barefoot across grass, concrete, asphalt, and dry sand on a hot summer day. These materials all receive the same amount of sunlight, but they are warmed to different temperatures. As a result, on a calm, sunlit day, wiggly currents of air can be observed rising over a hot asphalt street. Cooler, more dense air moves from a nearby cooler surface, such as grass. Figure 5.16 shows the movement of air from a patch of cool grass to a warmer asphalt street. The cooler, more dense air moves from the grass to the asphalt. It pushes the warm, less dense air upward. This process of warmer air being pushed upward is called **convection.** If the warmer air rises then moves back down onto the cooler surface, we say that a **convection cell** has been formed.

Figure 5.16 *above:* A visual model of a convection cell; blue represents cool, dense air; red represents warm, less dense air. Cool air pushes warm air upward.

Figure 5.17 *below:* Unusual images result when objects are viewed through convection currents rising from hot ground.

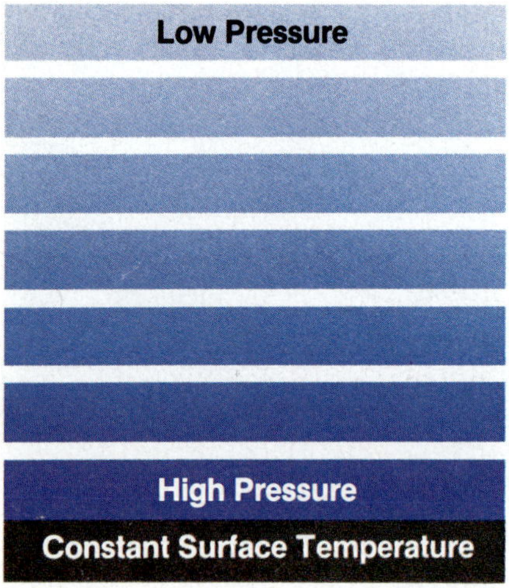

Figure 5.18 Pressure lines are straight when over a surface of constant temperature. The lines represent air pressure that is equal at equal altitudes. No air movement occurs.

Figure 5.19 When surfaces have different temperatures, air pressure varies. Equal pressure lines occur at different altitudes. Wind blows as air moves from areas of high to low pressure.

Earlier you learned that the air pressure decreases as you go higher into the atmosphere. You may expect the air pressure to be the same at the same altitude everywhere above the surface. If so, an air pressure model would look like Figure 5.18. The white lines you see represent equal pressure. Everywhere along the same line, the pressure is the same. Notice that in this model the pressure lines are straight. They will be straight if the entire surface is one temperature.

How does temperature change affect this air pressure model? Pressure is lower in warm air than in cold air. This is because air molecules move faster in warm air and are spaced farther apart than they are in cold air.

Look at Figure 5.19. Notice how the equal pressure lines are curved. The lines are closer together over the cold surface where the pressure is high. The lines are farther apart over the warm surface where the pressure is low. This arrangement of lines means that the air pressures are not equal at the same heights everywhere above the surface. Instead, equal air pressures occur at different heights when surface temperatures vary. Wind

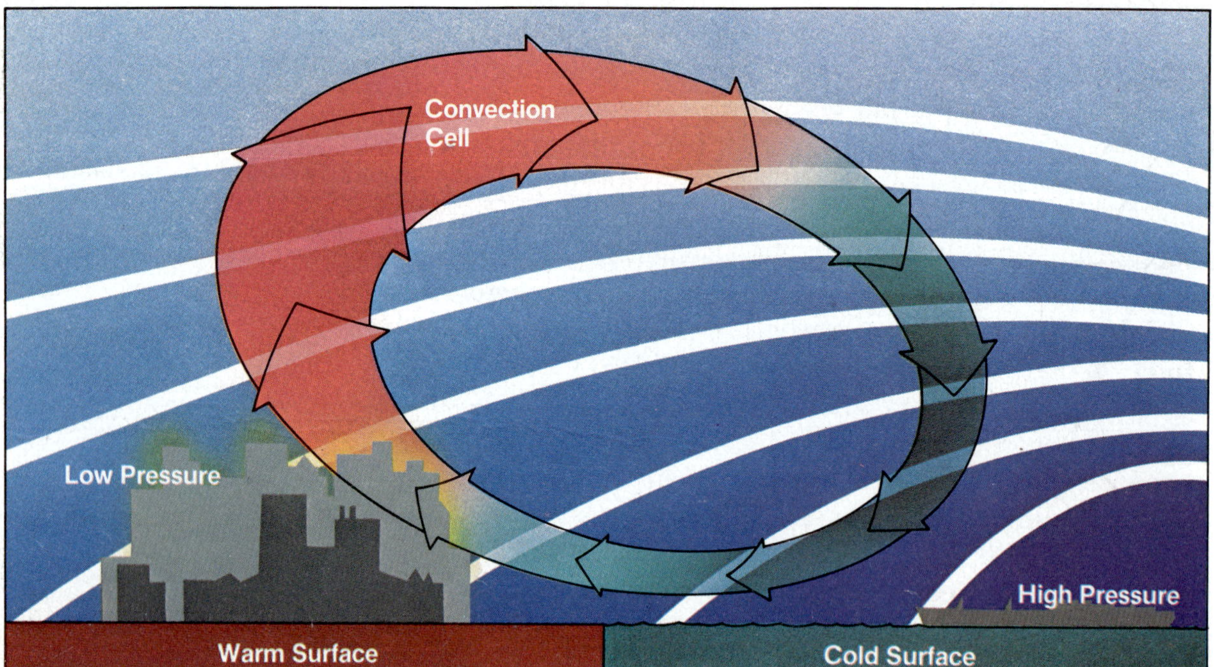

ACTIVITY

What Makes Air Move?

Materials

plastic food wrap
masking tape
cardboard box
knife
pan of ice cubes
black paper
smoking punk

Procedure

1. Make a smoke box as shown.

2. Position a pan of ice, a smoking punk, and a sheet of black paper in direct sunlight. Place the box over them so the sunlight shines directly on the paper and ice.

3. Observe how the air moves in the box for 10 to 15 minutes.

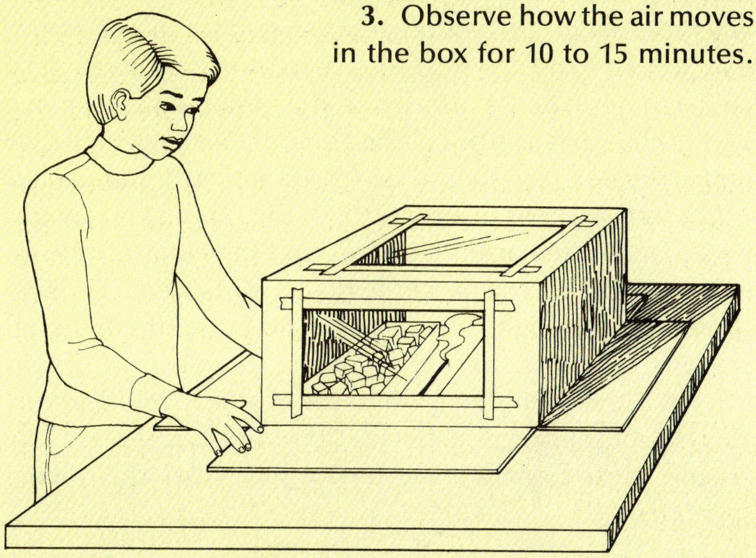

Questions

1. Do you see any evidence that the air is moving?

2. Where does the air move upward?

3. Why did the air move as observed?

4. Explain an afternoon, on-shore breeze at the beach. Use a diagram to aid your explanation.

blows because air moves from areas of high pressure to areas of low pressure. Factors such as temperature and altitude can cause pressure changes and so produce wind.

Study Questions for 5.3

1. Is warm air more or less dense than cold air?
2. What is a convection cell? Diagram one.
3. What makes the wind blow?

5.4 Moving the Atmosphere

You will find out

- why the equator is warmer than the poles;
- why the earth's large convection system appears to be deflected;
- that the whole earth has large-scale air movements.

Figure 5.20 Indirect sunlight at the poles results in cold temperatures.

The sun is shining on half the earth at all times. Different regions of the sunlit earth, however, have very different temperatures. For example, the equator receives more direct sunlight than the poles, so it is much warmer. Since the equator is warmer than the poles, you would expect a huge convection cell to form. The cooler, more dense air from the poles should move across the surface toward the equator. Warm, less dense air at the equator should be forced upward and should move in the upper atmosphere to the poles. You should find a belt of lower pressure around the equator and an area of higher pressure at the poles. Measurements have shown that these pressure areas do exist. If this is all there is to the model, the wind on the surface should always blow from the north in the northern hemisphere. In the southern hemisphere the wind on the surface should blow from the south. But how does this model test out?

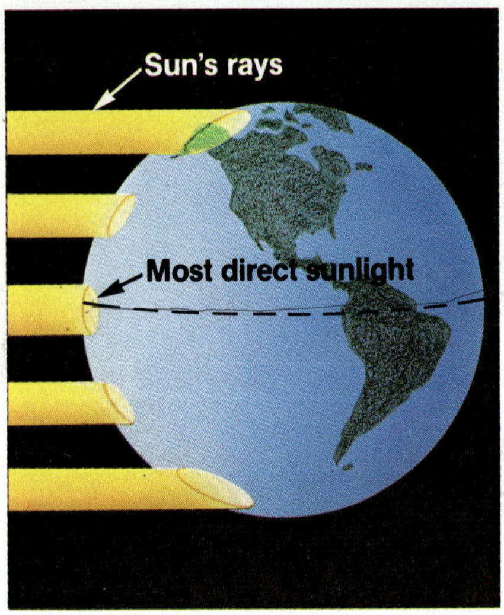

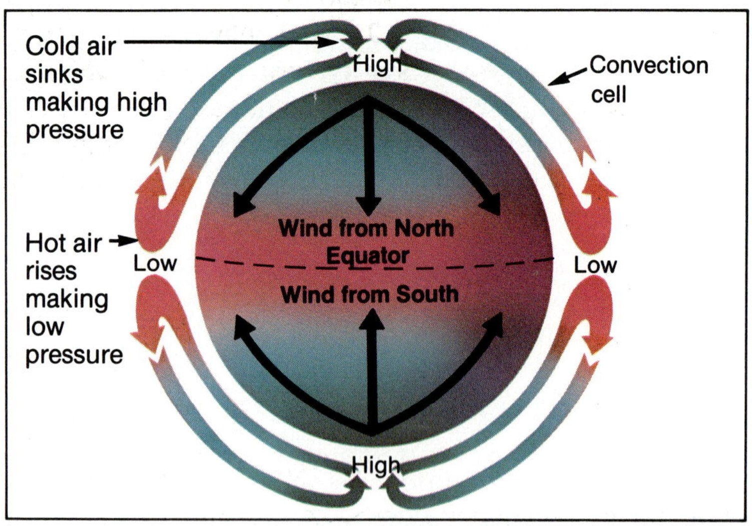

Figure 5.21 *above:* The sun's rays striking Earth's surface; *right:* If the earth did not rotate, convection currents, or winds, would behave like this model.

For one thing you know that the wind does not always blow from the north in the northern hemisphere. You also know that the wind does not always blow from the south in the southern hemisphere. It would if the earth were not spinning on its axis. The rotation of the earth has an important effect on large-scale movements of air. Except at the equator all objects moving in the northern hemisphere tend to turn or curve to the right as seen by a ground observer. The moving object may actually be moving in a straight line but appears to be turning because the earth is moving under it. The turning is called the **Coriolis** [kawr-ee-OH-lihs] **effect.** When viewed from a polar position, moving things in the northern hemisphere appear to turn to the right when moving away from you. In the southern hemisphere they appear to turn to the left.

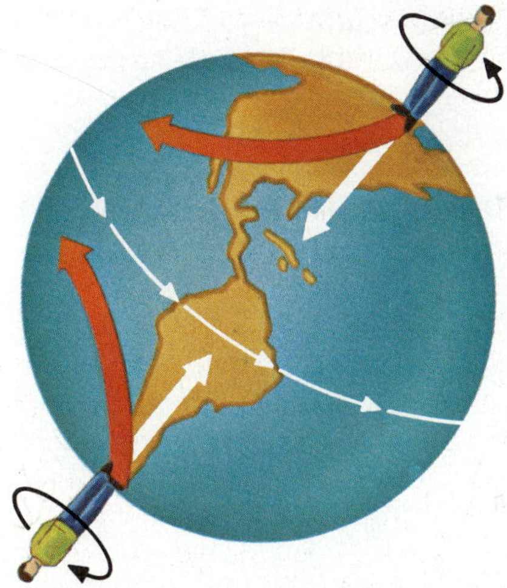

Figure 5.22 People at opposite poles experience the Coriolis effect differently.

ACTIVITY

Materials
globe
paper
pencil

Can You Go Straight?

Procedure

1. Place the paper on the globe to cover the route from St. Louis to New Orleans.
2. Practice making an air flight with your pencil point from St. Louis straight down to New Orleans.
3. Now have your partner slowly turn the globe counter-clockwise (looking down on the North Pole) as you move the pencil straight south from St. Louis.

Questions

1. Did you land in New Orleans? To the east or to the west of it?
2. Did your air flight path turn to the right or left (remember, you are inside the aircraft)?
3. Why does the Coriolis effect appear to make moving objects drift to the right or left?

Do You Know?
Global wind belts push against the surface, actually slowing the earth's rotation rate. Scientists have measured these tiny effects and found the length of a day varies less than one millisecond because of the winds.

Air is heated at the equator, becomes less dense, and moves upward and poleward above the earth. This forms a low-pressure belt around the equator. As the poleward moving air gets to 10° latitude north or south, it is influenced by the Coriolis effect. By the time it reaches 30° latitude north or south of the equator it tends to sink and pile up. Some of the air sinks toward the earth's surface. The piling up of air forms a warm high-pressure belt. Air moving away from the high-pressure belt produces two global wind belts, the **trade winds** and the **prevailing westerlies.** These wind belts are near the surface. Some of the upper air continues on to the poles. Over the poles it sinks toward the surface. It then moves toward the equator. This produces a third global wind belt, the **polar easterlies.**

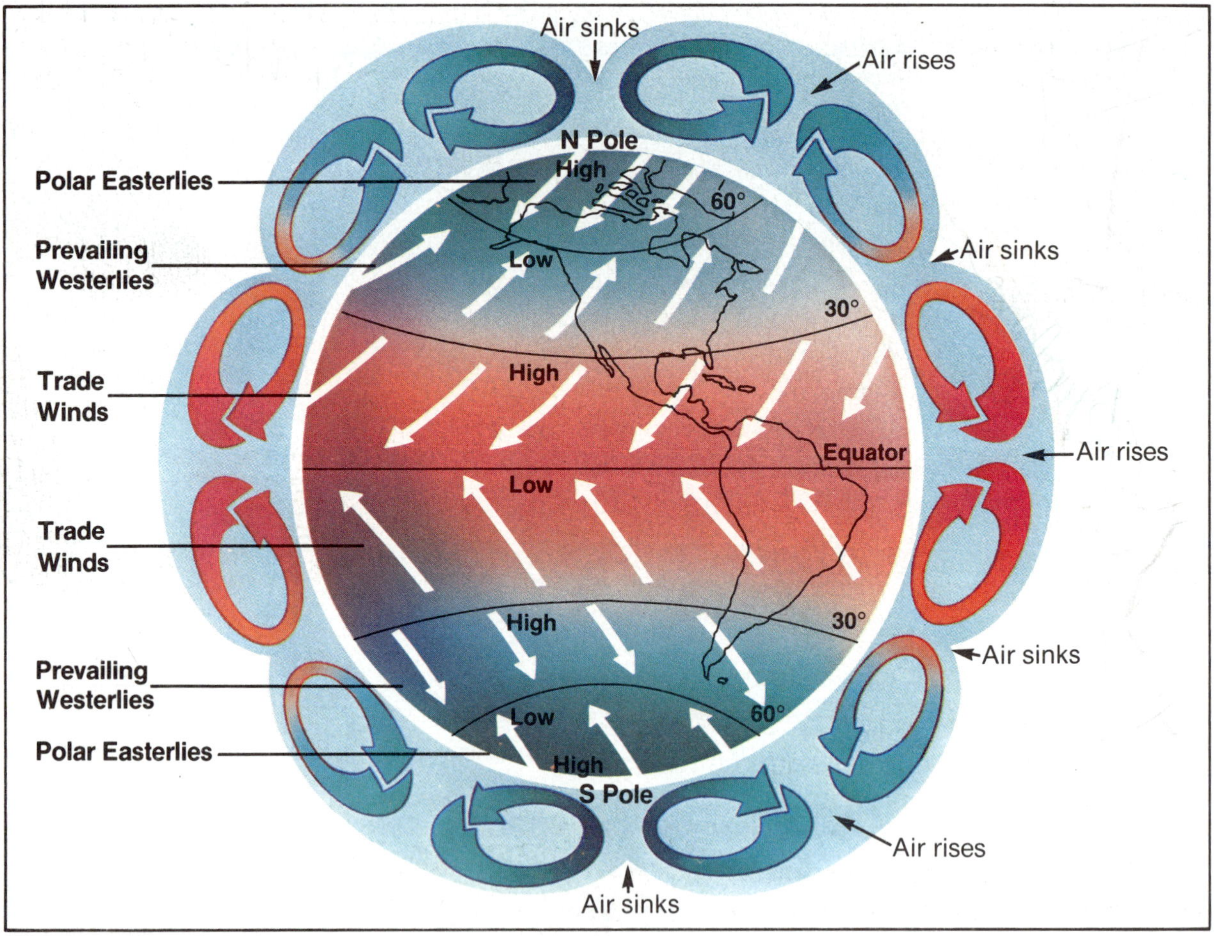

Figure 5.23 A model of the global winds that result from the Coriolis effect and convection

This is a simplified model of atmospheric air movement. It shows a zone of high pressure to the north and a zone of high pressure to the south of the United States. Since air moves from high pressure to low pressure, we would expect to find cold air moving from the north. We would expect warm air to move from the south. Most of the United States is in a "battle zone," where warm and cold air encounter each other and push back and forth. It is in this battle zone that rapid and frequent weather changes occur.

Figure 5.24 High-pressure zones at 30° and at the polar region cause air movement toward the region of low pressure near 60°. These opposing wind belts create the battle zone of changing weather patterns over North America.

Study Questions for 5.4

1. Why is the equator warmer than the poles?
2. What is the Coriolis effect?
3. Why is most of the United States in a battle zone between cold air from the north and warm air from the south?

What Did You Learn?

- The atmosphere is made up of nitrogen, oxygen, argon, carbon dioxide, small amounts of other gases, and water vapor.
- The atmosphere exerts a pressure as the result of molecular bombardment. This atmospheric pressure is measured with a barometer.
- Temperature is a measure of the average energy of motion of molecules.
- Radiant energy is energy that travels through space. Hotter objects give off shorter wavelength radiant energy than cooler objects.
- The lower atmosphere is warmed from below because of the greenhouse effect.
- Warming air causes it to expand, becoming less dense. Cool air is more dense than warm air.
- Cool, more dense air moves under and pushes warmer, less dense air upward. The movement is called convection.
- A convection cell is formed when air moves upward over a warmer surface and down onto a cool surface.
- Horizontal movement of air is called wind.
- The equator receives more direct sunlight than the poles, so it is warmer.
- The atmosphere and objects moving through it tend to turn because of the Coriolis effect. When viewed in a direction moving away from you, they turn to their right in the northern hemisphere. They turn to their left in the southern hemisphere.
- The atmosphere of the earth moves as a huge convection cell that is turned by the Coriolis effect. This produces high- and low-pressure zones and three global wind belts.

Key Terms

atmosphere
molecules
atmospheric pressure
barometer
temperature
radiant energy
greenhouse effect
wind
convection
convection cell
Coriolis effect
trade winds
prevailing westerlies
polar easterlies

Biography

Evangelista Torricelli (1608-1647)

Evangelista Torricelli was born October 15, 1608, in Italy. His father, who worked with textiles, sent him to his uncle for an education. When the boy was 17, he had done so well in his classes that his uncle sent him to Rome for further education. A teacher in the school in Rome recognized him as an outstanding student. The teacher hired Torricelli when he finished school, and he taught some classes for his teacher.

Many of the classes that Torricelli taught were on science topics. In one lesson on wind, he said that the current theory on what made wind blow was wrong. The theory at the time was that wind was produced by vapors escaping from damp earth. Torricelli thought that wind was produced when different air temperatures produced air of different densities.

Torricelli is famous for his experiments with the barometer. Once he was trying to settle an argument about how a siphon worked. He tried replacing the siphon with a 1-m glass tube. He filled the tube with mercury, closed the top end with one finger, and turned it into a bowl of mercury. He had invented the mercury barometer. However, he must have known about changes in atmospheric pressure before his experiment. In a letter to a friend, he wrote that he wished "to make an instrument that will show the changes of air, now heavier and denser, now lighter and thinner." Soon after the invention of the barometer, it was used in studying the atmosphere. One scientist wrote that the invention of the barometer changed weather science as much as the invention of the telescope changed astronomy. The barometer helped us to learn about our invisible sea of air, our atmosphere.

TO THINK ABOUT AND DO

Vocabulary

Number a piece of paper from 1 through 8. Write the correct word or words that fill the appropriate blank after each number. After each word or words write the letter of the related picture.

1. The sea of air around the earth is called the ___?___.
2. The bombardment of air molecules produces a(an) ___?___ ___?___.
3. The higher the ___?___, the faster the molecules move about.
4. A device that measures atmospheric pressure is a(an) ___?___.
5. Three global wind belts are the ___?___ ___?___, ___?___ ___?___, and ___?___ ___?___.
6. The increase of the temperature of the air caused by the radiation of longer wavelengths is called the ___?___ effect.
7. The process of warm air being moved is called ___?___.
8. Objects moving through the air appear to turn because of the ___?___ effect.

A

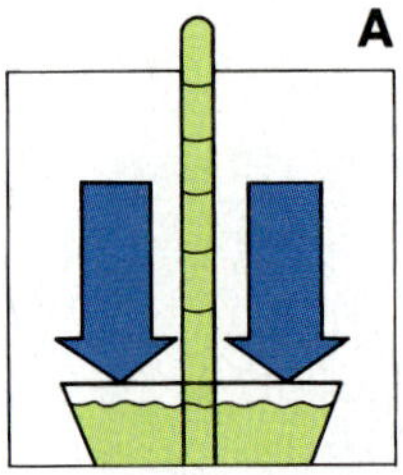

B

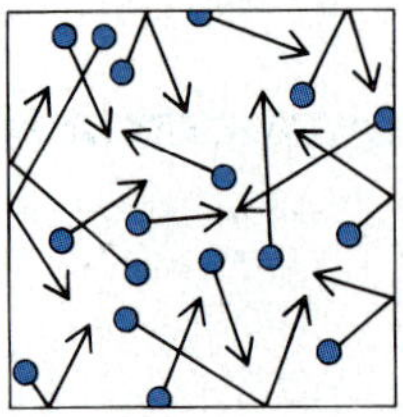

C

D

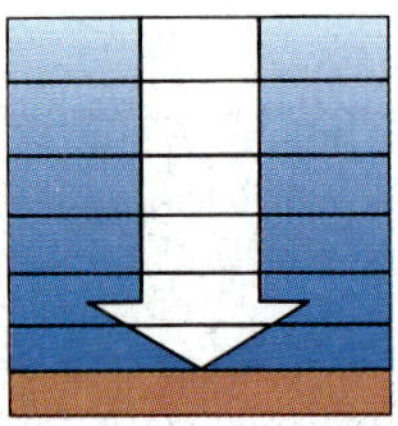

E

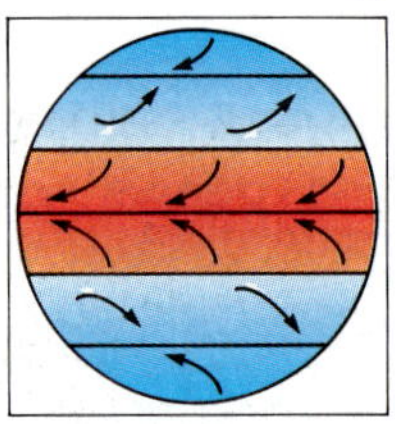

F

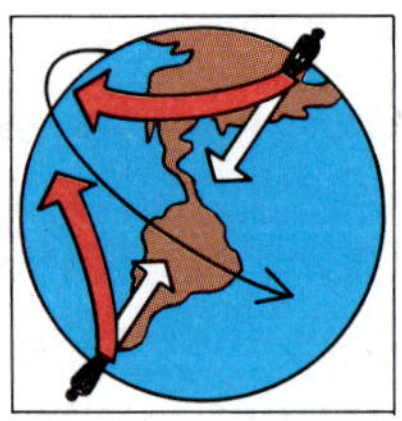

G

H

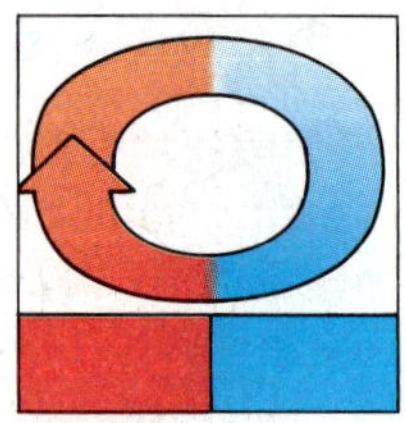

What Do You Remember?

The following questions are multiple choice. Only one of the choices for each answer is correct.

1. Atmospheric pressure is measured with (a) a barometer, (b) a thermometer, (c) a speedometer, (d) an odometer.

2. Radiant energy is energy that travels through (a) molecules, (b) bases, (c) space, (d) water.
3. Cooler air is (a) more dense than warm air, (b) less dense than warm air, (c) the same density as warm air, (d) none of these are correct.
4. Convection occurs because (a) warm air pushes up cool air, (b) cool air pushes up warm air, (c) hot air pushes up warm air, (d) water vapor causes air to move.
5. The whole earth has convection cells that are turned by the Coriolis effect. This produces (a) one global wind belt, (b) two global wind belts, (c) three global wind belts, (d) four global wind belts.

Applying What You Have Learned

1. Suppose you are growing fruit trees that are sensitive to cold air. Would it be better to plant the trees in a valley or on the slopes leading to the valley? Why?
2. Sometimes people open a window and let cold air in a room when trying to start a fire in a fireplace. How would this help start a fire?
3. Does the Coriolis effect apply only to air that is moving? Would it affect an airplane? A rocket?

Research and Investigation

Investigate wind patterns where you live. Obtain helium-filled balloons. Attach a stamped postcard that is addressed to you to each balloon. Write when and where the balloon was released and ask the person finding the card to write when and where he or she found it and to mail the card to you. Note on a map when and where each balloon was found. Prepare a report for your class.

6

Water in the Atmosphere

When will it rain? What causes storms? When will a storm occur? People have tried to answer questions like these throughout history. A long time ago people had to rely on their memories and experiences for answers. They sometimes told others of their experiences through weather proverbs. Here are some of the proverbs:

Mackerel sky and mare's tails
Make tall ships carry low sails.

A rainbow in the morning is the shepherds' warning;
a rainbow in the afternoon is the shepherds' delight.

When the grass is dry at morning light,
Look for rain before the night.
When dew is on the grass,
Rain will never come to pass.

These proverbs all refer to a sequence of weather events. All are related to some form of water. This chapter is about water in the atmosphere. After you finish the chapter, return to these proverbs. Will they help you predict when it will rain?

Figure 6.1 *opposite:* A mackerel sky, named for its speckled texture of cirrocumulus clouds

6.1 Water on the Move

You will find out

- how water moves into and out of the atmosphere;
- how the amount of water in the atmosphere is measured;
- how the amount of water in the atmosphere can be increased.

Figure 6.2 Some water molecules in the waterfall have enough energy of motion to escape into the air as vapor.

You know about water. You use it all the time. You use it as a liquid when you wash in it or drink it. You use it as a solid when you skate on ice or eat a snow cone. Water also occurs as a gas called **water vapor.** Water is different from most substances on the earth. It is different because it can occur naturally as a solid (ice), a liquid (water), and a gas (water vapor).

Water is always on the move. Liquid water moves across the land in streams and rivers. Water moves through the atmosphere as tiny droplets called clouds. It also moves through the atmosphere as invisible vapor. There is more water vapor moving through the atmosphere than there is water in all the rivers on land.

Water is also moving in other ways. It moves back and forth between liquid water and water vapor. Before discussing this movement, you need a model. How are water vapor and liquid water different? For your model you can think of water vapor as single molecules of water moving about. They move about freely in the air. Liquid water can then be thought of as many molecules being held together. They are held together by forces between the molecules. These forces are like springs. They let the water molecules move back and forth but hold them in the liquid form.

Water molecules on the earth are always moving. They never stop. How fast they move depends on the temperature. As the temperature goes up, they move about more rapidly. As the temperature goes down, they slow down. Molecules have more energy of motion at higher temperatures.

Water molecules in liquid form can obtain enough energy of motion to escape into the air. This happens when the energy of motion is great enough to break the force holding the molecules together. If more molecules are leaving the liquid state than are returning, it is said that the water is **evaporating** [ee-VAP-uh-rayt-ihng]. These water molecules going into the air are water vapor. Water vapor goes into the air through evaporation. Evaporation from the mirror in your bathroom after your shower is a good example of this.

Water vapor molecules in the air slow down when cooling occurs. If they slow down enough so that molecular forces grab them, they again form liquid water. When more molecules are returning to the liquid state than are leaving, it is said that water vapor is **condensing** [kahn-DENS-ihng]. Water vapor leaves the air through condensation. A good example is water on a kitchen window on a cold winter day.

As you can see, there is a relationship between evaporation, condensation, and the temperature. If the temperature is decreased, the average energy of motion of the molecule is decreased. Water vapor molecules condense into the liquid state when they slow down enough so that molecular forces grab them. Faster molecules are therefore less likely to be captured. As the temperature increases, there is less tendency for water vapor to return to the liquid state.

Figure 6.3 *top:* Water vapor molecules condense on the cool mirror. *middle to bottom:* Evaporation gradually returns water molecules to the air.

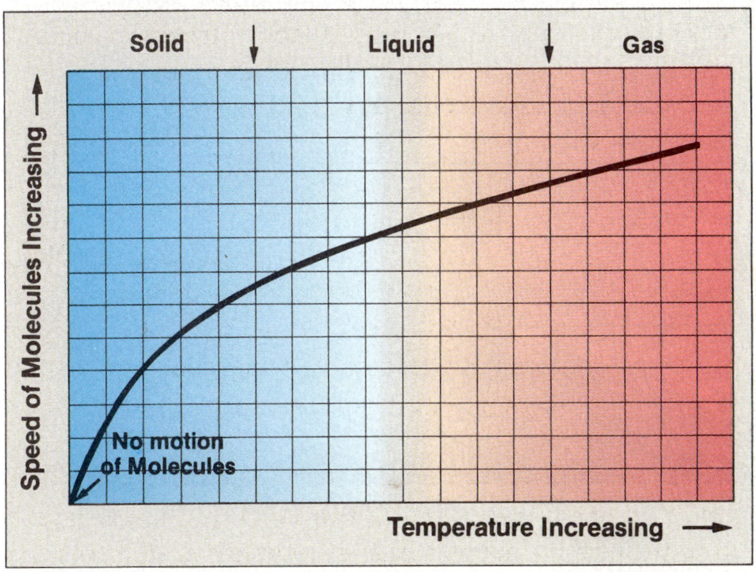

Figure 6.4 Average energy of motion of molecules increases at higher temperatures.

Warm air can therefore hold more water vapor than cooler air. In fact, warm air on a summer day can hold five times as much water vapor as cold air on a winter day. The amount of water vapor in the air is called **humidity** [hyoo-MIHD-uh-tee].

Air at the same temperature can have different amounts of water vapor in it. The amount of water vapor in the air may be referred to as low humidity or high humidity. Humidity, or the amount of water vapor in the air at a certain temperature, can be stated as a percentage.

One hundred percent humidity means that the air has all the water vapor it can hold at that temperature. Fifty percent humidity means that the air is holding half of the water vapor it can at that temperature. These percentages of humidity readings are called the **relative humidity.** The relative humidity tells you how much water vapor is in the air compared to the total that could be in the air at a certain temperature.

Relative humidity can be increased two ways. Adding more water vapor to the air increases the relative humidity because it increases how much water is in the air. Cooling increases the relative humidity because it

Figure 6.5 *top:* It's not the heat; it's the humidity! *below:* This bar graph illustrates that air at higher temperatures contains more water when at 100-percent humidity than air at lower temperatures. At the same temperatures, air at different humidities will feel different.

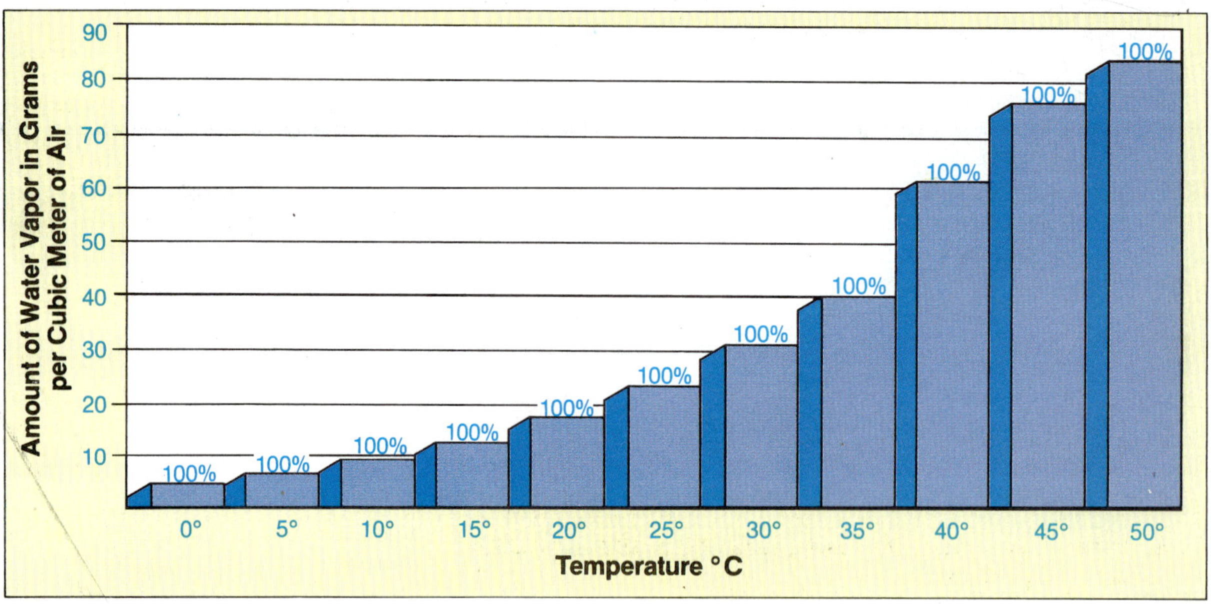

reduces the amount of water vapor that can be in the air. Have you heard a weather report lately? What is happening to the relative humidity?

Do You Know?
Cold air cannot hold much water vapor so the saying "It's too cold to snow" has meaning.

ACTIVITY

How Wet Is the Air?

Materials
cotton shoelace
small card
two thermometers

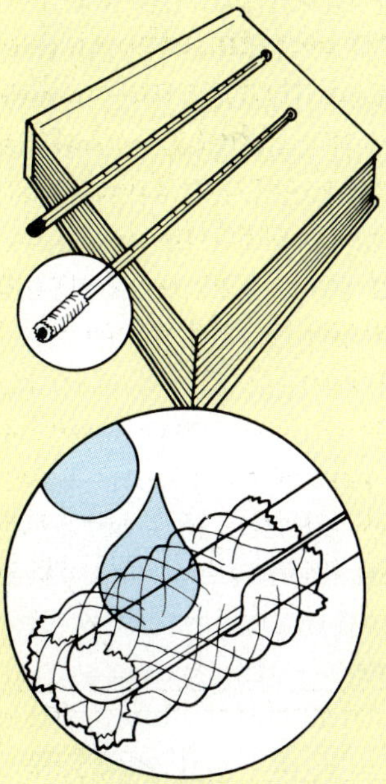

Procedure

1. Make a wet-bulb thermometer. Slip a short piece (3 cm) of shoelace over the bulb of a thermometer. Wet the shoelace with water.
2. Place the wet-bulb thermometer beside a dry-bulb thermometer on a book as shown.
3. Fan the bulb end of the two thermometers with a small card. Note the temperature of the wet-bulb thermometer.
4. Continue to fan the thermometers until the wet-bulb thermometer registers its lowest reading. At this time, record the wet-bulb and the dry-bulb temperatures.
5. Subtract the wet-bulb reading from the dry-bulb reading. Use this difference and the dry-bulb temperature on a relative humidity chart (page 513) to find the relative humidity.
6. Measure the humidity in your classroom, outside, and in other locations.

Questions

1. What data did you obtain about the humidity in the different locations?
2. What would explain any differences you may have found?

Study Questions for 6.1

1. Describe how water moves into and out of the atmosphere.
2. What is meant by the "humidity is 50 percent"?
3. What are two different ways that the humidity can increase?

6.2 Condensation

You will find out

- how condensation occurs;
- the relationship between relative humidity and condensation;
- how to predict when dew or frost will form.

Figure 6.6 Water vapor exhaled in warm breath condenses in cold air.

Have you ever seen your breath on a cold day? You can see your breath because your exhaled air is warm and contains much water vapor. Your breath forms a small pocket of air with a high amount of water vapor. As this pocket of air is cooled, its relative humidity increases. The humidity reaches 100 percent and condensation takes place, forming tiny water droplets. What you see is a small cloud with thousands of tiny water droplets. As the condensed droplets spread into the surrounding air with lower humidity, they evaporate and disappear.

In the atmosphere large amounts of air can also be cooled. Air near the ground cools during the night by contact with the cooling earth surface. It sometimes cools until the relative humidity reaches 100 percent. Further cooling then results in condensation. The temperature at which condensation occurs is called the **dew point.** If the dew point is above 0°C, the water vapor will condense on surfaces as a liquid, called **dew.** If the temperature is at or below 0°C, the vapor will condense on surfaces as a solid, called **frost.**

Figure 6.7 A spider's web looks like a jeweled necklace as dew condenses on it.

Dew or frost forms on *C* nights—clear, calm, cool nights. Let us see why dew or frost forms on *C* nights. Here is how it works. The earth is warmed by radiant energy from the sun and gives off infrared radiation (page 91). The surface warms during the day as more energy is received than is given off. The surface cools during the night because no energy is received, but the surface continues to give off infrared radiation. This infrared radiation passes back into space. If there are clouds they serve as blankets, keeping the radiation in. This keeps the temperature above the dew point. *C* nights favor dew or frost formation because there is nothing to keep the heat from the radiation in the area.

ACTIVITY

What Is the Dew Point?

Materials
shiny metal can
crushed ice
thermometer

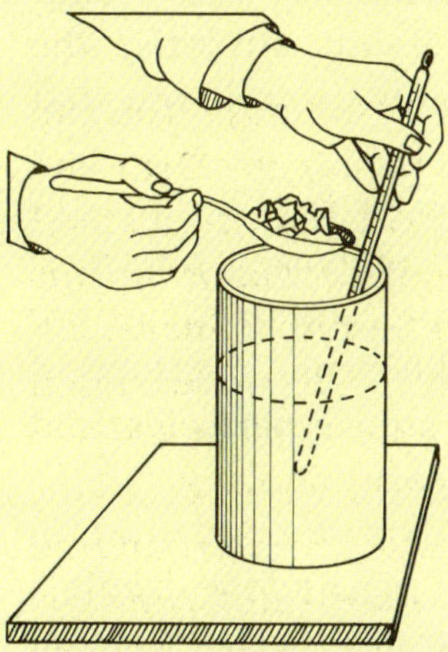

Procedure

1. Fill a shiny metal can about half full with tap water.

2. Place a thermometer in the water. Add a *small* amount of ice. Watch the sides of the can as you stir the water.

3. Continue adding small amounts of ice and stirring until dew begins to form on the can. The dew point of the air will be the temperature of the water in the can when dew begins to form on the can.

4. Repeat steps 1 through 3 outdoors. Compare the dew point of the air in your classroom to the dew point of the air outside.

Questions

1. What is the dew point inside your classroom?

2. What is the dew point outside your classroom?

3. Find out the predicted low temperature for tonight. Will the dew point be reached?

Fruit growers have developed special means to prevent frost in their orchards. Stack heaters are sometimes used to warm the air in the orchard. Other times large fans are used to mix the air being cooled near the surface with warmer air above the surface as another way of preventing frost.

When conditions are favorable, condensation occurs on surfaces as the dew point is reached. Note that dew or frost forms on, under, and on the sides of objects. People used to believe that frost and dew fell out of the air. But frost and dew really condense on cool objects, just as water vapor condenses on a glass of ice water.

Figure 6.8 Stack heaters protect orchard trees from damaging frost.

Study Questions for 6.2

1. The weather reporter states that "the dew point is 10°C." What does this mean?
2. What is the relative humidity when the dew point is reached?
3. Under what conditions does dew form?
4. Under what conditions does frost form?

6.3 Fog and Clouds

You will find out
- how a cloud is formed;
- the three basic cloud shapes.

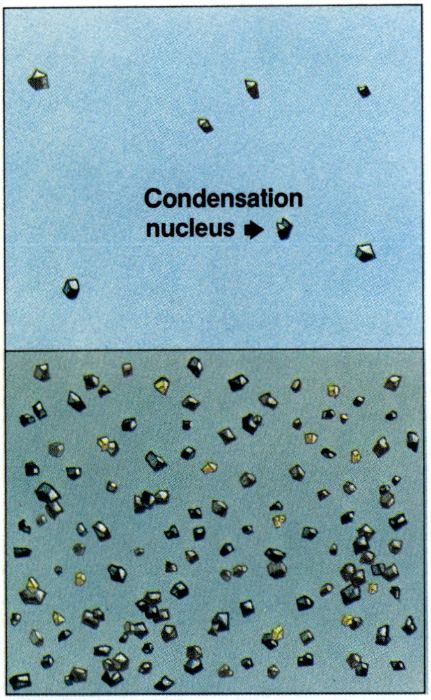

Figure 6.9 *top:* A clear sky has few condensation nuclei. *bottom:* A dirty sky has many.

Are there clouds where you are now? Some people watch clouds for clues about the weather. They call clouds "billboards of the sky" because clouds advertise the coming weather. Do you think clouds can tell you that much about coming weather?

What are clouds? Clouds are mostly tiny droplets of water. These droplets have been condensed on small particles that are present in the air. The particles are called **condensation nuclei** [NOO-klee-eye]. Particles of salt (from evaporating sea spray), smoke particles, and dust in the air all serve as condensation nuclei.

After water vapor molecules stick to a condensation nucleus, other water molecules will join and the droplets will grow larger. It may take several minutes to form the tiny water droplets that make up a cloud. It may take several hours for drizzle droplets to form. Compare the size of the water droplets in Figure 6.10.

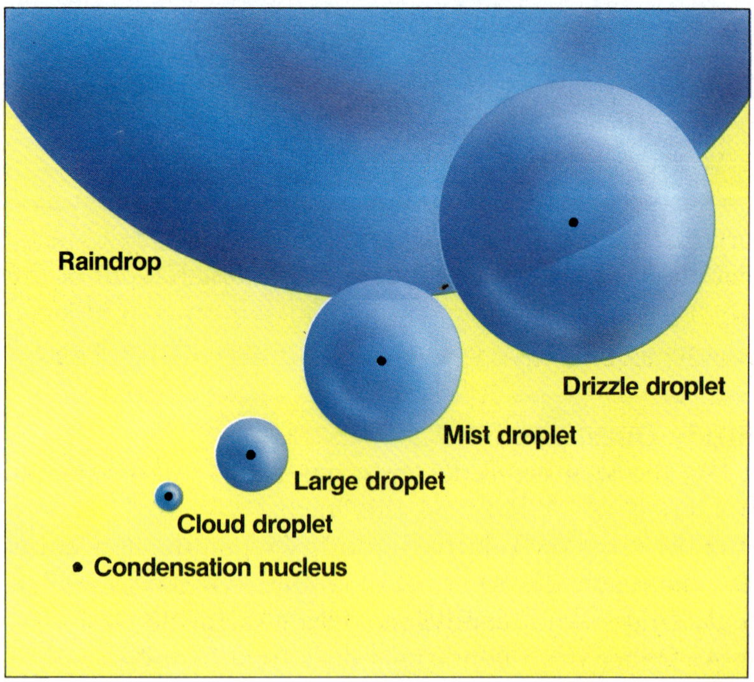

Figure 6.10 Relative sizes, enlarged 500 times, of water droplets in a cloud

Do You Know?
Volcanic dust travels all the way around the earth and can stay in the atmosphere for years.

Three things are needed for a cloud to form. Water vapor and condensation nuclei are needed, of course, but cooling must also take place. For example, convection may create a cloud. A pocket of air near the ground is moved upward by a convection current. The pocket of air cools as it travels upward. It soon reaches a height where it is cooled to the dew point. Condensation then takes place on condensation nuclei. Water droplets are formed, making a cloud.

Do You Know?
Water vapor from the engines of high-flying jet airplanes sometimes forms clouds. These clouds may be affecting the weather.

Figure 6.11 The development of cumulus clouds

The water droplets making up a cloud are very small. They fall to the earth very slowly. In calm air, droplets may fall less than 0.5 m per second. A very slight upward movement of the air will overcome this small falling speed. This is why clouds appear to float in the sky.

Fog is also a cloud. It's at or near the earth's surface. It is made up of very small water droplets that fall very slowly. Fog often forms under the same conditions favorable for dew or frost to form—*C* nights. Coastal fogs are usually fogs that form over the ocean and then move inland.

Figure 6.12 Ground fog looks like a cloud upon the earth.

Clouds are named for their shape and for the height at which they occur. There are three basic cloud shapes: **cumulus** [KYOOM-yuh-luhs] (piled up), **stratus** [STRAT-uhs] (spread out), and **cirrus** [SIHR-uhs] (hair like fibers). Usually you do not see pure examples of these three basic cloud shapes but combinations of the shapes.

Figure 6.13 Cloud types

When moist air rises rapidly over a small area, heaps of condensed moisture form. These heaps are called cumulus clouds. Occurring at an altitude of from 300 to 1,000 m, cumulus clouds are usually seen during fair weather. However, a cumulus cloud sometimes piles up to an imposing mountain of condensed water vapor of up to 9 km high. These clouds are called **cumulonimbus** [kyoo-myuh-loh-NIHM-buhs] (*nimbus* means "stormy"). Severe weather is often produced by the cumulonimbus clouds. This includes tornadoes, heavy rain, thunder, lightning, and hail.

Sometimes the air rises gently over a large area and cools slowly. Then the water vapor often condenses into spread-out clouds that look like gray sheets. If these clouds are at low levels, they are called stratus clouds. If stratus clouds touch the ground, they are called ground fog. Stratus clouds that occur between 2 and 7 km are called **altostratus** clouds. These sometimes block out much sunlight, and the sun is visible only as a bright spot. Stratus clouds above 7 km are made of tiny ice crystals. They are called **cirrostratus** clouds. When the sun or moon shine through cirrostratus clouds the light is scattered by ice crystals. This causes a halo around the sun or moon.

The highest of the common clouds look like white, hairlike fibers. These are cirrus clouds. Because of the low temperature at their heights, they form when water vapor condenses directly into ice crystals. Cirrus clouds sometimes have a curled feathery look that is called mare's tails.

The color of a cloud depends on its relative position to you and the sun. A cloud will appear white if you are between the cloud and the sun. This cloud reflects sunlight back to you. The same cloud might appear gray or black if it is between you and the sun. This cloud is blocking the sunlight.

Figure 6.14 Cirrocumulus clouds

Figure 6.15 Can you see the halo formed around the sun by the cirrostratus clouds?

Study Questions for 6.3

1. What is a cloud?
2. What is fog?
3. What three things are necessary for water vapor to condense into water droplets to make a cloud?
4. Name, describe, and sketch the three basic cloud shapes.

6.4 Precipitation and the Water Cycle

You will find out
- how water falls from clouds;
- how water cycles through the atmosphere.

Water that falls to the surface of the earth—either in solid or liquid form—is called **precipitation** [pree-sihp-uh-TAY-shuhn]. Dew and frost are not called precipitation. They form directly on the surface.

Much has yet to be learned about how precipitation forms in clouds. One way it may form is by droplets of water sticking to other droplets. The droplets in a cloud are not all the same size. As the air moves about, the larger droplets collide and combine with smaller ones. This would make large, heavy raindrops that would fall to the surface of the earth.

A second way that precipitation may form is by droplets forming ice. The tops of clouds are often made of ice crystals. Air movement can mix this ice with water droplets from lower layers. Water molecules can escape from ice less easily than they can from water. Ice crystals grow at the expense of water droplets. As the ice crystals grow, they become heavier, falling earthward. If they fall through cold air, they will reach the surface as snow. If they fall through warm air, they will melt and reach the surface as rain. If they fall through warm air and melt and then fall through cold air and freeze again, they arrive as sleet.

Water evaporates. Water vapor condenses. Clouds form. Precipitation falls to the land. Then the water evaporates again. As you can see, the process goes on as a never-ending series of events. A series of events that keeps happening over and over is called a **cycle** [SY-kuhl]. The water cycle is considered to have three main events. The first main event is that water vapor evaporates from the oceans into the atmosphere. Next the atmosphere transports the vapor over the land where it condenses into clouds and then falls as precipitation. The last main event takes place when streams carry the water back to the oceans where the cycle begins all over again.

Figure 6.16 When clouds let moisture go, the result is precipitation.

Do You Know?
The wettest place in the United States is Mount Waialeahe, Hawaii. It receives about 1,168 cm of precipitation a year.

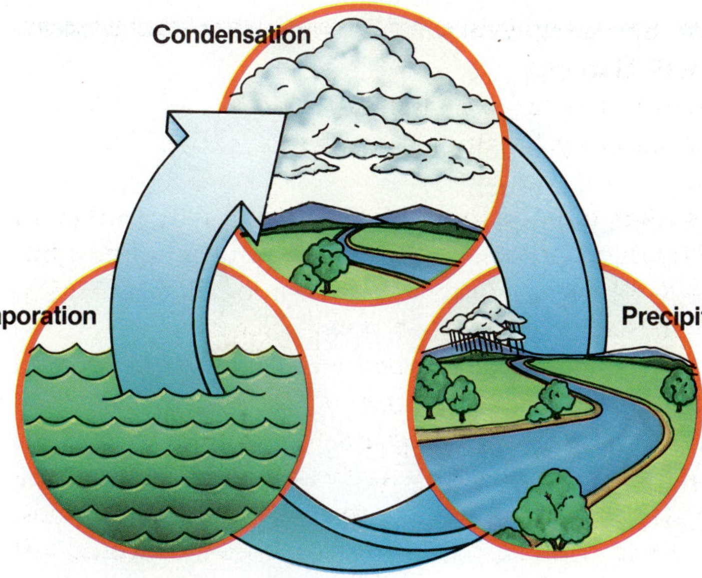

Do You Know?
The driest place in the United States is Death Valley, California. It receives about 4 cm of precipitation a year.

Figure 6.17 The water cycle

Figure 6.18 During droughts, mud flats can form where riverbeds and lakebeds dry.

This is a simplified description of the water cycle because it does not all happen this way. You know, for example, that water evaporates from the moist land too. This vapor may condense and fall on the land without ever returning to the ocean. This is a subcycle within the water cycle. There are many small exchanges of vapor and precipitation in the overall water cycle.

Due to changes in the atmosphere the water cycle at times does not seem to really be a cycle. Sometimes there is too much rain in a place, and there is a flood. Sometimes there is not enough rain, and there is a **drought** [drout]. Droughts may last a season, a year, or many years. In the next chapter you will look at why water from the water cycle is at times distributed unevenly.

Study Questions for 6.4

1. A newspaper article states that there has not been enough precipitation this year. What does *precipitation* mean?
2. How does precipitation form from the water droplets in a cloud?
3. What are the main parts of the water cycle? Why is it called a cycle?

What Did You Learn?

- Water enters the atmosphere through evaporation and leaves by condensation.
- Relative humidity is the percentage of how much water is in the atmosphere compared to how much could be in the atmosphere at a particular temperature.
- The dew point is the temperature at which water vapor condenses on a surface.
- Dew or frost forms on *C* nights—clear, calm, cool nights.
- Fog and clouds are made of tiny droplets of water. Water vapor condenses on condensation nuclei to form these droplets when the air has 100-percent humidity and is cooled.
- The basic cloud shapes are cirrus, cumulus, and stratus.
- Precipitation may form from small cloud droplets of water colliding and joining together to form large ice crystals.
- Precipitation can take the form of a solid or a liquid that falls to the surface of the earth.
- The earth's water cycle, which keeps occurring, is evaporation, condensation, and precipitation.

Key Terms

water vapor
evaporating
condensing
humidity
relative humidity
dew point
dew
frost
condensation nuclei
fog
cirrus
cumulus
stratus
cumulonimbus
altostratus
cirrostratus
precipitation
cycle
drought

Career

Meteorologist

People often need to know future weather conditions. Builders need to know if it is going to rain before they pour concrete. Farmers need to know future weather conditions to plan crop planting and harvesting. You need to know about future weather to plan what you will wear to school or will do during the weekend. Information about future weather conditions comes from a meteorologist, a scientist who studies the weather.

June Bacon-Bercey is a meteorologist. As a meteorologist, she has been a weather analyst, a radar meteorologist, and then a chief forecaster, predicting the weather for New York, Connecticut, Pennsylvania, and New Jersey, including Long Island Sound and Cape May to Block Island. She has also been a television weather caster (WGR-TV in Buffalo) and an aviation operations meteorologist. She is presently a public information specialist with the National Oceanic and Atmospheric Administration (NOAA). As you can see, a meteorologist can do more than gather information about temperature, wind speed, pressure, and humidity.

June Bacon-Bercey became the first woman recipient of the Seal of Approval of the American Meteorological Society for television weather casting. She has received many awards for her work. Her present job is to inform people about weather. She makes films to prepare people for weather disasters. These films are shown on television. She says that the United States needs more and better scientists and a public that understands science. She says that in her present job she can encourage people to consider careers in meteorology and earth science. She means what she says. In 1977 she won $64,000 on a TV quiz show. She used it to set up a scholarship assistance program for women wanting a career in meteorology.

TO THINK ABOUT AND DO

Vocabulary

Write the numbers 1 through 10 on a sheet of paper. Read the numbered phrases. Write the word or words in the list that best fit the meaning for each phrase beside the appropriate number.

fog	dew
water vapor	frost
evaporation	cirrus
relative humidity	cumulus
precipitation	dew point

1. the cloud shape that is hairlike
2. a percent telling how much water vapor is in the air compared to how much could be at that temperature
3. a cloud at or near the ground
4. water in the form of a gas
5. temperature at which condensation forms
6. what happens when more water molecules are leaving the liquid state than are returning
7. the cloud shape that is piled up
8. water vapor that condenses on the surface as a liquid
9. water that falls to the surface of the earth either in solid or liquid form
10. water vapor that condenses on the surface as a solid

What Do You Remember?

Number your paper from 1 to 10 and write *true* or *false* for each statement. If the statement is false, change the underlined word (or words) to make it correct.

1. Single molecules of water that move through the atmosphere are called <u>liquid water</u>.
2. Water vapor enters the atmosphere through <u>evaporation</u>.
3. Water vapor returns to the liquid state through <u>condensation</u>.
4. Cooler air has molecules with <u>less</u> energy of motion than warmer air.

5. Warmer air can hold less water vapor than cooler air.
6. The relative humidity can be increased by adding more water vapor or by cooling the air.
7. The temperature at which water evaporates is called the dew point.
8. Dew and frost most likely form on clear, calm, cool nights.
9. A piled-up cloud shape is a cumulus cloud.
10. A cloud shape of hairlike fibers is called a serious cloud.

Applying What You Have Learned

1. Under what conditions would you expect dew to form?
2. What is wrong with the statement "A heavy dew fell last night"?
3. Why is it that dew and frost usually form at night?
4. Explain why relative humidity readings are often lower in the afternoon than they were in the morning.
5. Why does warm air hold more water vapor than cool air?

Research and Investigation

1. Set up a weather watch. See if you can find relationships between cloud types, wind speed, wind direction, humidity, and coming weather. Will identifying cloud types alone enable you to predict coming weather?
2. How large is a raindrop? Fasten some cloth material such as nylon hose over the opening of a small box. Smooth flour over the cloth. Hold the cloth horizontally for a few seconds in a rainfall. Drops passing through the cloth will leave their mark, which can be measured. What is the diameter of an average raindrop? Find out how to figure the volume of an average raindrop. Report your findings to your class.

As water vapor emerges from hot jet engines, it condenses to form water droplets called contrails.

7 Weather

You're tired of clear blue skies. You want to vote for a deep snow. But others want to go for a bike ride, and they vote for more of the same clear weather.

Voting for weather? Will this be possible in the future? If it is, who will decide the kind of weather you should have? Perhaps it will be you pushing a button to vote for the kind of weather you want. Do you think it will ever be possible to change the weather that much? Probably not until there is a better understanding of the weather. But understanding the weather does not mean that you will be able to change it. Think about who may be affected by the weather you may choose.

Meteorologists understand that most weather moves across the country. Today's rain in Atlanta may be in Boston tomorrow. Meteorologists are good at predicting tomorrow's or the next day's weather. But predicting the weather beyond two or three days requires a lot of educated guesswork.

Do you know what tomorrow's weather will be? Do you know why you will have that weather? In this chapter you will learn what causes changes in the weather.

Figure 7.1 *opposite:* These people are more than just fair-weather fans.

7.1 Air Masses

You will find out

- what an air mass is;
- that air moves across the country;
- the three things an air mass depends on.

Figure 7.2 These toes feel the draft of cold air flowing downward.

Every time you open the refrigerator door, cold air pours out onto the floor. The cold air that leaves the refrigerator is replaced by warm air from the kitchen. When you close the door, the warm air is cooled by the cooling coils of the refrigerator. The next time you open the door cold air will again pour out.

Much the same thing happens to air outdoors. Large sections of the atmosphere near the ground become "air conditioned." That is, the air takes on the same temperature as the earth's surface it is over. It also becomes almost as moist or dry as the surface. When such a body of air over thousands of square kilometers has uniform temperature and moisture, it is called an **air mass.**

How do air masses form? They form mostly in high-pressure belts. Recall the model developed in Chapter 5 of air movement of the earth's atmosphere. There are two zones of high pressure in the northern hemisphere. In both places air piles up, sinks, and spreads out slowly across the surface. The air slowly becomes changed by its surroundings. An air mass is formed. Air masses can be very large, thousands of square kilometers across. But they may be only a few kilometers deep.

Figure 7.3 Air movement patterns in the earth's atmosphere

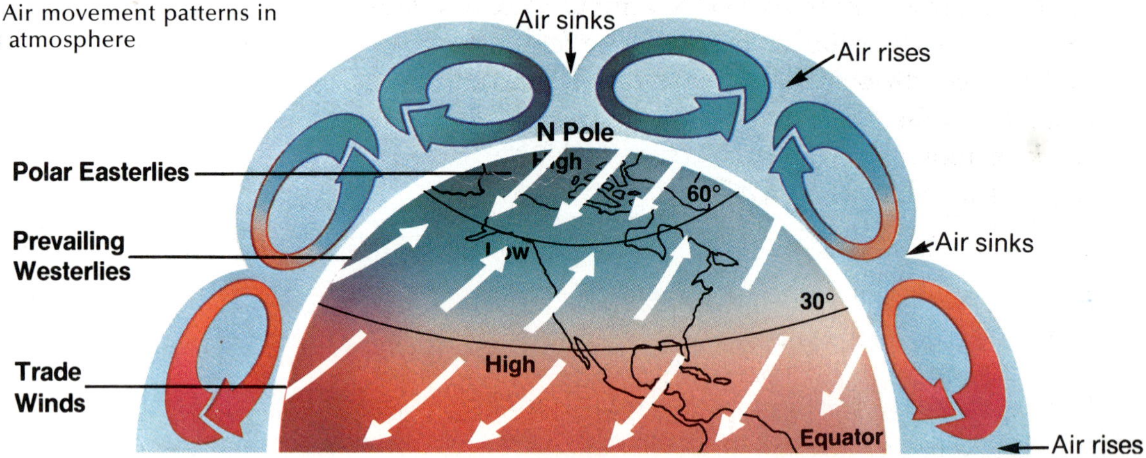

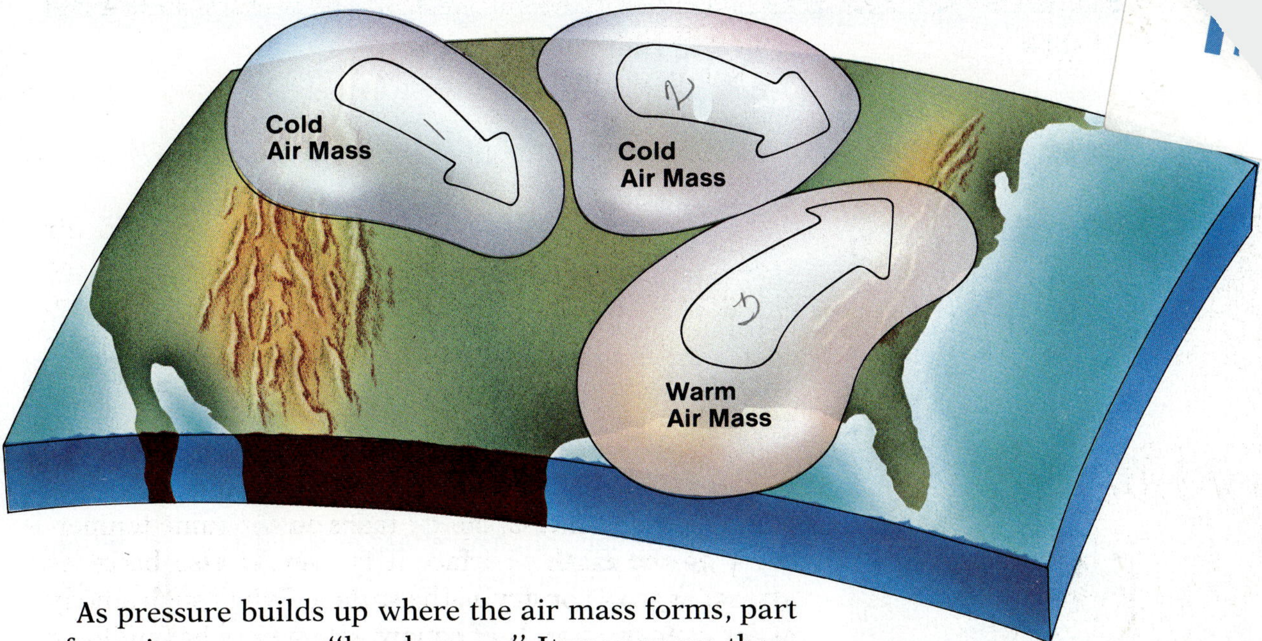

Figure 7.4 Air masses as they may be located over the continental United States

As pressure builds up where the air mass forms, part of an air mass may "break away." It may move thousands of kilometers. A moving air mass is like a large, flattened bubble of air. As the air mass moves across the surface, it may keep nearly the same temperature and moisture conditions. Since this bubble is so large, days may be required for it to move past your location. During this time, your weather is controlled by the conditions of the air mass that you are currently in.

If the moving air mass begins to slow or stop over your area, however, the conditions of the air you are in begin to change. The weather in the air mass depends on three things. These are (1) the temperature and moisture conditions of the air mass, (2) the temperature and moisture conditions of the area over which it has moved, and (3) how fast it has moved. Meteorologists keep close track of moving air masses. This tracking gives them important information for weather predictions.

Study Questions for 7.1

1. What is an air mass?
2. How does an air mass form?
3. What three things does the weather in an air mass depend on?

7.2 Weather

You will find out

• how air masses affect the weather.

The weather can be full of surprises for you. One day can be hot and humid with drizzling rain. The very next day can be clear and dry with a fresh clean smell in the air. It is as if something changed all the outside air. In fact, that is exactly what happens. A new air mass arrives and pushes the hot, humid air away. In its place comes a cool, dry air mass. But air masses must come from some place. In fact, air masses develop in a number of places.

Figure 7.5 Four types of air masses

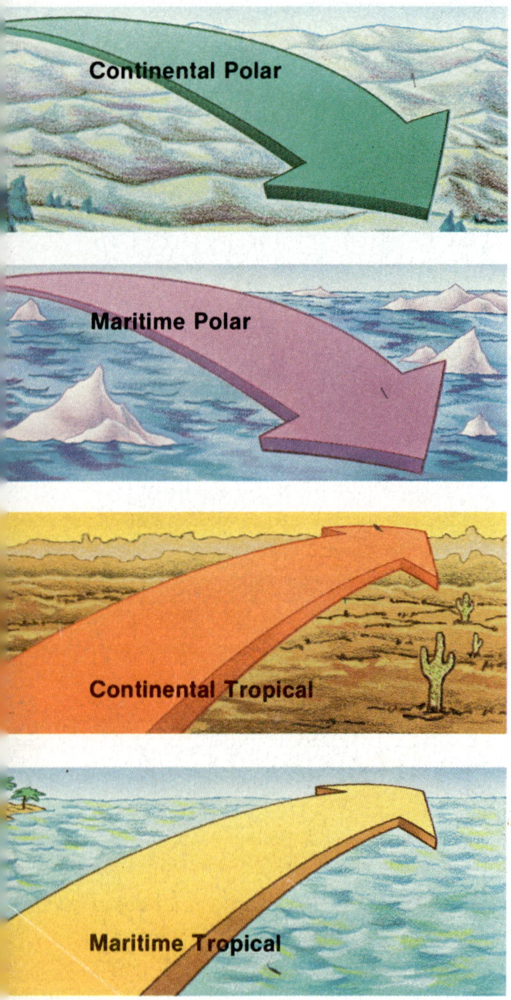

The birthplace of an air mass is called a **source region.** One source region includes almost all of Canada and part of the northern United States in winter. Air masses formed here are cold and dry. They are called **continental polar** air masses. When a continental polar air mass arrives in the winter, the meteorologist reports a cold wave. The arrival of a continental polar air mass in the summer often provides relief from a heat wave. After a continental polar air mass arrives, your day may begin with crisp, clear air and blue skies. As the sun warms the surface, you see cumulus clouds start to form. By afternoon the sky may be covered with cumulus clouds. After sunset the clouds evaporate. Another source region is the hot and dry area of southwest United States. This air mass is called **continental tropical** and brings hot dry weather with it.

Another source region includes the mid Atlantic, the Caribbean, and the Pacific near the high-pressure areas found at about 30° north latitude. Air masses formed here are called **maritime tropical.** A maritime tropical air mass is warm and moist because it formed over subtropical waters. Maritime tropical air masses typically bring hot, humid weather. Clouds are more of the stratus type.

Still another source region for air masses is the cool northern Atlantic and northern Pacific oceans. What will these air masses be like? The moist, cool air masses formed here are called **maritime polar** air masses. Maritime polar air masses that form over the northern Atlantic Ocean may move into eastern Canada and the northeastern United States. They may be part of a storm called a "nor'easter" (northeaster). Maritime polar air also forms over the Pacific Ocean. It brings precipitation to the western states and to the Rocky Mountains.

Air masses are moved across the United States by the prevailing westerlies which you learned about in Chapter 5. Higher pressure to the north may send a bubble of continental polar air south into the United States. At other times pressure differences send a bubble of maritime tropical air northward into the United States and Canada. The different air masses tend to alternate and are moved eastward by the prevailing winds.

Figure 7.6 Air masses moving across North America create changing weather patterns.

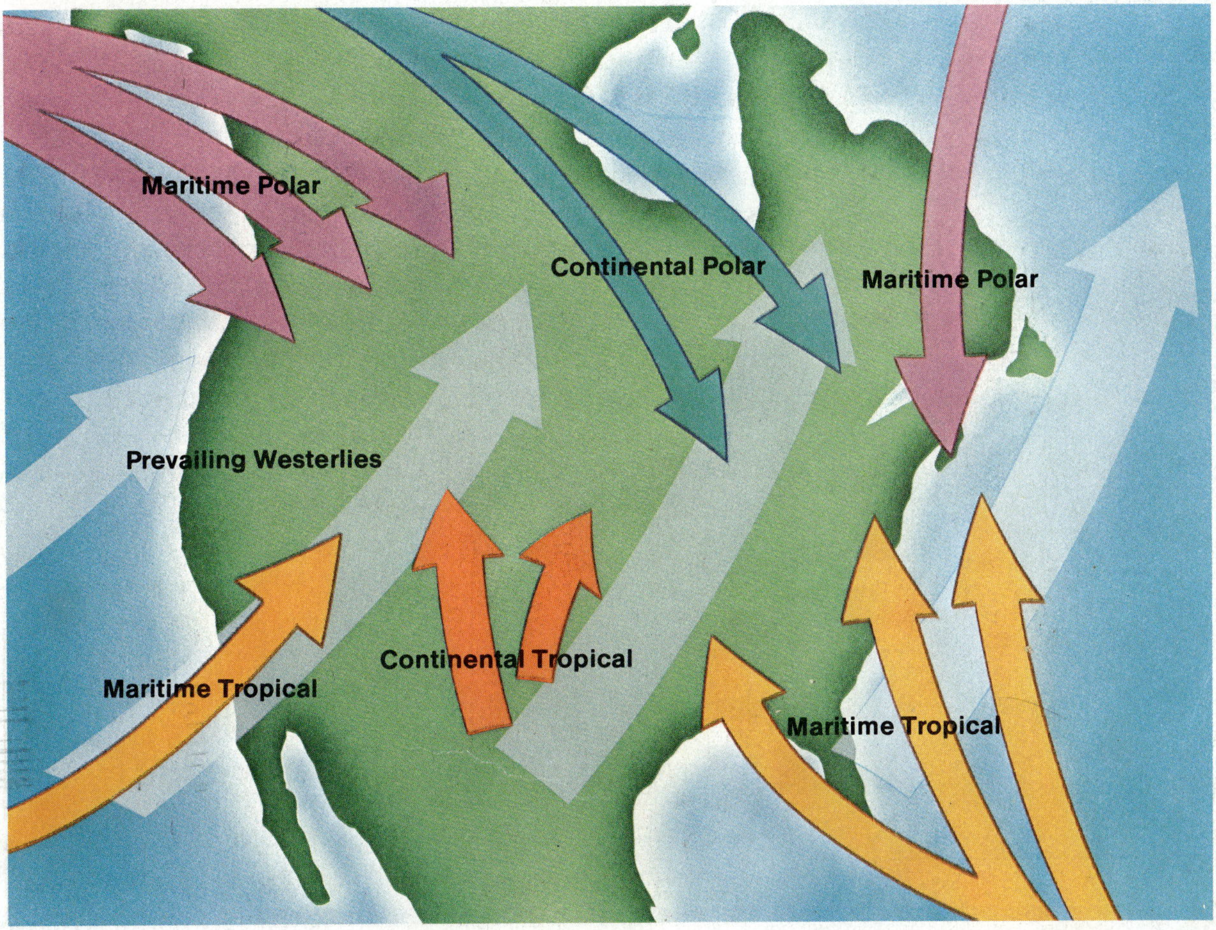

Because air masses are so large, it may take days for one to move past one location. During this time you have **air mass weather.** Air mass weather is about the same from day to day with slow, gradual changes. Weather changes usually occur as one air mass is being moved out and replaced by the leading edge of another air mass.

The leading edge of a moving air mass is actually a zone from 5 to 80 km thick. This zone is called a **front.** There are different kinds of fronts. A **cold front** is at the front of a moving cold air mass. A **warm front** is at the

ACTIVITY

How Does an Air Mass Change the Weather?

Materials

A series of at least six newspaper weather maps. The maps should show a moving front passing some city on the second day.

Procedure

1. Study six weather maps that show a moving front. Identify a city that the front passes on the second day.

2. For a six-day period, make a list of the daily high temperatures, the low temperatures, any precipitation, wind speed, and direction.

Questions

1. Describe the weather before and after the front moved past the city. How did the low and high temperature change? Was there precipitation? What else changed?

2. What kind of air mass moved over the city?

3. What was the air mass weather like on the fourth, fifth, and sixth day? How did the low and high temperatures change? Was there precipitation? Were there any other changes?

Some Commonly Used Weather Symbols

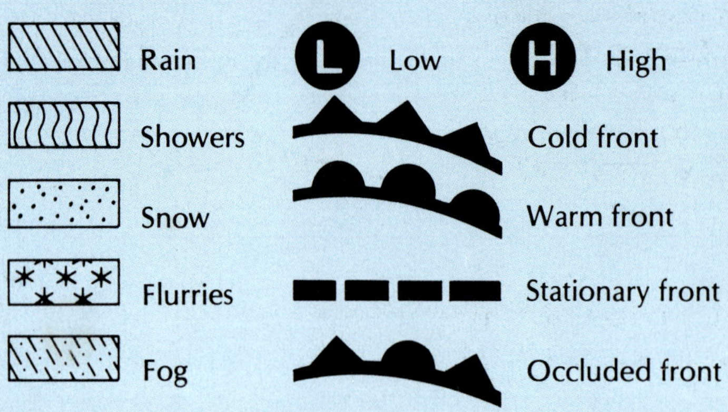

front of a moving warm air mass. Sometimes air masses stop and do not move. An unmoving zone or boundary between two air masses is called a **stationary front.**

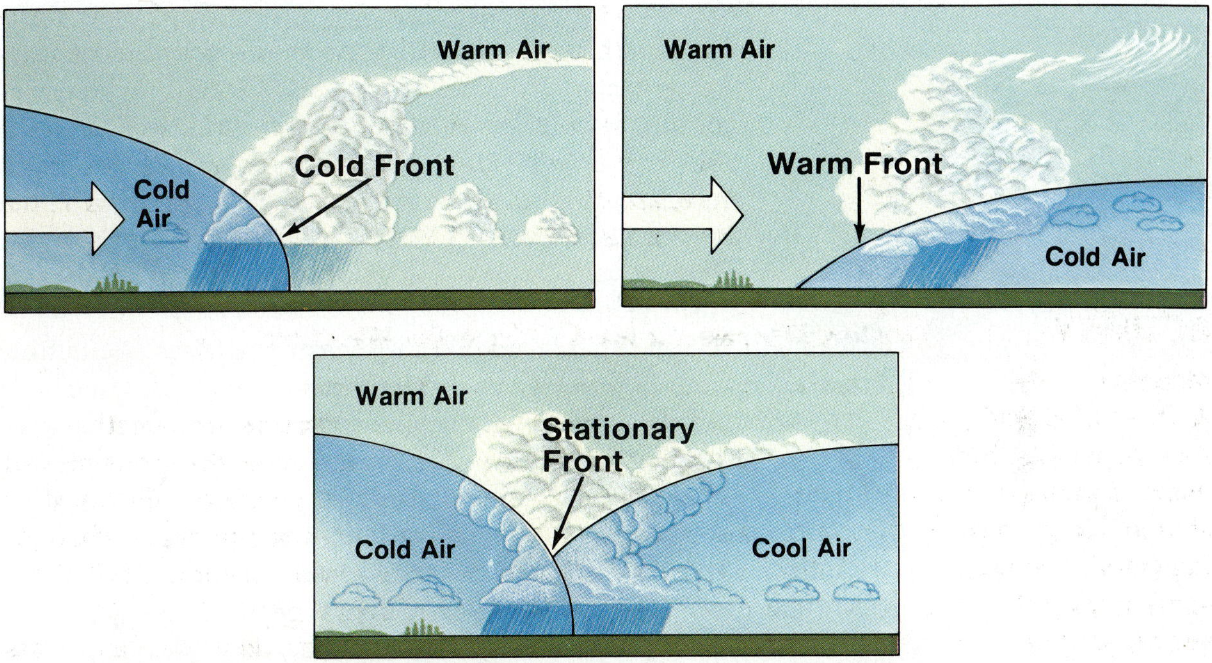

Figure 7.7 Weather fronts bring changes in local weather as they pass.

Fronts are shown as lines on a weather map. Different symbols are used to represent the different kinds of fronts. A symbol represents the place where the leading edge of a huge, flattened bubble of air touches the surface. When a front moves by, it may cause rapid and major weather changes. The changes caused by a front moving by are called **frontal weather.** Look at Figure 7.7 to see how a cold and warm front may change the weather. What type of weather are you having right now—air mass weather, cold front, or warm front?

Study Questions for 7.2

1. Name the source regions that produce the air masses that generally affect continental United States weather.
2. What is a cold front and what kind of weather follows it? A warm front?
3. What are the characteristics of "frontal weather"?

7.3 Forecasting

You will find out

- what is meant by highs and lows on a weather map;
- how a weather forecast is made.

Figure 7.8 Being prepared for the weather sometimes means guessing.

You may have looked outside this morning before leaving for school. You looked to see if it was sunny, if it was raining, or if it was windy. Perhaps you listened to the radio to see what to expect later in the day. If so, you were listening to a weather **forecast.** It tells you about the coming weather. Did you use the forecast to decide what to wear? You could also use it to plan what to do after school.

Sometimes a weather forecaster will say something about high pressure or low pressure. Or sometimes just the words *high* and *low* are used. When barometric pressure is marked on a map, areas of high and low pressure are found. These pressure areas are important to meteorologists because certain kinds of weather can be expected with each kind of pressure area.

Figure 7.9 This map shows one possible pattern of high- and low-pressure areas over the United States.

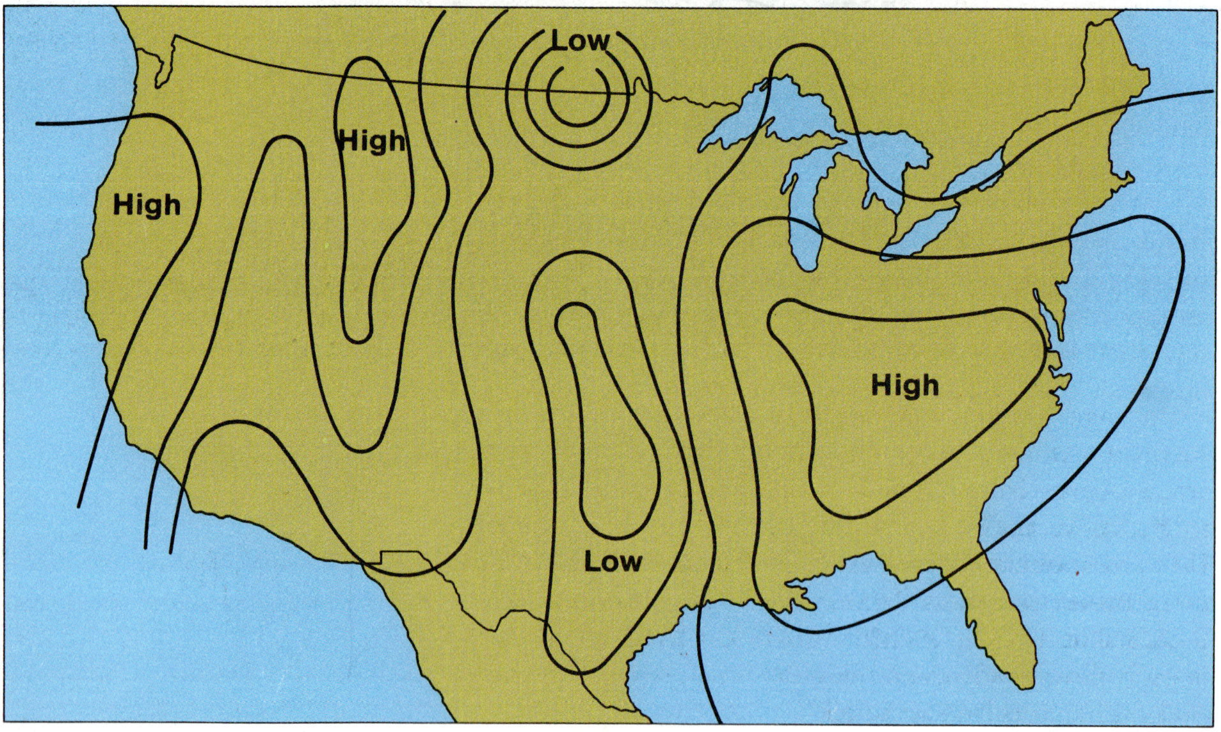

Suppose that a cold air mass moved over where you are. If you were watching a barometer, you would notice that the pressure goes up. Usually you are in a high-pressure area when you are in a cold air mass. The cold air produces a higher pressure because it is more dense than the warm air. The dense colder air sinks and spreads out. The downward movement of air is not very fast. It is usually not more than a few centimeters per second. It is enough, however, to warm the air some. This makes a clear sky. The air mass weather in a high-pressure area is usually fair. Most of the time high pressure means no clouds and fair weather.

Figure 7.10 *left:* The effect of a low-pressure air mass; *right:* The same location in a high-pressure air mass

On page 130 you learned that when a cold air mass meets a warm air mass you have a front. Sometimes a wave occurs in the front. As a result, some cold air might move under part of the warm air, lifting it up. Upward moving air makes a lower pressure. The upward movement cools the air due to its expansion. This makes clouds. The weather in a low-pressure area is usually cloudy with precipitation. Most of the time, low pressure means bad weather.

Figure 7.11 The prevailing westerlies help to push low-pressure areas east over the country.

Low-pressure areas and their associated bad weather usually follow certain tracks across the country. You can see some of the major tracks in Figure 7.11. According to Figure 7.11, where should you look to find out what kind of weather is moving toward Kansas?

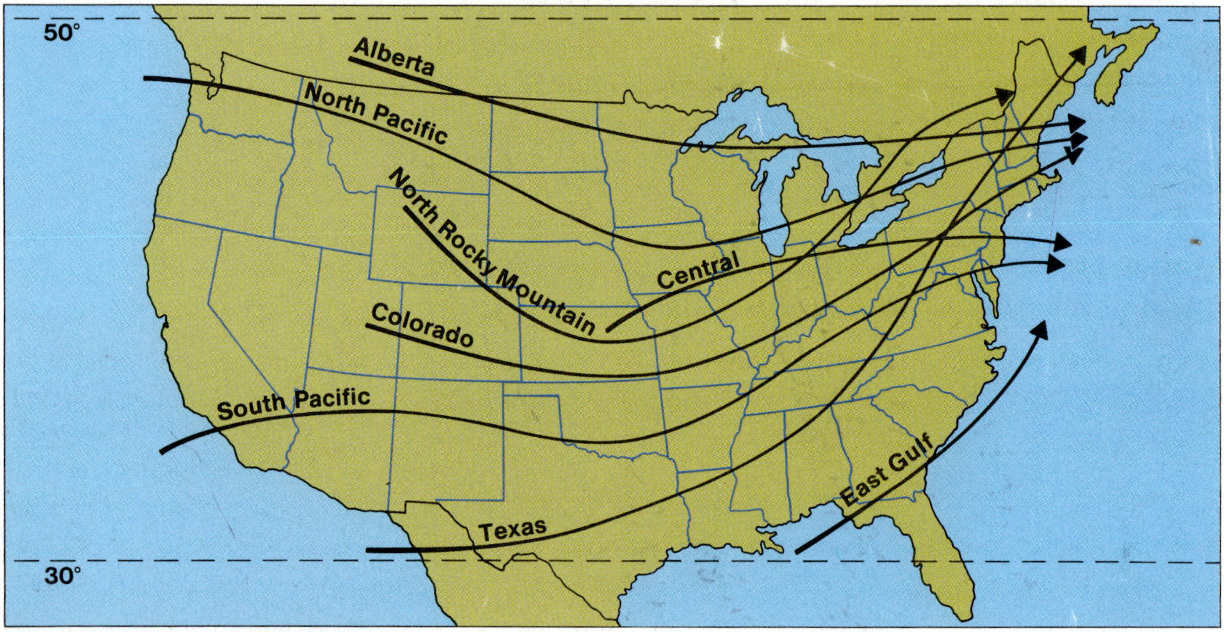

Figure 7.12 This meteorologist is recording data that are coming from other weather stations.

Weather forecasts begin with reports from the various weather stations. There are thousands of weather stations all across the country. People at the stations observe, measure, and report weather conditions. They describe the kinds of clouds over their stations. There are instruments to measure the temperature, relative humidity, pressure, wind, and precipitation. These observations and measurements are sent to regional offices of the National Weather Service. Meteorologists use the weather station information and information from weather satellites, radar, and their own instruments. They make a map that shows the location, size, and strength of high- and low-pressure areas. Fronts are then located and drawn on the map. Such a map and charts are used to determine the present state of the atmosphere. After all this is done, the meteorologists then must decide where the high- and low-pressure areas, air masses, and fronts will be tomorrow. From this, they can forecast where the clouds will be and if there will be precipitation. They also decide what the temperature will be and the kind of precipitation, if any. With all this information they then make a forecast. Such a forecast may state that there is a "30 percent chance of rain tomorrow." This means that when the conditions (air mass, front, pressure, etc.) are like they believe will occur, it is likely to rain three times out of ten. Three times out of ten is 30 percent. So the forecast is for a 30 percent chance of rain. With such a low chance for rain you can plan to have your picnic.

Do You Know?
The pressure difference between a high-pressure area and a low-pressure area is usually less than the pressure difference between the top floor of a skyscraper and the ground.

ACTIVITY

How Do Highs and Lows Help the Forecaster?

Materials
daily newspaper weather maps

Procedure

1. Study daily weather maps showing a high-pressure area and a low-pressure area.

2. Read the daily weather forecasts for the days before and after each pressure area goes by.

3. Study the daily weather maps until it looks like a high- or low- pressure area is approaching where you are. Make your own forecast.

Questions

1. Describe the National Weather Service forecast that was made after the high-pressure area arrived.

2. Describe the National Weather Service forecast that was made as the low-pressure area arrived.

3. Describe the forecast you made. Was your forecast correct?

Study Questions for 7.3

1. A weather forecast mentions high- and low-pressure areas. What are these and what do they mean about the weather?

2. A weather forecast states that there is a "70 percent chance for rain tomorrow." What does this mean?

7.4 Major Storms

You will find out

- what the characteristics of a thunderstorm are;
- what is considered to be the most violent kind of storm;
- what a hurricane is and the major kinds of damage it causes.

Figure 7.13 Cumulonimbus clouds such as these are also known as thunderheads.

The passage of many air masses, fronts, and pressure areas may go unnoticed. But they are sometimes accompanied by rapid and violent weather changes which are called **storms.** A snowstorm, for example, is a rapid change that a low-pressure area may bring. The three most violent kinds of storms, however, are thunderstorms, tornadoes, and hurricanes.

The cumulonimbus cloud that you studied in Chapter 6 is also called a **thunderstorm.** What is a thunderstorm? As the name suggests, it is a storm with thunder. The storm is an intense rainstorm with thunder, lightning, often strong winds, and sometimes hail.

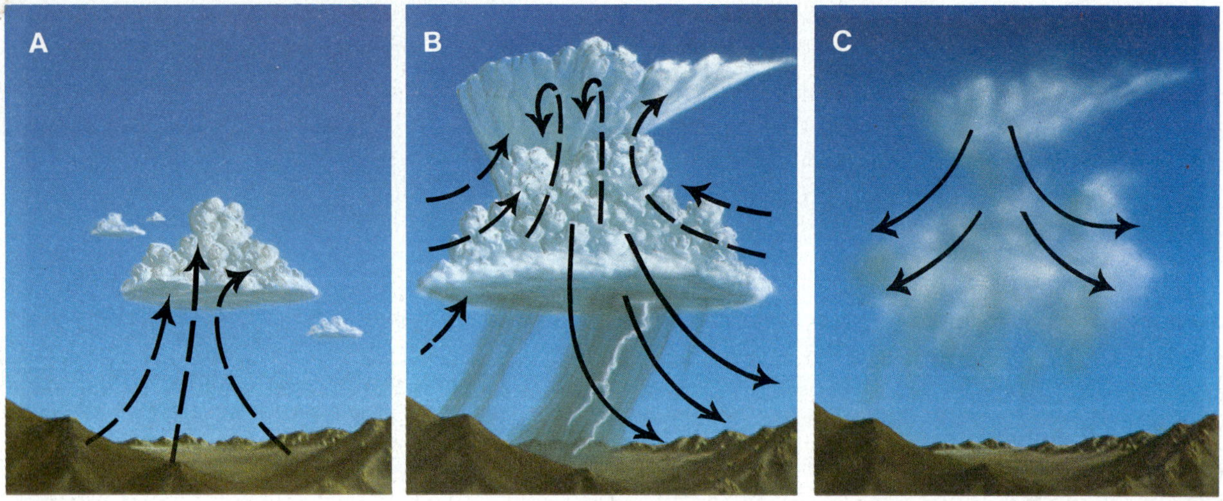

Figure 7.14 These are the stages of development of a thunderstorm cell. Dotted arrows show updrafts, and solid arrows show downdrafts. In *A*, updrafts form cumulonimbus clouds. In *B*, updrafts and downdrafts cause rainfall. In *C*, downdrafts mostly occur as the cloud disappears.

The updrafts and downdrafts in thunderstorm cells are responsible for hailstones and intense rainfall with large drops. The updrafts keep returning the hailstones and raindrops upward above the freezing level. When the rain and hail finally become too heavy for the updrafts, they fall to the ground. The result is a downdraft

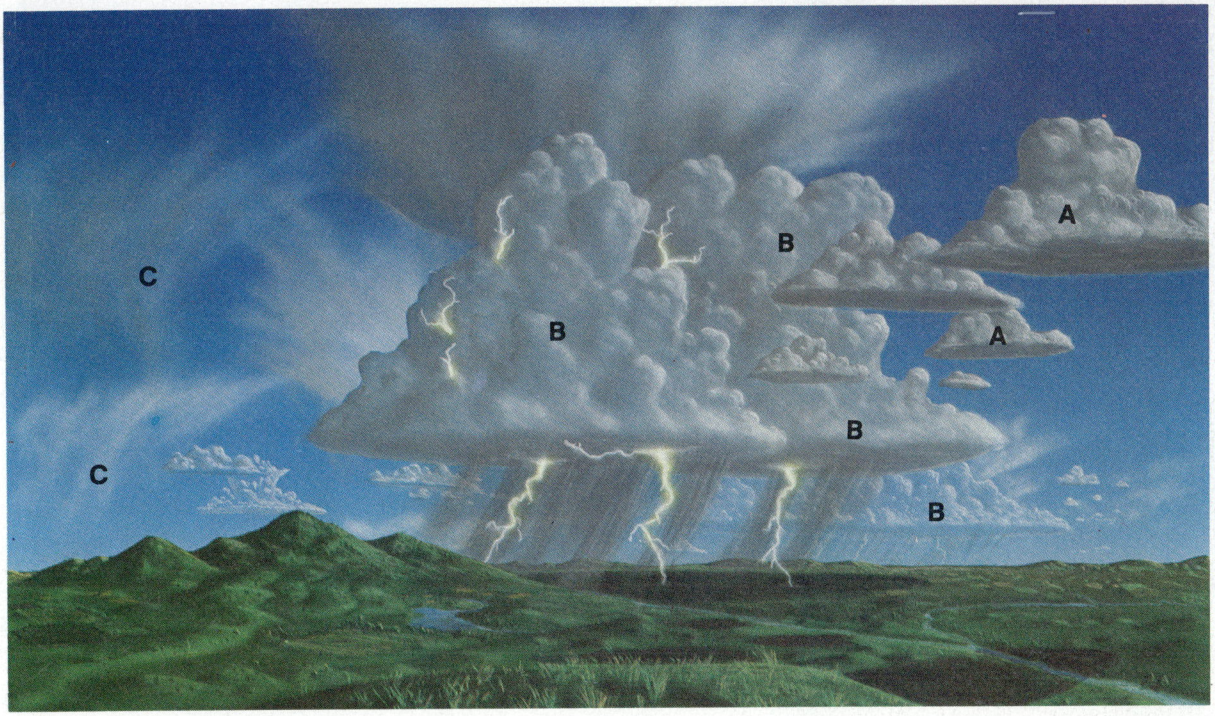

that produces a strong, cool wind that often arrives before the storm.

A thunderstorm is made up of one or more cells. Each cell is about 2 to 8 km in diameter. In Figure 7.14 you can see the three stages in the life of each cell. It usually takes less than an hour for a cell to go through all three stages. A thunderstorm, however, may last longer than an hour because new cells are formed as old ones die.

Air mass thunderstorms are the most common. More air mass thunderstorms occur in July than in any other month. These develop in warm, moist air from convection currents. They form mostly in the afternoon. Frontal thunderstorms can occur at any time of the day during any month. These develop as warm, moist air is lifted by a cold front as the front moves through.

Damage from a thunderstorm is usually caused by its lightning, wind, or hail. Lightning is really a giant electrical spark. The spark jumps between clouds, from a cloud to the ground, or from the ground to a cloud. Lightning heats the air, expands it, and produces a sudden pressure wave which you hear as thunder.

Figure 7.15 A thunderstorm moves over the land. The cells shown by letters here correspond to the stages of thunderstorm development in Figure 7.14.

Figure 7.16 Contrary to popular belief, lightning can strike the same place any number of times.

Figure 7.17 Luckily this tornado is over open land. Sometimes tornadoes are called twisters.

A **tornado** is the most violent storm that occurs on the earth. It is smaller but more violent than a thunderstorm or hurricane. A tornado often looks like a funnel or rope-shaped cloud that dips down from a cumulonimbus cloud. The diameter of the tornado is usually less than 400 m. It is a small low-pressure area around which air swirls violently up to speeds of 650 km per hour or more. If this swirling low-pressure area contacts the ground, it acts like a huge vacuum cleaner and causes considerable damage. It moves across the countryside at about 40 to 65 km per hour toward the northeast, and it may last up to an hour.

Most tornadoes occur in the afternoons of late spring or early summer. They are common in Kansas, Iowa, Arkansas, Texas, Mississippi, and Oklahoma. About 150 tornadoes occur in the United States each year.

A **hurricane** is a large, violent circular storm that is born over tropical bodies of water between the latitudes of 5 and 15 degrees. Hurricanes form from colonies of cumulonimbus clouds during late summer and fall when the bodies of water are the warmest. Once formed, they usually move to the northwest. If they reach north of 30° latitude, they usually start moving toward the northeast. Why do they move in this manner? How do the winds generally move north and south of 30° latitude?

Do You Know?
Hurricanes are given names that alternate between boys' and girls' names. The first hurricane of the season receives a name that begins with the letter *A*, the second hurricane one with a *B*, and so forth.

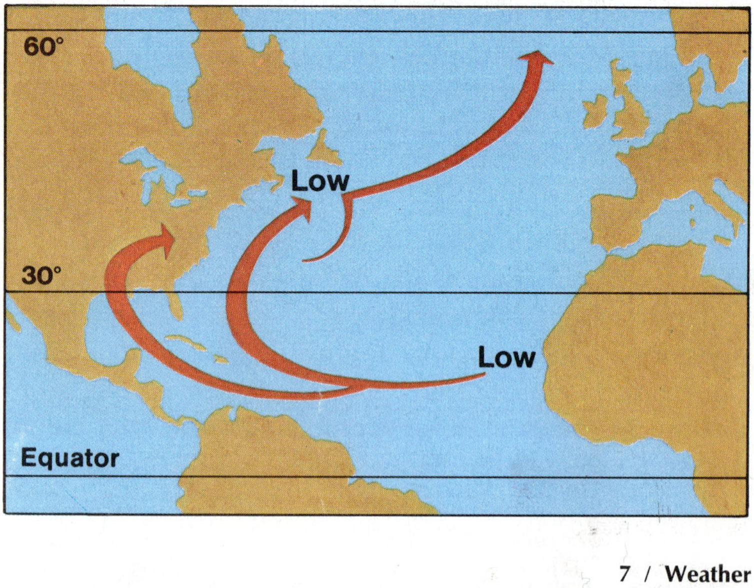

Figure 7.18 Common paths followed by hurricanes in the Caribbean Sea area

Figure 7.19 This man is watching the track of a hurricane with radar.

Figure 7.20 This is a computer radar picture of a hurricane. The calm blue area at the center is called the eye.

A hurricane may be from 150 to 650 km wide. In the northern hemisphere the wind blows around the center of the storm counterclockwise. These winds are greater than 120 km per hour and may reach 320 km per hour. The center or core of the system is called the eye of the storm. Little wind occurs in the eye and few clouds are present. The eye is usually about 15 to 30 km wide. As the eye of the storm passes an area it is calm for a short period of time.

Hurricanes may last up to nearly two weeks as they move across the ocean. They move at a rate of 15 to 50 km per hour. Most of the damage from hurricanes results from winds and flooding. Strong winds cause a "pileup" of water along the coast. Increases in sea level can be 4 to 6 m above normal. When storm waves of 10 to 15 m pile up above this raised sea level, large inland areas can be flooded. Hurricanes have caused extensive property damage and loss of life.

Now the National Weather Service can provide early hurricane warnings. Based on information from weather satellites, these early warnings have greatly reduced loss of life from hurricanes. Property damage, however, can be very large depending on the track of the storms. Hurricanes sometimes have thunderstorms with tornadoes that add to the overall damage.

Study Questions for 7.4

1. List the major characteristics of a thunderstorm.
2. What are tornadoes? Where do they form?
3. Describe a hurricane. How does one cause damage?

What Did You Learn?

- An air mass is a large part of the atmosphere that has much the same temperature and moisture conditions as the earth's surface it is over.
- Air masses form where air sinks and spreads out slowly across the surface.
- Air masses form in source regions where they acquire conditions of temperature and moisture similar to the region.
- Air masses can leave where they form and move across the surface of the earth.
- Air mass weather is what we experience while inside an air mass. Changes are slow and gradual.
- The area where two air masses come in contact is called a front.
- The weather change that happens when a front moves by is called frontal weather.
- Fronts are called cold fronts, warm fronts, or stationary fronts.
- Areas of high and low atmospheric pressure are associated with certain types of weather.
- High pressure is associated with dry air masses and fair weather.
- Low pressure is usually associated with fronts and cloudy weather with precipitation.
- Weather forecasts are made from information about air mass movement, fronts, and high- and low-pressure areas.
- Storms are rapid and violent weather changes.
- A thunderstorm is a cumulonimbus cloud with accompanying thunder, lightning, strong winds, heavy rain, and sometimes hail.
- The tornado is a funnel-shaped cloud that dips down from a violent cumulonimbus cloud. It can be the most violent of all storms.
- A hurricane is a large circular storm that forms over tropical waters and moves westward and then northward in the northern hemisphere. Winds in a hurricane may cause coastal flooding in addition to wind damage.

Key Terms

air mass
source region
continental polar
maritime tropical
maritime polar
air mass weather
front
cold front
warm front
stationary front
frontal weather
forecast
storm
thunderstorm
tornado
hurricane

Biography

Carl-Gustaf Rossby (1898-1957)

Carl-Gustaf Rossby, an assistant chief of the United States Weather Bureau in the 1940's, was born in Stockholm, Sweden, in 1898. After studying mathematics and astronomy Rossby began to work as a meteorologist in Stockholm. He came to the Weather Bureau in the United States in 1926 on a fellowship and later became a United States citizen. Rossby was an early pioneer in providing a weather service for aviation.

In 1941 he was appointed chairman of a new department of meteorology at the University of Chicago. Here, he organized a program to teach the military about meteorology.

In addition to organizing and promoting the study of weather, Rossby found time to make his own studies. While studying air masses, Rossby designed a graphical method to identify them. The graphical method is now called the Rossby diagram. This diagram helps explain how air masses form and what changes them.

Rossby also studied the worldwide movement of the atmosphere. He investigated waves in the atmosphere (now called Rossby waves) and found they influenced our weather. He then developed a model and a formula to measure their speed. The formula is famous and is called the Rossby equation.

Rossby's work made numerical forecasts possible. In 1940 he and his co-workers made the first numerical forecast. His work and the introduction of computers in the 1940's set the stage for the forecasting techniques of today.

TO THINK ABOUT AND DO

Vocabulary

Some of the words in the following sentences have scrambled letters. Unscramble each word and write the correct word on a sheet of paper.

1. A large part of the atmosphere with temperature and moisture like the surface is a(an) RIA ASMS.
2. The birthplace of an air mass is called a(an) ROUSCE GIREON.
3. The NEITCANOTLN OPRAL air mass is cold and dry.
4. The RIMTAEIM APOICRLT air mass is warm and moist.
5. The TAMEIRIM APLOR air mass is cold and moist.
6. RAI SASM HEWERAT has slow, gradual changes.
7. RATOFLN weather has rapid, major changes.
8. The thin zone between two air masses is a(an) TORNF.
9. A moving cold air mass makes a(an) ODLC ONRFT.
10. A moving warm air mass makes a(an) RAMW RONTF.
11. A(An) NOTRAYTAIS NOFTR separates two air masses that are not moving.
12. A prediction about the weather is a(an) ARTHEEW ARTCOFES.
13. The most violent of all weather changes is a(an) TOMRS.
14. The most violent of all storms is a(an) TONRDOA.

What Do You Remember?

Number your paper from 1 to 8. Write the letter of the correct answer beside the appropriate number.

1. A moving cold air mass makes a (a) cold front, (b) warm front, (c) stationary front.
2. Fair weather is associated with (a) low-pressure areas, (b) high-pressure areas, (c) zero-pressure areas, (d) medium-pressure areas.
3. Air masses (a) always stay where they form, (b) can move across the surface.
4. A continental polar air mass is (a) cold, (b) moist, (c) dry, (d) both cold and dry.

5. A maritime polar air mass is (a) cold, (b) moist, (c) warm, (d) both cold and moist.
6. A maritime tropical air mass is (a) cold and dry, (b) moist, (c) warm and moist, (d) dry.
7. The leading edge of a moving air mass is called a (a) front, (b) middle, (c) back, (d) side.
8. Rapid weather changes occur in (a) air mass weather, (b) frontal weather.

Applying What You Have Learned

1. Why do you suppose that hailstones fall only from thunderstorms?
2. Why do tornadoes do so much damage?
3. How do hurricanes do their damage?
4. Compare how the weather changes *as* a cold air mass arrives where you are and *after* the air mass arrives.
5. Explain how a weather forecast is made.
6. What general direction do low-pressure areas take in crossing the continental United States? Why do they follow these tracks?

Research and Investigation

1. Track a thunderstorm. Obtain a compass, a city map, and straight pins. Use the compass to determine the direction of a lightning flash. Judge the distance of the flash by timing how long it takes for the thunder to reach you after you see the lightning. Count "1001, 1002, 1003," etc., at a normal speed to count seconds. It takes about three seconds for sound to travel one km. Stick pins in the map to trace the path of the thunderstorm. Prepare a bulletin-board report showing your map.
2. Plan a field trip to a weather observing station. What instruments are used? Find out what happens to the measurements that are made.

This is an example of the destruction left by a tornado. Can you see the path the tornado followed?

8

Climate

You know that the weather may change from day to day. The weather this week is probably different from the weather last week. Some months are colder than other months. But suppose you look at the temperature averages for a location. You will find a pattern each year. This pattern does not change much. The location where oranges grow, for example, has warm average monthly temperatures throughout the year.

Different monthly temperature and precipitation patterns occur in different locations. These determine the kinds of plants and animals that live in the locations. The patterns also determine what kinds of houses are built. The patterns even determine what people do for recreation. Consider skiing, for example. Would you expect to find a ski resort close to the area where orange trees grow? What kinds of weather patterns would you expect in places where a ski resort is located?

This chapter is about weather patterns that occur in different locations. You will find out what causes these patterns.

Figure 8.1 Citrus groves in Florida

8.1 Sunlight and Precipitation

You will find out

- the difference between weather and climate;
- how sunlight makes temperature patterns;
- how atmospheric movement makes precipitation patterns.

Figure 8.2 What tells you that the moose, the forest, and the rafters share the same climate? The clue is the abundant amount of water.

Suppose two of your best friends move to two different places. Before they leave, both promise to write and send you pictures. Several months later, their letters arrive at the same time. Someone opened the letters, however, and the pictures were mixed up. Which of the pictures go together?

It looks as if your friends moved to very different places. How are the places different? Would you say that they have different weather? That might be true, but on any given day the weather in both places could be the same. The major difference in the two places is their average weather measured over a long period of time. The average weather over a long period of time is called **climate.** The two places have a different climate.

Do You Know?
Weather may change
from day to day.
The climate we have
is here to stay.

Weather changes from day to day and from week to week. The weather this winter may be very different from the weather last winter. But the climate of a location does not change so much. Climate describes the averages, totals, and the extremes for weather factors over long periods of time. Table 8-1 compares some of these weather factors for the places where your two friends moved. How would you describe the climate of the places where your friends live?

As you can see, climate is really weather considered over a long period of time. The two most important climate factors are temperature and precipitation. Let's see what determines these two climate-making factors. First, let's find out about temperature.

Table 8-1 Climatic Data for Two Locations

Temperature	Site 1	Site 2	Precipitation	Site 1	Site 2
Average January High Temperature	18 °C	-3.3°C	Average Rainfall – Wettest month	2.5 cm	14.4 cm
Average January Low Temperature	1.6°C	-14.4°C	Average Rainfall – Driest month	0.2 cm	6.6 cm
Average July High Temperature	41 °C	30 °C	Average Rainfall for Year	18.2 cm	124.9 cm
Average July Low Temperature	24 °C	13 °C	Seasonal Average for Snow and Sleet	0.0	190.5
Average Temperature for Year	21 °C	6 °C	Percent of Yearly Possible Sunshine	84%	50%
Record Highest Temperature	47.8°C	37 °C	Number of Days with Thunderstorms per Year	26	43
Record Lowest Temperature	-8.8°C	-49 °C	Number of Days with Fog per Year	1	21

Temperature, as you know, is determined by how much sunlight is received. You learned in Chapter 5 (page 98) that different parts of the earth receive different amounts of sunlight. The greatest amount is received near the equator. The least amount is received at the poles. The yearly average temperature should therefore decrease as you move from the equator to the poles. In general, the temperature does decrease in this way.

Figure 8.3 The locations of the three major climate regions on the earth

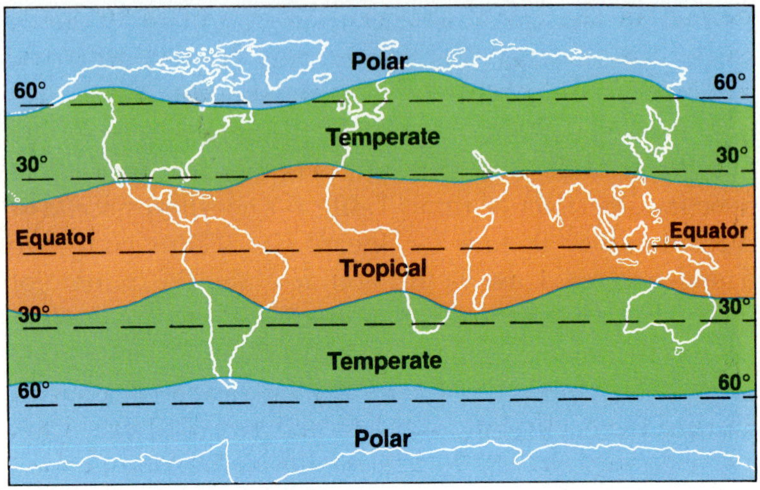

Figure 8.4 Vegetation in the three types of climate varies greatly. *top:* Polar; *middle:* Temperate; *bottom:* Tropical

There are three major groups of climates that are based on yearly temperature averages. These occur in broad zones, or regions, as shown in Figure 8.3. They are (1) the **tropical** climate zone, (2) the **polar** climate zone, and (3) the **temperate** climate zone. There are subdivisions that occur within each of the three major zones. These will be discussed later.

The tropical climates are near the equator and receive the most direct sunlight all year. The average temperature during the year is hot, staying above 18°C.

The polar climates are near the poles. The poles receive a lot of sunlight in the summer. In fact, the sun never sets during part of the summer. However, it never rises during part of the winter. Overall, the polar climates are cold. The average temperature during the year is 10°C or less.

The temperate climates are between the polar and tropical climates. Temperate climates are neither very cold nor very hot. The average temperature during the year is between 10°C and 18°C.

The distance from the equator also determines precipitation patterns. See the model of atmospheric pressure air movement on page 101. The model has air moving upward near the equator. This moist air cools and forms clouds, which result in rain. On the other hand, air is slowly sinking near 30° latitude north or south, which makes it dry all season. Most of the great deserts of the world are near 30° north or 30° south latitude for this reason. There is another wet zone near 60° latitude and another dry zone near the poles.

These wet and dry zones move north, then south during the year. The wet and dry zones move in this way because the sun's rays hit the earth at different angles during different seasons. The sun's rays are received at the most direct angle in the northern hemisphere on June 21 or 22. They are received at the most direct angle in the southern hemisphere on December 21 or 22. After December 21 or 22 the wet and dry zones are moving toward the north. Then, after June 21 or 22, they start to move toward the south. This results in different precipitation patterns in each season. Figure 8.6 shows where the wet and dry zones are in winter and in summer. Imagine these zones moving north, then south, between two places. You see how the atmospheric movement produces precipitation patterns that change with the seasons.

Figure 8.5 The angle of the sun's rays striking Earth changes throughout the year. The northern and southern hemispheres have opposite seasons.

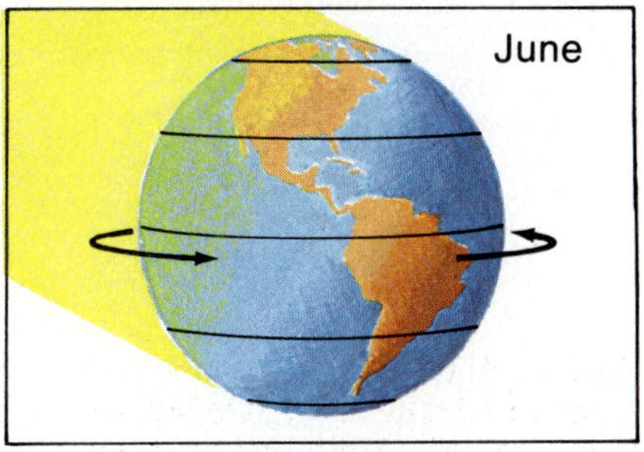

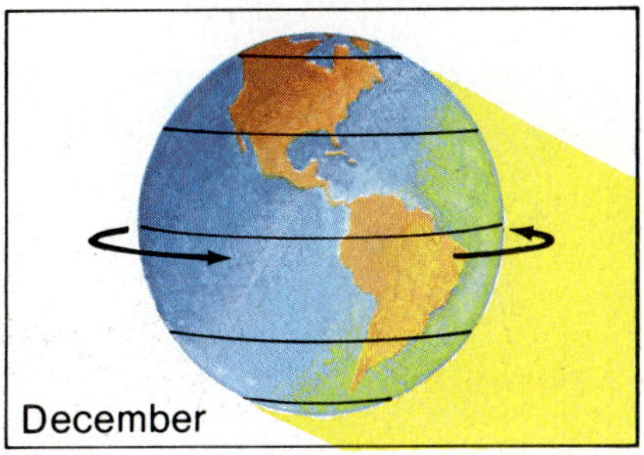

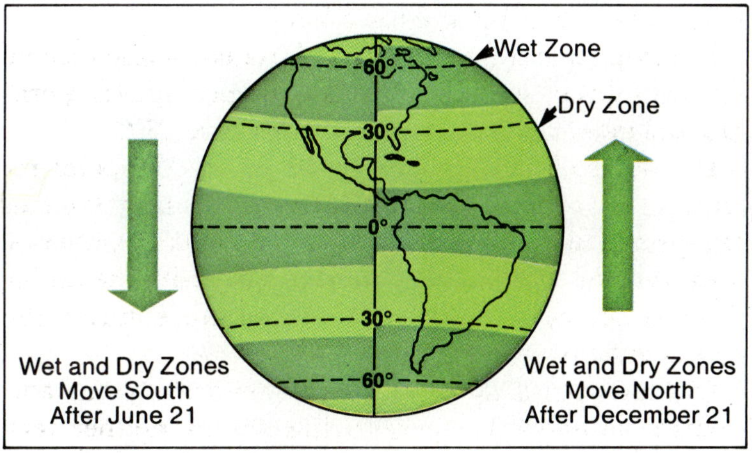

Figure 8.6 Compare the positions of the wet and dry zones at different times of the year.

Study Questions for 8.1

1. What is climate? How is it different from weather?
2. Name three major groups of climates that are based on yearly temperature averages.
3. Why would you not expect to find a tropical climate that is dry all year?
4. Why would you not expect to find a polar climate that is wet all year?

8.2 Climate: Local Influence

You will find out

- why mountains are climate makers;
- how climate changes with altitude;
- why large bodies of water are climate makers;
- how ocean currents influence climate.

So far you have looked at the kinds of climates caused by large zonal factors. These included such factors as tropical areas, polar areas, wet areas, dry areas. There are also four local influences that affect nearby climates. These are (1) altitude, (2) mountains, (3) large bodies of water, and (4) ocean currents. Let's find out why.

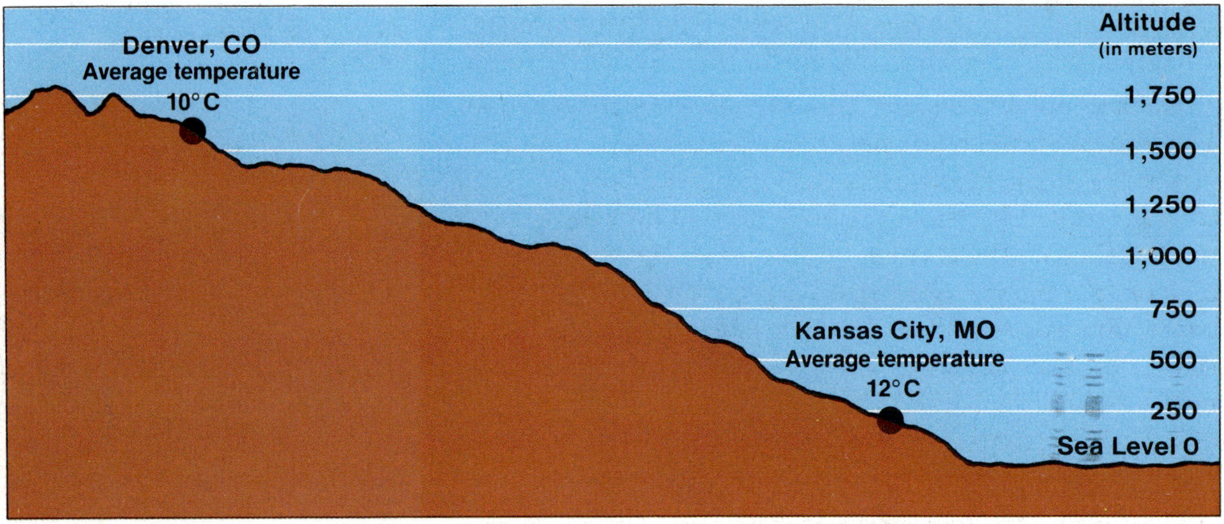

Figure 8.7 The difference in altitude between Kansas City and Denver is one reason their climates are different.

The first of the four local climate makers is altitude. As you go to higher altitudes, places will have lower average temperatures. Denver, Colorado, and Kansas City, Missouri, for example, are both in the temperate climate zone. They are almost at the same latitude. You would expect the two cities to have about the same average temperature. Denver, however, has an altitude of 1,609 m, and the altitude of Kansas City is 228 m. The average yearly temperature for Denver is about 10°C and for Kansas City it is over 12°C. Altitude makes the difference. In fact the temperature decreases about 3°C for each 500 m rise in altitude.

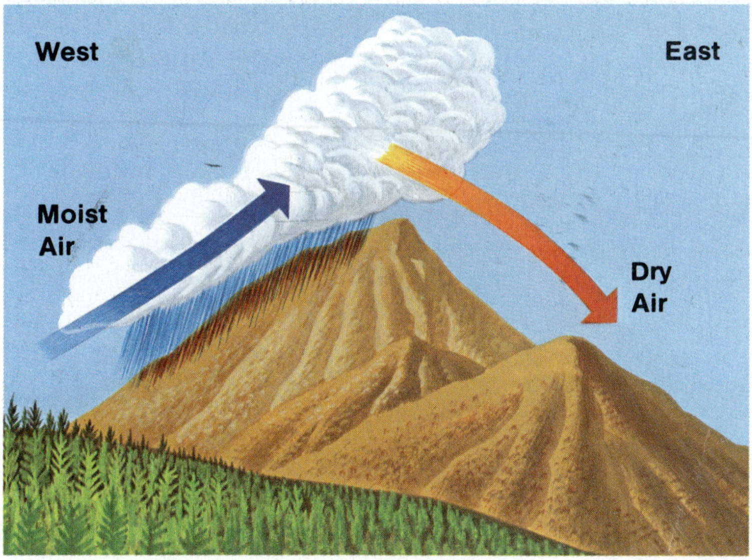

Figure 8.8 *left:* Wet air rising up the western side of the mountain produces rain. There is little moisture left in the air as it moves down the eastern slope.

Figure 8.9 *top:* Pear orchards grow well on the moist, western side of mountains in Oregon. *bottom:* The eastern slope has only desert grasslands.

The second of the local climate makers is mountains. You can be in the snow on a mountaintop even though you are near the equator. In addition to the temperature change caused by the altitude of the mountain, mountains influence the climate another way. Look at a map of North America. There are mountains on the west coast that run north and south. The western side of these mountains receives much precipitation. There are farms and fruit trees there. On the eastern side there is a desert. Only plants such as cactus can live on the eastern side.

Why is this so? You learned earlier that the earth has prevailing wind belts. The United States is in the prevailing westerly wind belt. These westerly winds move maritime air inland. This causes the west coast to have cooler summers and warmer winters than the east coast. The winds also bring moisture from the Pacific Ocean.

Moist air from the Pacific is forced up the mountain slopes by the prevailing westerlies. As the air moves up the slopes, it expands and cools. Water vapor condenses, clouds form, and precipitation forms. The precipitation falls on the western side of the mountains, the side that the wind pushes against. Air moving down the eastern slope is compressed. It becomes warmer and dryer, and deserts are found here.

Figure 8.10 *top:* San Diego; *bottom:* Dallas

Figure 8.11 Major ocean currents that influence climate in North America.

The third of the local climate makers is a large body of water. The climate of a seacoast is influenced by a large body of water, the sea. Large bodies of water increase the humidity. Nearness to a large body of water influences the temperature as well. If it is hot inland, the coast will not be so hot. If it is cold inland, the coast usually will not be so cold. San Diego, California, and Dallas, Texas, for example, are at about the same latitude. Dallas, however, is inland, and San Diego is at the seacoast. The average summer temperature is 7°C cooler in San Diego. San Diego has 5°C warmer average winter temperature than Dallas. Nearness to water keeps the temperature more even at San Diego. In general, nearness to large bodies of water tends to prevent extreme or rapid changes in weather.

The fourth of the local climate makers is ocean currents. Ocean currents can transport warm water northward or cool water southward. These currents will influence the temperatures of the air mass over the area. As this air mass moves inland, it also affects the areas near the seacoast. Look at Figure 8.11. What ocean currents are influencing the east and west coasts of North America?

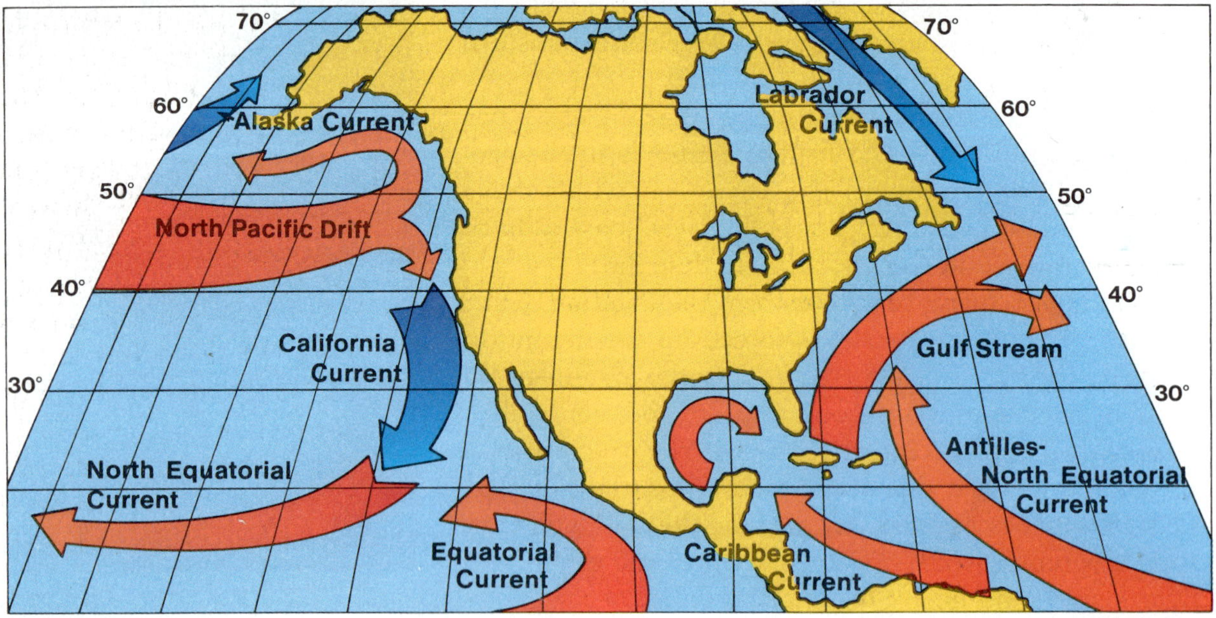

ACTIVITY

What Influences Your Climate?

Materials

map of the United States
World Almanac or *Information Please Almanac*

Procedure

1. Find where you live (or a nearby large city) on a United States map. Estimate and record the latitude.

2. Identify another large city at about the same latitude that has the opposite climate-making influences to your location. For example, if you live near a seacoast, identify a city that is inland. Also find an opposite city for (1) northern United States and southern United States and (2) high and low altitude.

3. Find (1) the average yearly temperature, (2) the average high temperature in the summer, and (3) the average low temperature in the winter for each location. Record the information in a table like the one shown.

Questions

1. What is the latitude where you live?

2. Does altitude influence the climate where you live? What evidence can you give for your answer?

3. Does nearness to a large body of water influence the climate where you live? What evidence can you give?

City	Latitude	Average yearly temperature	Average high temperature	Average low temperature	Altitude	Nearest large body of water

Study Questions for 8.2

1. What happens to the climate at higher altitudes?
2. Describe how mountains are climate makers.
3. What are two influences that a large body of water has on climate?
4. How do ocean currents modify climates?

8.3 Describing Climates

You will find out

- how to describe climates;
- what description to give the climate where you live;
- what a microclimate is.

How can you describe the climate where you live? First, you have learned that there are three major climate zones on the earth. The polar, temperate, and tropical zones are based on temperature. Second, you have learned that certain precipitation patterns occur because of atmospheric movement. Some places are wet all year, and others are dry because of the movement. Third, you studied the additional influence of mountains, large bodies of water, altitude, and ocean currents. Now, let's see how all this information is used to describe a climate.

Figure 8.12 Which of these climate types appeals to you? Look at the chart below.

Climate Type	Description	Location in North America
Polar	very cold, long winters, cold summers – dry	Northern Canada, Alaska, Greenland
Humid Continental (Sub-arctic)	long cold winters, cool summers – moderate precipitation	Most latitudes of Alaska, Canada
Humid Continental (Middle Latitudes)	cold winters, moderate summers – moderate precipitation	East of the Rockies in Northern U.S., south Canada
Humid Continental (Low Latitudes)	mild winters, hot summers – moderate precipitation	Inland U.S., east of the Rockies to Atlantic Coast
Humid Sub-tropical	short, mild winters, humid summers – moderate precipitation	Southeastern U.S.
Tropical Wet	hot and humid all year – heavy precipitation	Hawaii, eastern Mexico
Tropical Wet/Dry (sub-tropical)	hot all year – alternating wet and dry seasons	Hawaii, south Florida, coasts of Mexico
Semi-Arid	varying temperatures – low rainfall	U.S. Great Plains, parts of Canada
Desert	hot with cold nights – arid	Inland southwestern U.S.
Marine	mild winters, rainy summers, moderate temperatures	Northwestern coast U.S., western coast of Canada
Mediterranean	short, mild, moist winters, hot dry summers	Southwestern coastal U.S.
Highlands	temperature and precipitation vary with altitude and latitude	Mountainous regions in U.S., Canada, Mexico

In addition to the three major climate zones of polar, temperate, and tropical, certain areas between these zones are classified as subpolar or subtropical. A subpolar zone is between the temperate and polar zones. A subtropical zone is between the tropical and temperate zones.

There are subdivisions within each of the three major world climate zones. Within a major zone, for example, there may be an area near the ocean. This area may be influenced by air masses from the oceans. If so, the area has a **marine** climate. As you learned earlier, the ocean makes temperatures more even. Areas with a marine climate are cooler in the summer and warmer in the winter than farther inland.

Also within a major zone there may be an area that is far from an ocean. This area may be influenced mostly by air masses from large land areas. If so, the air has a **continental** climate. The land heats and cools rapidly. Summers are hot and winters are cold. A continental climate does not have as even temperatures as marine climates.

Climates can also be classified as either **arid,** meaning "dry" or **humid,** meaning "moist." Arid climates receive less than 25 cm of precipitation a year. Humid climates receive 50 cm or more of precipitation a year. If an area receives between 25 and 50 cm of precipitation, it is called **semiarid.**

Figure 8.13 Examples of vegetation types in the four major climate areas of North America

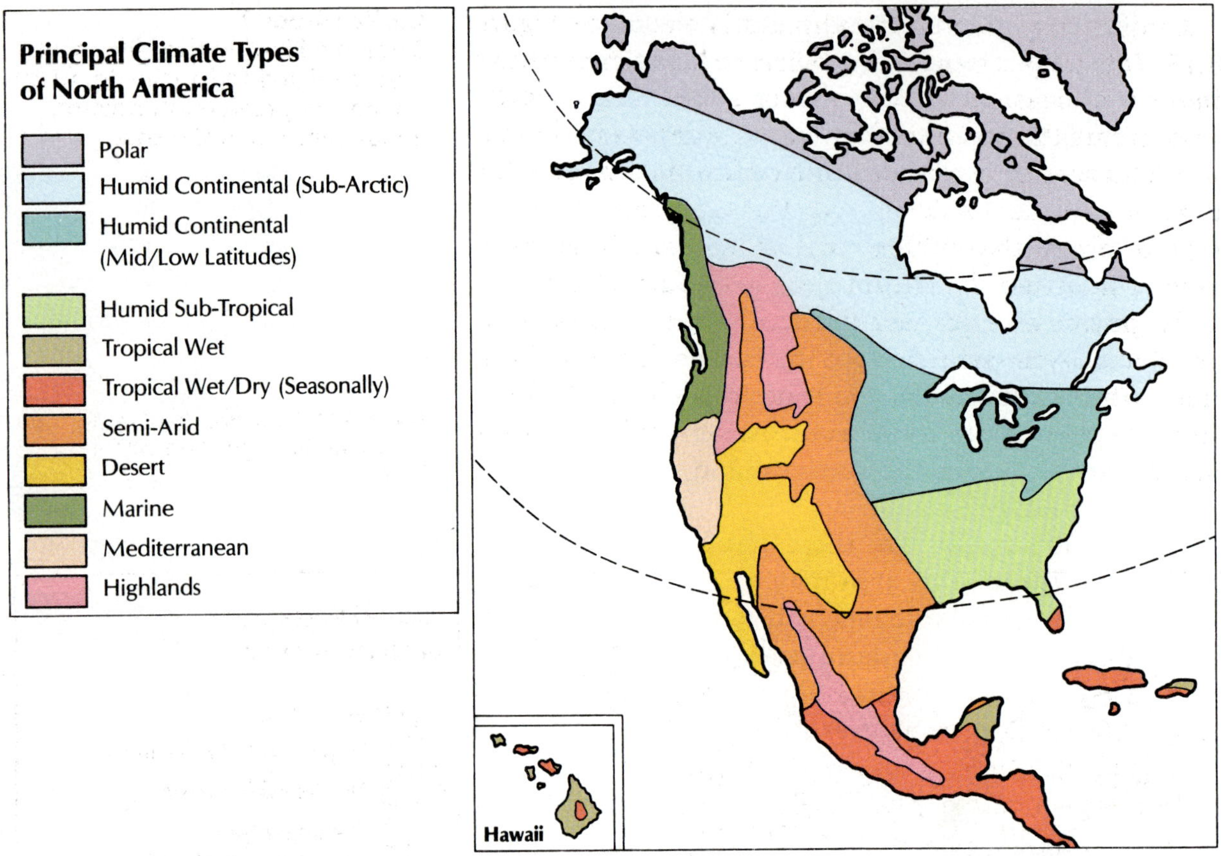

Figure 8.14 Refer to Figure 8.12 for descriptions of these climate types.

The map in Figure 8.14 shows the different climate areas of the United States. This is a generalized picture of the different climate areas according to temperature and precipitation patterns as determined by latitude and local influences. The areas are said to be generalized because some other feature close by may change the local climate. Also the climate really doesn't change all of a sudden if you move across one of the lines. The climates really blend into each other. Find your location on the map. What is the climate called where you live? Whatever the climate is called in your area, remember that the two basic determiners of climate are the amount of precipitation and the average temperature. The name given to the climate generally relates to conditions that influence one or both of these. For example, nearness to an ocean, an inland location, altitude, latitude, or mountains all have an influence on climate.

Another way to describe climate is shown in Figure 8.15. This map is based on a **moisture index.** A moisture index is a measure of the moisture gained by precipitation minus the moisture lost by evaporation and plant use. This way of describing climate has meaning for the type of soil that is developed, the landscape, and the type of plants that will grow. Forests grow in areas of greater moisture than those areas where grasslands are. Only grasses and desert plants can grow in areas that are dry or semi-arid. Study Figure 8.15 for your location. What kinds of plants are shown where you live? How do they relate to the amount of rainfall in your area?

Do You Know?
Since the kind of plant life in an area is related to its climate, fossil pollen can provide information about climates of the past.

Figure 8.15 How does this map represent the vegetation where you live?

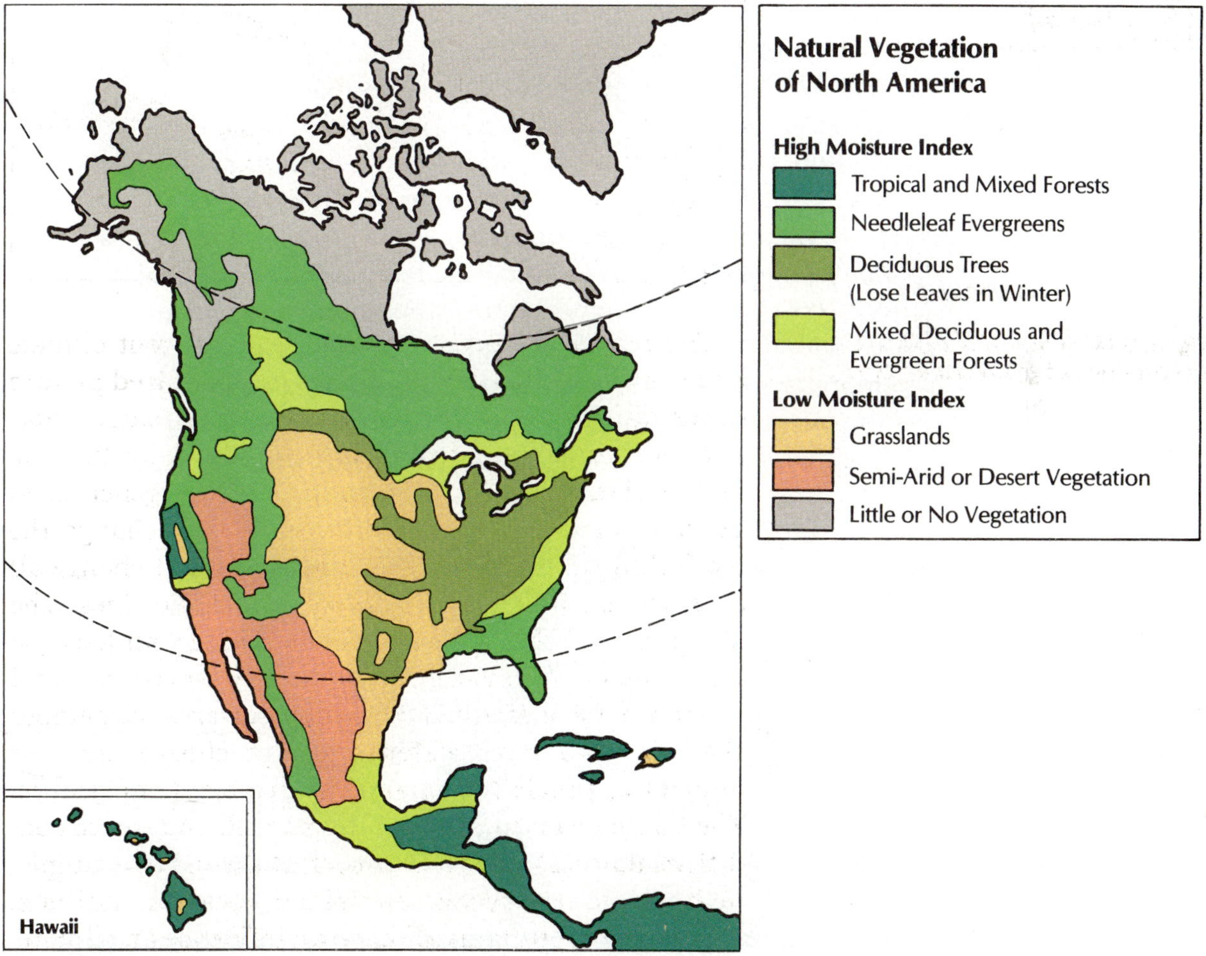

Do You Know?
The 1883 volcanic eruption of Krakatoa threw dust in the atmosphere which circled the earth with the result that summers were cooler for about five years afterward.

Once you describe the climate of a region, do you think the climate would be the same throughout the region? In general your answer should be yes, but the climate of any specific location in a region may change. New types of vegetation, buildings, roads, and water reservoirs can change the local climate. Local patterns of climate are called **microclimates.** A microclimate usually exists, for example, between the north and south side of a building. Certain plants may grow in one microclimate but not in the other. On a larger scale,

ACTIVITY

Where Is a Microclimate?

Materials

thermometer
wet bulb thermometer
water

Procedure

1. Choose two areas in which you can make measurements. One area should have nearby buildings, asphalt streets, and parking lots. The other area should have grass, trees, and plants, such as a park.

2. Measure and record the temperature and relative humidity (see page 110) in both locations. Report and record representative measurements on five separate days. Remember that representative measurements require that you follow the same procedure each day.

Questions

1. What were the temperature and relative humidity in the two areas on each day?

2. What were the average temperature and relative humidity for the two areas?

3. What did your data tell you about the two areas?

4. What would cause any differences you may have found?

Day	Grassy Area		Building Area	
	Temperature	Relative Humidity	Temperature	Relative Humidity
1				
2				
3				
4				
5				
Total				
Average				

trees are sometimes planted to break the force of the wind, called a windbreak. This makes a microclimate. The trees not only slow the wind but also may make the land warmer in the winter and cooler in the summer. They also increase the relative humidity. Many people are now considering microclimates when they build new buildings. How would people benefit from considering microclimates?

Figure 8.16 *left:* The tree is an example of a natural microclimate in which moss can grow. *right:* This microclimate was made by people.

Have you read or heard a weather forecast that predicts a different temperature for a city and for its suburbs? Cities create microclimates. The asphalt, concrete, and bricks of a city absorb and radiate energy differently from soil and plants. Temperatures are therefore usually higher in cities than in the country.

Cities also change the climate in other ways. Rainfall collects and runs off sidewalks, buildings, and streets. In the country, the rainfall soaks into the ground. Cities have less grass and trees. The humidity is usually lower in cities than in the country.

Figure 8.17 Smog is an artifical microclimate that is unwanted.

Factories and automobiles pour gases and small particles into the atmosphere. These particles serve as condensation nuclei, increasing the amount of fog that forms. When the fog combines with exhaust gases and smoke, it is called **smog.** Smog helps to change the climate of a city. It makes the city warmer at night. It also reflects radiation during the day, making the high temperatures cooler. This makes the smog last longer.

Study Questions for 8.3

1. What are the two basic determiners of climate?
2. What is the climate where you live?
3. What is a microclimate? What causes it?

What Did You Learn?

- Climate is the average of the many kinds of weather that occur at a given place over a period of time.
- The two most important factors that determine climate are temperature and precipitation.
- The major groups of climates of the earth are tropical, polar, and temperate.
- Worldwide atmospheric movement makes precipitation patterns at different latitudes.
- Climates are modified by mountains, large bodies of water, ocean currents, and altitudes.
- Climate can be described by seasonal temperatures and precipitation patterns.
- Climate can also be described in terms of precipitation and moisture lost by evaporation and plant use.
- Regional climates may be changed by local influences. These patterns are called microclimates.
- Cities can create microclimates. A city is usually warmer, less humid, and has more fog than the country.

Key Terms

climate
tropical
polar
temperate
marine
continental
arid
humid
semiarid
moisture index
microclimate
smog

Career

Climatologist

Climate is weather considered over a long period of time. But there is more to climate than average temperature, rainfall, and snow depth measurements. There are questions about the climate. Is the climate changing? Is the earth's climate becoming warmer? If so, by how much? And if it is changing, what is causing the changes? These are some of the questions that climatologists, scientists who study climates, try to answer.

Kristina Katsaros is a climatologist with a university. As a research professor, she can choose the questions about the climate that she wants to investigate. She then tries to obtain the money to pay for the equipment, supplies, and computer time that she needs to investigate her question. Then she works with her students to collect the information that will help to answer her question about the climate. Kristina Katsaros says it is exciting to think up a question, conduct an experiment, and then find out answers. She says that her work never becomes dull and is always changing.

One of her questions is about how sunlight, the atmosphere, and the oceans interact. She uses ships, airplanes, and satellites to measure sea surface temperature, wind speed, and humidity. This work deals with the energy transfers which move both the atmosphere and the ocean. What she finds out, for example, will make it possible for other scientists to verify or reject theories about climates.

Kristina Katsaros says, that as a working scientist, there are prices to be paid in sacrifices and discipline, but the rewards are worth it.

TO THINK ABOUT AND DO

Vocabulary

Write the numbers one through seven on a sheet of paper. Read the numbered phrases. Write the word or words in the list that best fit the meaning for each phrase beside the appropriate number.

climate	continental	polar	humid
tropical	arid	semiarid	

1. climates near the equator where the temperature averages above 18°C
2. the average weather of a location over a long period of time
3. a term meaning that a location is moist and receives more than 50 cm of precipitation a year
4. climates near the poles where the temperature averages 10°C or less
5. a term meaning that a location receives between 25 and 50 cm of precipitation a year
6. a term meaning that a location is dry and receives less than 25 cm of precipitation a year
7. the climate of an area that is influenced mostly by air masses from large landmasses

What Do You Remember?

Number your paper from one to eight and write the letter of the correct answer next to the appropriate number.

1. The patterns of such things as temperature, precipitation, and humidity over a long period of time are called (a) weather, (b) climate, (c) records, (d) forecasts.
2. An important factor that determines climate is (a) temperature, (b) clouds, (c) wind, (d) fog.
3. The greatest amount of sunlight is received (a) at the poles, (b) at about 30° latitude, (c) at the equator, (d) equally over the earth.
4. The average temperature for a tropical climate is (a) below 10°C, (b) above 18°C, (c) between 10°C and 18°C, (d) unknown.

5. A tropical climate has precipitation all year because air near the equator is (a) rising, (b) sinking, (c) moving sideways, (d) not moving.
6. A humid continental climate is changed by air moving from the oceans. It is changed to a (a) marine climate, (b) tropical climate, (c) subpolar climate, (d) arid climate.

Applying What You Have Learned

Consult encyclopedias, almanacs, and maps for information.

1. Oklahoma City, Oklahoma, averages 72 days each year with temperatures over 32°C. New Orleans is farther south but only averages 57 days each year with temperatures over 32°C. What would explain this climate difference?
2. The humidity at 7:30 P.M. in July is usually about 70 percent in Atlanta, Georgia, but only 20 percent in Phoenix, Arizona. What would explain this climate difference?
3. Buffalo, New York, averages about 190 cm of snow each year, but Chicago averages less than 90 cm. What is influencing the climate in Buffalo, New York? Why doesn't it affect Chicago, Illinois?

Research and Investigation

1. Report on how the climate influences gardening in your location. Find out the date of the last killing frost in the spring. How long is the growing season? What can be grown during this period? Consult weather data from a meteorological office, a county extension agent, or seed stores.
2. Investigate the climate data for your location. Find the average maximum and minimum temperatures for each month. What are the record lows and highs? What is average for rainfall, snow, and sleet? How does the humidity change daily and monthly? What are the average wind speeds for different months?

Unit Three

The Waters of the Earth

CHAPTER 9
Fresh Water

CHAPTER 10
The Oceans

9

Fresh Water

What do you think about when you look at a stream? Do you wonder where it came from or where it's going? Will it run out of water?

Not all people look at a stream and see the same thing. A park manager might see the stream as a place for boating. A city engineer might see the stream as a source of water for drinking, cooking, and washing. A farmer might see the stream as a source of water for thirsty crops. An electric utility planner might see the stream as a source of power. Many streams can provide all these things and have even other uses. The water in a stream is an important resource that can be used many ways.

Will there be enough water for all these uses? If there is not enough water for all these uses, which uses are more important? In this chapter you will learn where water comes from, where it goes, and some new ways to add to the available water supply.

Figure 9.1 *opposite:* A freshwater river is many things to many people.

9.1 Earth's Water

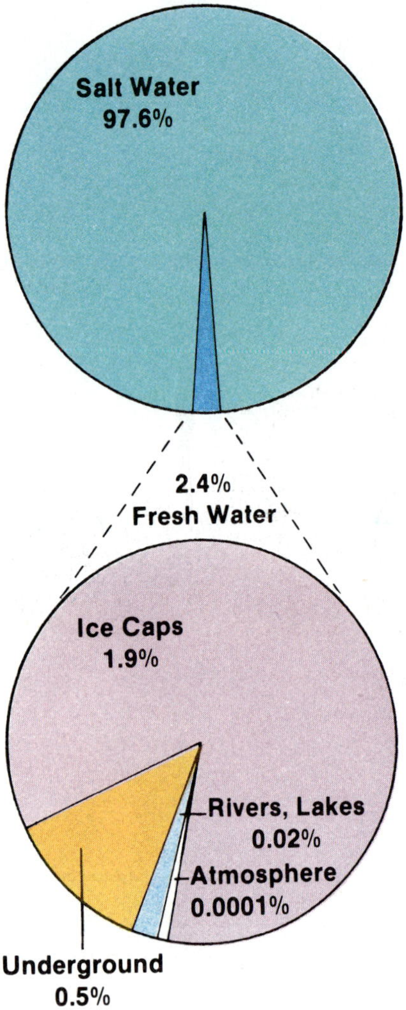

Figure 9.2 *top:* Compare the amounts of salt water and fresh water found on the earth. *bottom:* The distribution of the 2.4% of fresh water on the earth

You will find out

- how much of the earth's water is fresh water;
- why only one third of the earth's water is available for use;
- how the water cycle replaces fresh water.

The earth appears to have plenty of water. In fact, astronauts looking at the earth from space have called it the blue planet. One reason it appears blue is the large amount of water on it. The oceans cover about 70 percent of the earth's surface. The oceans also contain more than 97 percent of all the water on the earth. However, ocean water is salty. You cannot drink it.

This leaves only about 3 percent of all the water found on the earth that you can drink. This water, which is not salty, is called **fresh water.** About two thirds of the world's fresh water is locked up in ice, mainly in Greenland and the Antarctic. What remains, then, is less than 1 percent of all the water on the earth as the available fresh water. This fresh water is found underground, in rivers, in lakes, in ponds, and in the atmosphere. Most plants and animals that live on land need fresh water to survive. Since such a small percentage of the earth's water is fresh, you may think there is really not enough fresh water. Most places have enough, however, because the supply is frequently replaced.

The atmosphere doesn't hold much water at any given time. Suppose all the water vapor in the atmosphere fell to the earth. How deep a layer of water would be made? The entire earth would be covered by only a few centimeters of water. But remember that water is continually evaporating, condensing, and falling as precipitation in what is called the water cycle. Because of the water cycle, enough precipitation falls each year to cover the entire surface of the earth 85 cm deep in water.

The amount of water that evaporates, condenses, and returns as precipitation is not the same over the oceans as over the land. This is due to the wind patterns. Look at Table 9-1.

Table 9-1 Worldwide Evaporation and Precipitation

Evaporation		Precipitation	
From Oceans	84%	On Oceans	77%
From Land	16%	On Land	23%
Total	100%	Total	100%

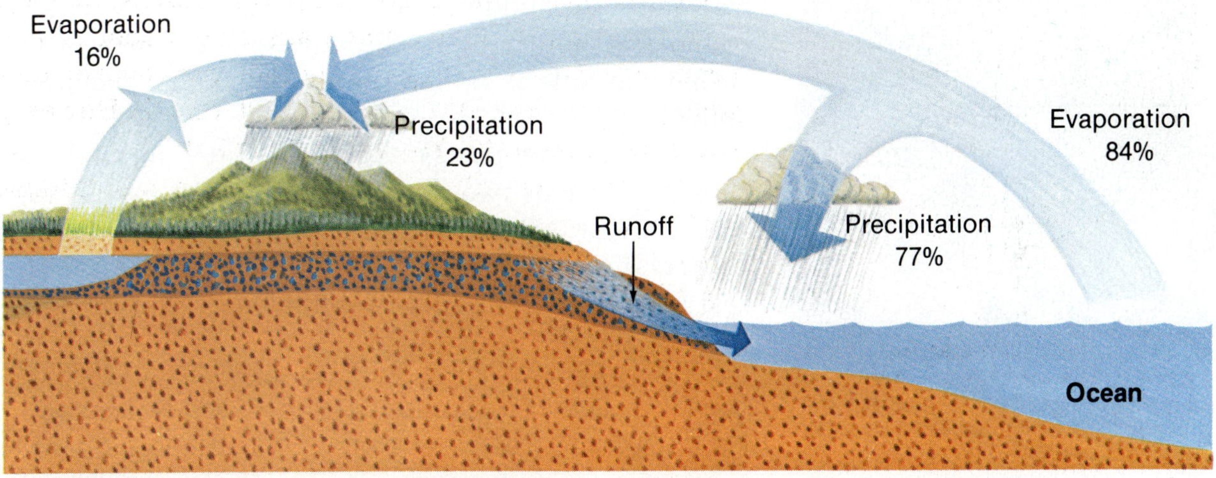

Figure 9.3 The water cycle

Notice that more water evaporates from the oceans than falls on the oceans as precipitation. Notice also that more water falls on land than evaporates from the land. Some of this rainwater soaks into the ground, and some of it flows across the land. Eventually the water returns to the oceans, balancing the cycle. As you can see, the water cycle is continuously replenishing the fresh water on land from the vast supply of water in the ocean.

Study Questions for 9.1

1. How much of the water on the earth is salt water?
2. Why is only one third of the earth's fresh water available for use?
3. Construct a pie graph to compare evaporation and precipitation over the oceans to evaporation and precipitation over the land. What happens to the difference?

9.2 Fresh Water on the Surface

You will find out

- what happens to water that falls on the land;
- what kind of soil holds more water;
- what surface runoff is.

What happens to rain after it falls? Rain falling on a sidewalk or street forms many puddles. If the rain is heavy enough, the street may become flooded. After the rain stops, the puddles gradually disappear as the water evaporates, returning to the atmosphere. Much of the water that falls as rain evaporates back into the atmosphere.

Basically three things happen to water that falls as rain. As much as two thirds of any rainfall may evaporate back into the atmosphere. Of the remaining one third of the rainfall, some soaks into the ground, and the rest flows across the surface of the land.

Figure 9.4 *above:* Puddles that remain after rain can be a minor inconvenience.

Figure 9.5 *right:* Flooding that results from heavy rain can be a major problem.

The amount of water that soaks into the ground depends on how dry the ground is and what kind of soil is there. Some kinds of soil have more space between the soil particles than others. This space between soil particles is called the **pore space.** The more pore space a soil has, the more water it will hold.

Do You Know?
Dry soil with a lot of pore space can soak up to 10 cm of water an hour.

Figure 9.6 When rain hits the soil's surface, water is absorbed into available pore space.

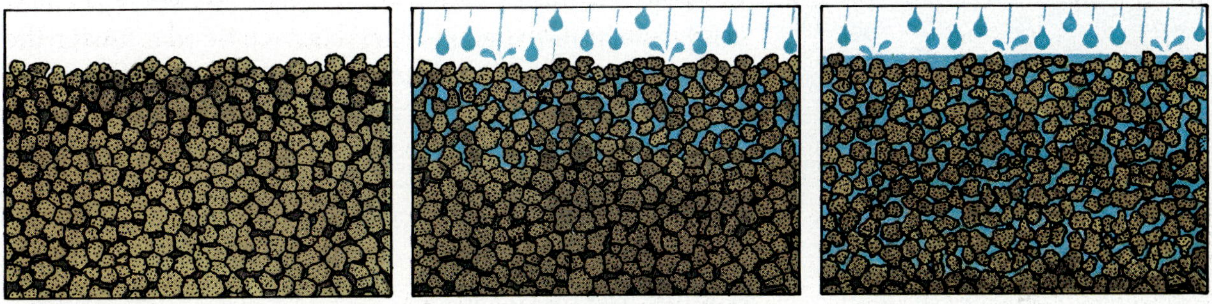

ACTIVITY — How Much Pore Space?

Materials

sand
clay
gravel
garden soil
glass containers
graduated cylinder
water

Procedure

1. Fill four glass containers with sand, clay, gravel, and garden soil.
2. Use a graduated cylinder to add measured amounts of water slowly to each container.
3. Record how much water each container of soil material will hold.

Questions

1. How much water were you able to add to each container?
2. What does this mean about the pore space in sand? Clay? Gravel? Garden soil?
3. Which would soak up the most rainfall?
4. Which would have the most surface water at first?

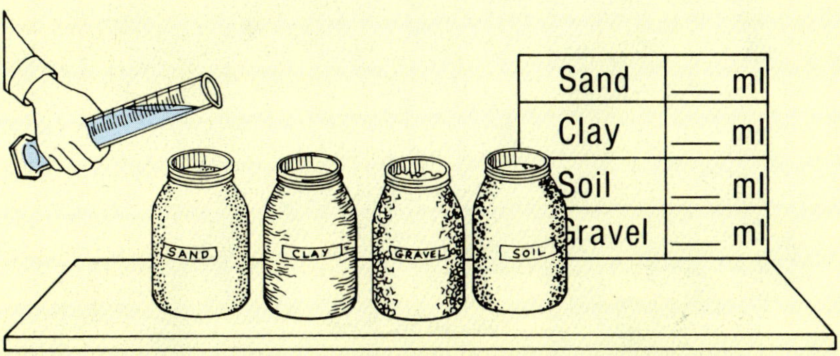

Figure 9.7 This runoff will eventually drain into a lake or a pond or a river that empties into the ocean.

The water moves down into the soil as the pore spaces fill. If rain falls faster than the water can move through the soil, the water begins to collect on the surface. This water may move downhill across the surface of the land. When water moves across the surface, it is called **surface runoff.**

Surface runoff moves on the land from high places to lower places. It collects in small streams that drain the land. The small streams carry the water to a few larger streams, and the larger streams drain into major rivers. The area of land drained by a river is known as the river's **watershed.** Figure 9.8 shows three large watersheds. The Columbia River watershed and the Colorado River watershed drain water into the Pacific Ocean. The Mississippi River watershed drains water into the Atlantic Ocean.

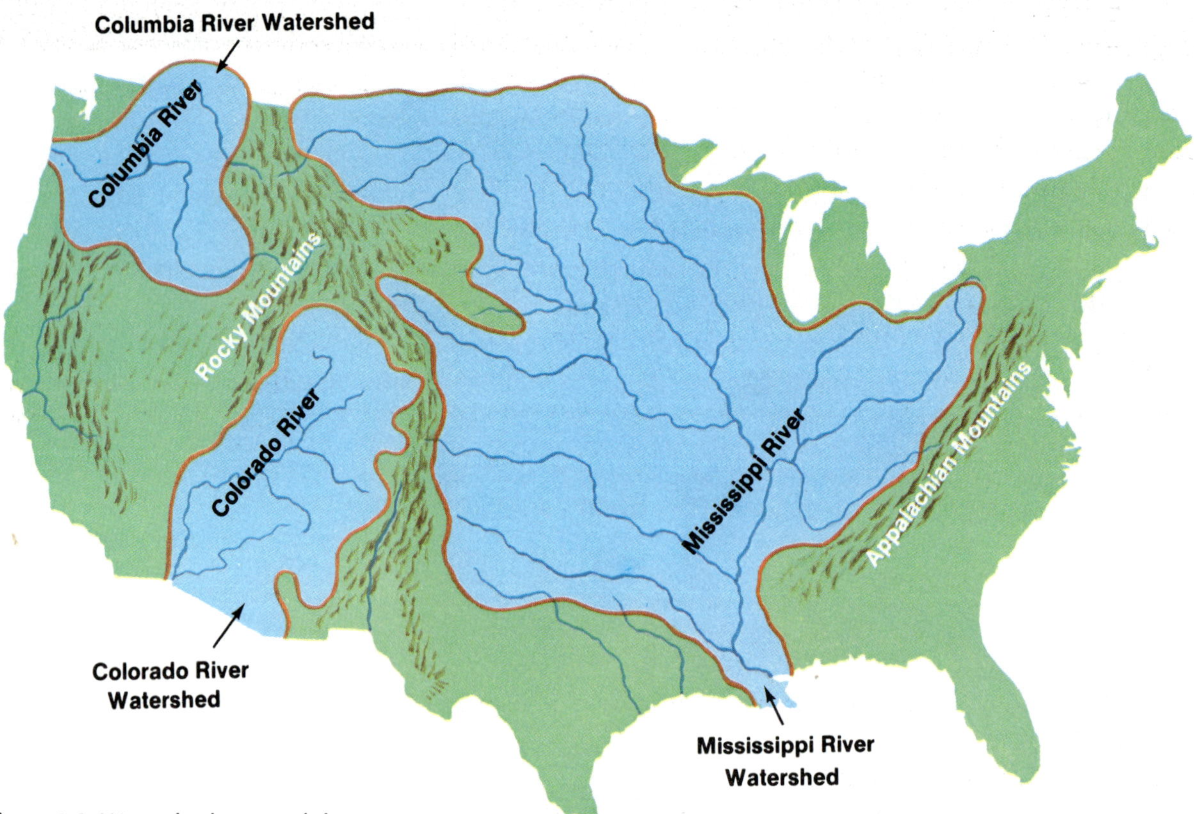

Figure 9.8 Watershed areas of the continental United States

Within a watershed, some of the runoff may be caught in a low place. This forms a standing body of fresh water. This body of water is called a **pond** or **lake,** depending on its size. To be called a lake, the water must be deep enough in some places so that sunlight does not reach the bottom. Some lakes and ponds, however, must have been named before people knew how deep they were. Sometimes lakes are called ponds and some ponds are called lakes. Do you know of any that are misnamed? The water found in all the rivers, streams, lakes, and ponds is also called surface water.

A lake can be created by building a dam on a stream or river. A lake formed in this way by a dam is called a **reservoir.** Dams are built and reservoirs are created for three different purposes: (1) water storage, (2) flood control, and (3) generating electricity. Water storage is used to make a new water supply or increase a supply that is already there. This supply can be used to provide water for cropland and cities. Flood-control reservoirs catch heavy runoff and hold the water, preventing flooding downstream. In many areas it is possible to store water in huge reservoirs for the generation of electricity. However, a reservoir cannot be used for all three purposes at the same time. For example, you cannot store water and release water to generate electricity at the same time. To generate electricity, water must be drawn from the reservoir to turn the turbines. What problem can you see if you tried to use a reservoir for both water storage and flood control? Consider where the flood water can go if the reservoir is already full.

Figure 9.9 Glen Canyon Dam in Arizona

Do You Know?
Straddling the equator and receiving 1.5 m to 2.5 m of rainfall every year, the Amazon River basin contains fully two thirds of all the river water on earth.

Study Questions for 9.2

1. Describe what happens to rain when it reaches the earth's surface.
2. What kind of soil holds more water?
3. Name one watershed. Describe how rain falling on this watershed ends up in the ocean.

9.3 Fresh Water Below the Surface

You will find out

- what the water table is;
- how water moves under the ground;
- what are some problems related to using groundwater.

Figure 9.10 Groundwater can be easily reached in some areas but not in others.

Have you ever visited a seashore or lakeshore? If so, perhaps you have dug a hole in the sand. You did not have to dig very deep before the lower part of the hole filled with water. You reached a depth where all the pore spaces in the soil were filled with water. Water from such a water-filled underground zone is called **groundwater.**

If you look down into the hole, you will see the surface of a tiny pond. Below the surface level of this tiny pond the surrounding ground is full of groundwater. The level of the water in the tiny pond and the surrounding ground is called the **water table.** A hole dug or drilled to the water table to get water is called a well. Near a large body of water the water table is not very deep, and wells can be quite shallow. In other places you may have to dig deeper, perhaps hundreds of meters deeper, to find the water table.

In a moist climate region the water table is close to the surface. In fact, it might even come right up to the surface and form a swamp, a lake, or a spring.

Figure 9.11 This illustrations shows some of the characteristics of the water table.

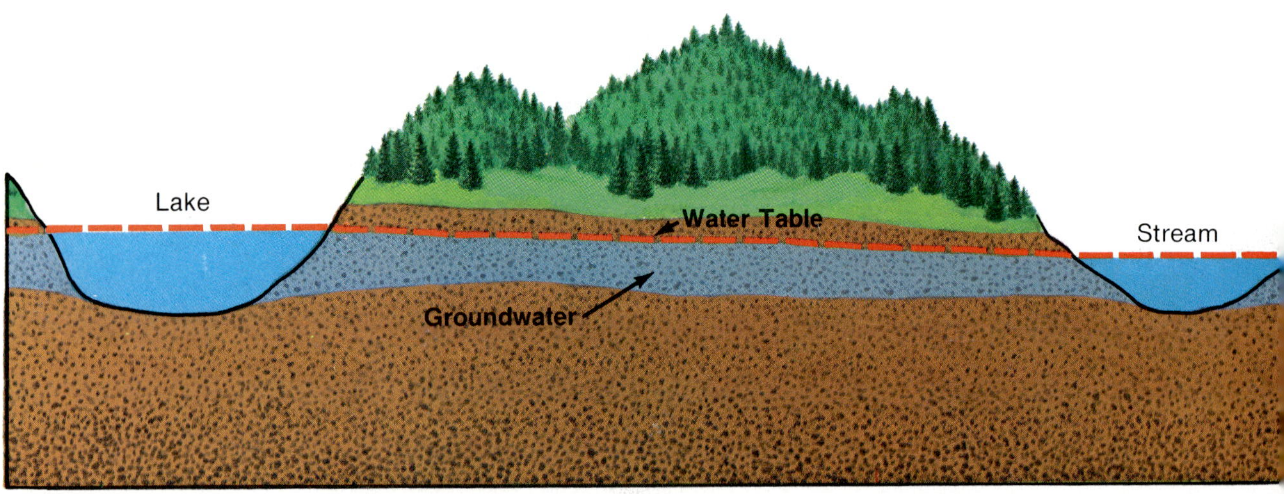

Figure 9.12 Rivers that flow through underground caverns can sometimes be reached by visitors.

Where does groundwater come from? Almost all of it was once precipitation. Some of the precipitation soaks into the soil and is used by plants. The rest of the precipitation continues to move slowly downward until it reaches the water table. Then it will generally move downward underground. It continues flowing underground until it finds an opening to the surface at a lower level. Such an opening forms a spring as in Figure 9.11. There the groundwater may join a stream and flow back to the ocean. This trip may be short, or it may take thousands of years. In some places groundwater moves only a few meters a year. In other places it may move several meters a day.

Do You Know?
There is 30 times more groundwater than water in all the streams, rivers, and lakes of the world. We can pump out a volume of this equal to the amount of precipitation that falls in 10 years.

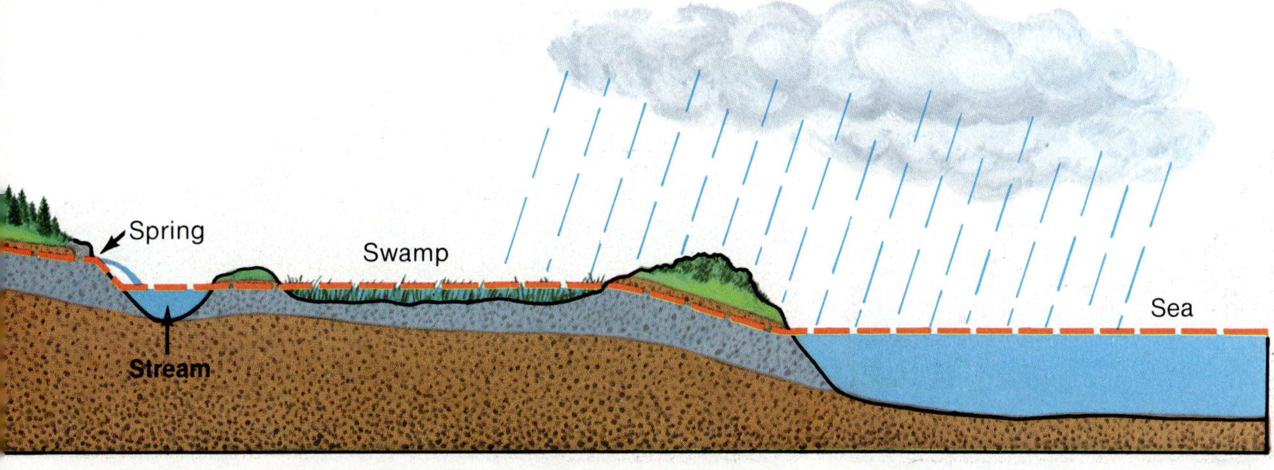

Groundwater also moves through rocks that have connected cracks and through rocks and materials that have openings through them. Rocks that allow water to move through them are said to be **permeable** [PUHR-mee-uh-buhl]. Sandstone is an example of permeable rock. Shale is an example of a nonpermeable rock. These rocks are usually found in layers. Often a layer of sandstone that contains water may be trapped between two layers of shale. In this case the sandstone with the groundwater in it is called an **aquifer** (AH-kwih-fuhr). An aquifer forms when any permeable rock containing water becomes trapped between two layers of nonpermeable rock. An aquifer may also form when soil saturated with groundwater is located above a layer of nonpermeable rock.

Figure 9.13 *top left:* Shale does not permit water to move through it. *top right:* An outcrop of shale

Figure 9.14 *bottom left:* Sandstone is permeable. *bottom right:* A sandstone landscape

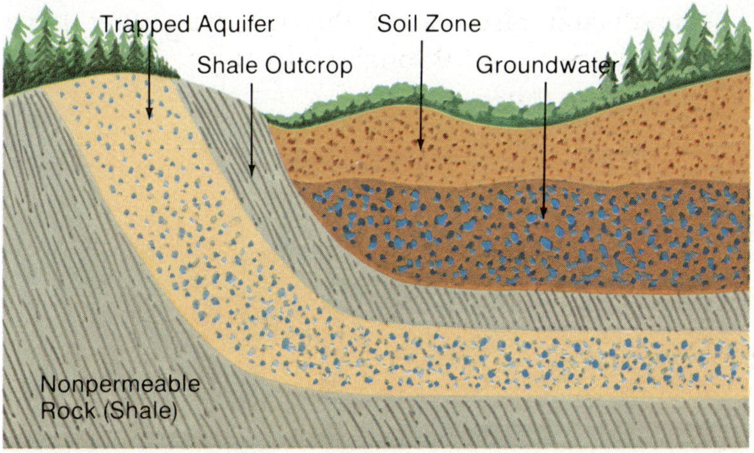

Do You Know?
It takes 200 years for a drop of water to move through one aquifer from Wisconsin to Chicago.

Figure 9.15 Water becomes trapped by layers of nonpermeable rock, creating an aquifer.

ACTIVITY

How Much Water Can a Rock Hold?

Materials

egg-sized pieces of sandstone, shale, and other rocks
water
containers
balance

Procedure

1. Find and record the mass of each rock sample.

2. Soak the rocks in a container of water overnight.

3. Remove the rocks from the water. Again find and record the mass of each rock sample.

Questions

1. What data did you obtain about the rocks?

2. Which rock would make the best aquifer?

3. How would this rock compare to sand as an aquifer (see data from the activity on page 171)?

Rock sample	Mass before soaking	Mass after soaking	Difference
1			
2			
3			

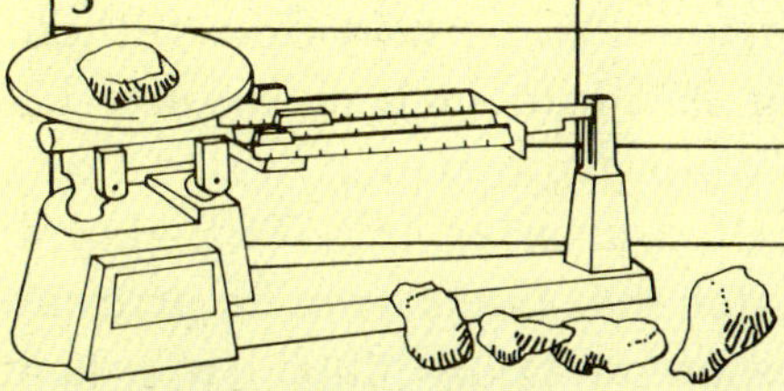

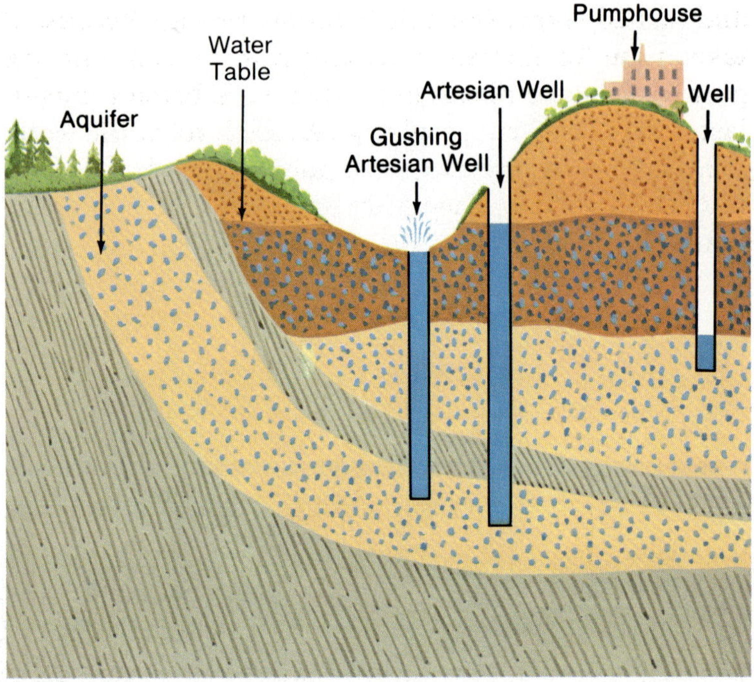

Figure 9.16 *right:* Underground water sources can be tapped for use through wells that bring the water to the surface.

Figure 9.17 *below:* A diesel pump brings underground water to the surface for irrigation.

Aquifers are a source of groundwater. A well can be drilled into an aquifer. Groundwater then moves into the hole forming a pool of water. The level of the pool of water is the same as the water table. As water is pumped from the well to the surface, more water moves into the hole through the aquifer. Since the water table rises in wet seasons and falls in dry seasons, the well hole often must go some distance below the water table to provide water all the time. About one third of all cities in the United States obtain water from wells. Does your city or town use surface water or groundwater?

In some places an aquifer carries water from a higher altitude to a lower altitude. If the aquifer has nonpermeable rock layers above and below it, pressure may build up at the lower altitude. Water may flow without pumping from a well drilled into such an aquifer because of the pressure. Wells from which water flows without pumping are called **artesian** [ahr-TEE-zhuhn] **wells.** The amount of water that flows from an artesian well depends on the pressure and how much water is in the aquifer.

Do You Know?
Water in some aquifers is in the same place it was 25,000 years ago.

Groundwater that is pumped out is replaced by rain and melted snow. But this is a slow process because it takes time for melted snow and rain to soak into the ground. In some places groundwater is being removed faster than it is being replaced. As the level of the water table is lowered by excessive pumping, wells must be drilled deeper and deeper. In some areas it is possible that the groundwater could be used up.

Do You Know?
If all the groundwater within 30 meters of the surface were placed on the surface, it would make a layer 5 meters deep across all of the United States.

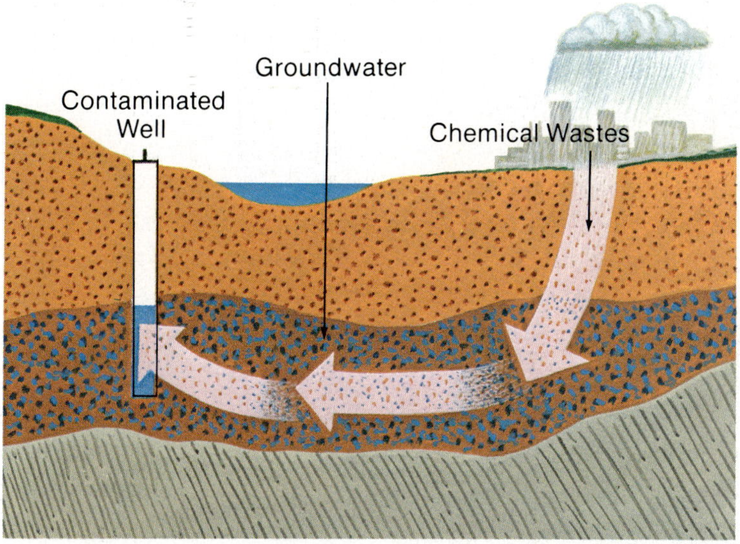

Figure 9.18 *left:* Chemical wastes from one area may contaminate groundwater that is then drawn to the surface somewhere else.

Figure 9.19 *below:* Contaminated water is a problem people must learn to solve.

Wastes from cities and industries can also affect groundwater. They can seep into the ground just like the water. In some areas groundwater has been polluted with waste chemicals. When this happens, people can no longer use this groundwater, and the wells have to be shut down.

Study Questions for 9.3

1. Where does groundwater come from and what is the upper level of groundwater called?
2. What is the property of a rock that allows water to move through it?
3. What are underground rock formations that hold water called?
4. What are two problems that can happen when using a well for a freshwater supply?

9.4 Fresh Water as a Limited Resource

You will find out

- why fresh water is a limited resource;
- possible new sources of fresh water.

All of the water you use comes from precipitation. If your supply is from surface water, the precipitation probably occurred within the last year. If your supply is from groundwater, the water may be from precipitation that occurred a long time ago. Whatever your supply, enough precipitation falls in the United States each year to supply 21,570 L of water a day for each person. Do you use that much water?

The average daily use of water per person in the United States is about 6,500 L per day. Of course, each person does not use that much water. Less than 10 percent of the water is used for your home water supply. Farmers and ranchers use about 40 percent of the supply for crops and animals. Industry uses the other 50 percent. Even with evaporation, it seems that there is plenty of water. All of the uses together consume an amount that is less than a third of the yearly average precipitation.

Figure 9.20 *above:* One of the many vital uses for water

Figure 9.21 *right:* The use of freshwater in the United States

Even though there appears to be enough water, problems do develop with the supply. One problem for users of surface water is that it depends on precipitation. If precipitation does not fall for a long period of time, a drought occurs. Rivers and lakes become low, and there is not enough water in that area. One solution to this problem is to build large reservoirs that will store enough water to be used during a drought. Another solution is to use other water sources.

Another problem relates to the climate. Because of climatic differences not all areas of the country share equally in the total precipitation. In some areas of the United States all of the surface water is already being used. Even the groundwater is being pumped from the ground faster than it is being replaced. In these areas water may be brought in from long distances.

As the population grows, more and more demands are placed on the limited water supply. New industries develop. People want more food. Should farmers give up their share of the water to the city dwellers? Should city dwellers give up their water so new industries can develop? Should an upstream city or irrigation project take so much water that downstream cities will not have any? Who has the right to use the water when there isn't enough for everyone? For people in some places these questions are very difficult to answer.

Do You Know?
Some parts of Arizona, California, Florida, and Texas are withdrawing groundwater up to 20 times faster than it is being replaced. This causes the ground to sink in these areas.

Figure 9.22 *above:* This aqueduct in California helps transport water from where it is stored to where it is used.

Figure 9.23 *left:* New aqueducts will bring Colorado River water into central Arizona.

Figure 9.24 Snowed in?

Water supply problems related to climate could be solved by increasing the water supply. Some possible ways to increase water supplies are (1) increasing the rainfall, (2) recycling waste water, (3) moving large icebergs from the polar oceans, and (4) desalting ocean water.

There have been attempts to increase precipitation in the Rocky Mountains. Devices were placed in the mountains and operated during the winter. The devices made billions and billions of snowflake-forming nuclei that were carried into clouds. These nuclei caused water droplets in the clouds to form snowflakes, which fell to the ground. Snowfall was increased 10 to 15 percent as a result. Many people who lived in mountain towns objected to the increased snowfall. They did not want more snow to shovel from walks and to drive through. Do you think they should have to put up with more snow so other people can have more water?

Another way to increase the water supply is to recycle waste water. This is already happening in some places. A nuclear power plant uses treated sewage water from Phoenix, Arizona. The power plant purifies the treated sewage water and then uses it for cooling. Part of the increased demand for water can be met by more recycling of water.

Figure 9.25 By using purified waste water, this power plant helps conserve fresh water for use elsewhere.

Remember that two thirds of the earth's fresh water is trapped in ice. In the polar regions, these huge masses of ice break away and float in the ocean as icebergs. They are made of fresh water, since they formed from snow. Some people have discussed using ships to push or tow these icebergs to coastal cities for use as a water supply. What problems do you think might develop when attempting to push icebergs to Los Angeles or New York City for a water supply?

Still another supply of fresh water could come directly from the oceans. Salt can be removed from ocean water. Desalting factories would produce a limitless supply of fresh water for areas near the ocean. But the cost of removing salt from seawater by evaporation or by chemicals is very high. So far it is cheaper to build pipelines that carry fresh water hundreds of kilometers than it is to build a factory to remove salt from ocean water.

Can you think of other new sources of fresh water? The greatest amount of water used is not by farming, industry, or people. A much greater amount of water is lost by evaporation from lakes and rivers. In fact, evaporation and growing green plants use up nearly three fourths of all precipitation that falls. Yet, few ways have been found to prevent this loss. Perhaps in the future you may be the one to find a new way to slow water loss by evaporation.

Figure 9.26 *top:* Do you think icebergs will be a source of fresh water in the future?

Figure 9.27 *bottom:* In the future, scientists may find a way to collect and use water that is released by growing plants.

Study Questions for 9.4

1. How does the amount of fresh water used in the United States each year compare to the yearly average precipitation?
2. What problems can develop in some places with the water supply?
3. Discuss four possible *new* sources of fresh water.

What Did You Learn?

- Of all the water on the earth, 97 percent is salt water in the oceans, and 3 percent is fresh water.
- About 2 percent of the earth's water is fresh water frozen in ice. Less than 1 percent of all the water on the earth is fresh water that is available for use.
- Of all the water on the earth, 97 percent is salt water in the oceans, and 3 percent is fresh water.
- Of all the precipitation falling on the land, most finds its way back into the atmosphere. About one third soaks into the ground or flows across the land.
- Accumulated water that moves across the surface is called surface runoff.
- The area of land drained by a river is called a watershed.
- Standing bodies of fresh water are called ponds or lakes, depending on their size. A lake formed by a dam is called a reservoir.
- Water moves down into the ground until it reaches an area where all the pore spaces are filled with water. This water is called groundwater. The upper level of the soaked area is called the water table.
- Permeable materials full of groundwater are called aquifers. A well is a hole drilled into an aquifer.
- A well that flows without pumping is called an artesian well.
- Problems may develop when groundwater is pumped out faster than it is replaced. Problems also develop from pollution of groundwater.
- The United States uses an amount of water equal to about a third of the yearly average precipitation. Some areas of the United States do not receive enough precipitation to meet their needs.
- Industry uses about 50 percent of the water supply, farmers and ranchers about 40 percent, and cities less than 10 percent.
- Possible new supplies of fresh water can be created by weather modification, recycling water, towing icebergs, and desalting ocean water.

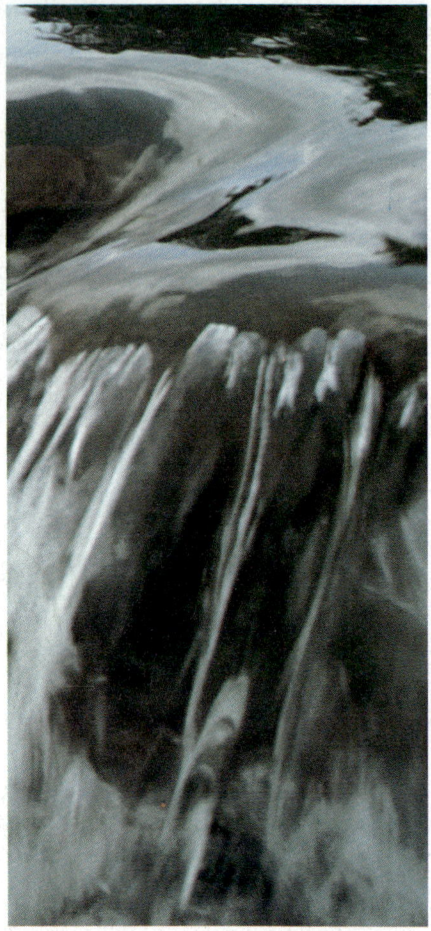

Key Terms

fresh water
pore space
surface runoff
watershed
pond
lake
reservoir
groundwater
water table
permeable
aquifer
artesian well

Career

Hydrologist

Some cities use streams and lakes as a source of water. Other cities use groundwater. How do the cities know about the water quality and how much water is available? They ask a hydrologist, a scientist who measures the amount and quality of fresh water.

Do you know anyone who started in one career but later changed to another one? Judy D. Fretwell, who started as an elementary school teacher, did that. She taught school for four years. She found doing research for science projects very exciting. But she also wanted to work outside, and she enjoyed writing.

Judy Fretwell left her teaching job and went back to college. There she studied science, mathematics, and computers. Two years later, she found work that included all the things she enjoyed doing. She was hired as a hydrologist by the U.S. Geological Survey. As a hydrologist, Judy Fretwell measures the water levels in lakes, streams, and wells. She also evaluates the water quality. After collecting this and other information, she writes reports. The reports tell other people about possible threats to their water supply.

Judy Fretwell wants other people to know about their surface water and groundwater supply and not to take it for granted. She states that water is our most valuable resource.

TO THINK ABOUT AND DO

Vocabulary

Write the numbers 1 through 10 on a sheet of paper. Select the word or phrase from the following list that matches each numbered definition. Write the appropriate word or phrase on your paper.

water table	groundwater
watershed	fresh water
reservoir	pond
pore space	surface runoff
lake	permeable

1. water that does not contain salt
2. space between soil particles
3. water that moves across the surface
4. land drained by a river
5. a body of fresh water that sunlight shines through to the bottom
6. lake created by a dam
7. a body of fresh water that sunlight does not shine through to the bottom
8. water from a water-filled zone below the surface
9. surface of a water-filled zone below ground
10. characterizing rocks that water can move through

What Do You Remember?

Write the numbers 1 through 10 on a sheet of paper. Write the letter of the best answer next to the number.

1. How much of the water on the earth is fresh water? (a) 97 percent (b) 3 percent (c) 2 percent (d) 1 percent
2. Approximately how much of the water on the earth is fresh water that is available for use? (a) 100 percent (b) 97 percent (c) 3 percent (d) 1 percent
3. If all the water in the atmosphere were to condense and fall to the earth, it would make a layer how deep? (a) 2 cm (b) 7 cm (c) 23 cm (d) 85 cm

4. Most of the precipitation that falls on the land (a) soaks in, (b) runs off, (c) returns to the atmosphere, (d) stays on the land.
5. The area of land drained by a river is a(n) (a) water table, (b) watershed, (c) reservoir, (d) aquifer.
6. A reservoir may be created for the purpose of (a) water storage, (b) flood control, (c) generating electricity, (d) any of the above.
7. Groundwater is water (a) on the ground, (b) at the bottom of the lake, (c) from the water-filled zone below ground, (d) all of the above.
8. A well is a hole drilled into the ground below (a) the water table, (b) the watershed, (c) a spring, (d) a reservoir.
9. Our use of water takes how much of our yearly average precipitation? (a) all of it (b) about half of it (c) about a third of it (d) less than a tenth of it.

Applying What You Have Learned

In answering the following questions, remember that surface runoff depends on precipitation and that the amount of groundwater that can be pumped out is equal to a ten-year supply of precipitation. Assume there is a river and a large reservoir where you live.

1. Suppose it does not rain for a full year. Where could you obtain a freshwater supply at the end of this year?
2. Suppose it does not rain for five years. Where could you obtain a freshwater supply at the end of five years?
3. Suppose it does not rain for 15 years. Where could you obtain fresh water at the end of 15 years?
4. Considering the precipitation that falls on the surface, how could you increase your water supply?

Research and Investigation

Plan a field trip to your local water treatment plant. What is the source of water? How much is available? How much is used? How is the water treated? How does it get to your house?

10 The Oceans

Have you ever walked along a sandy beach at the ocean? Perhaps you walked just out of reach of the waves that crashed onto the beach. Each wave pushes a surge of churning water and foam across the sand. The water hesitates and then rushes back toward the ocean only to be swallowed by the next crashing wave. What makes these waves? Where do ocean waves come from?

If you have ever explored a beach, you probably have found rocks and seashells of all shapes and colors. Where do these interesting, wondrous things come from? As you look out over the vast, watery surface of the ocean, do you wonder what lies beneath that surface?

People have always wanted to know more about the ocean. Some of the most exciting things about the earth are being discovered in the depths of the ocean. Scientists are studying the ocean in many different ways. In this chapter you will study the ocean and the strange but exciting world beneath the ocean surface.

Figure 10.1 *opposite:* A sandy beach on the island of Kauai, Hawaii

10.1 Waves and Currents

You will find out

- what makes waves on the ocean;
- what makes currents in the ocean.

On their first trip to the ocean many people quickly notice the vast, watery surface. Water as far as they can see! Then they notice something else. The surface of the ocean is rarely, if ever, still. On some days small waves tumble and surge on the shore. On other days large waves, topped with foam, thunder and crash onto the shore, threatening boats and buildings with disaster. Waves are in many ways the most destructive part of the ocean. They erode and shape the coast. They are always a threat to ships and buildings near the shore. One after another, the waves keep coming. What causes waves and where do they come from?

Figure 10.2 Wave action always brings change and sometimes causes destruction.

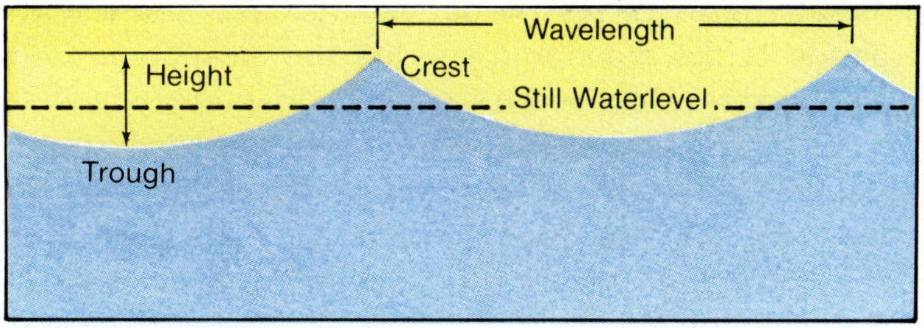

Figure 10.3 *left:* A profile of a wave

Figure 10.4 *below:* Rough seas are familiar to most ships. Would you like a ride such as this boat is getting?

You have made waves if you have thrown a rock into a pool of water. The rock made wrinkles on the surface that moved away from the disturbance. These moving wrinkles are called waves. A simplified sketch of a wave is shown in Figure 10.3. The highest part of a wave is the **crest.** The lowest part is the **trough** [trawf]. The horizontal distance between two crests is defined as the **wavelength.** The vertical distance between a crest and a trough is the **wave height.** A wave can have various wavelengths and wave heights. Ocean waves to a height of 165 m have been observed. The wavelength usually ranges from 65 m up to 6,500 m.

ACTIVITY

How Does a Wave Move?

Materials

rope about 15 meters long
10-cm piece of colored yarn or string

Procedure

1. Tie a 10-cm piece of colored yarn or string to the center part of a 15-m rope.

2. Have someone hold one end of the rope still while you shake the other end up and down.

3. Observe what happens to the yarn as waves move down the rope.

Questions

1. Make a sketch of the wave formed along the rope.

2. What was the motion of the colored yarn?

3. What evidence do you have that a wave moves forward, not the material that the wave is in?

Figure 10.5 *top:* The ocean surface is calm when there is little wind. *bottom:* Seas become rough when winds are strong.

Figure 10.6 *below:* This illustration represents the pattern of a swell as it reaches a gently sloping beach.

What causes waves? Wind, earthquakes, and the attraction of the moon and sun are the most important wave makers. Most ocean waves are made by winds.

When the wind begins to blow over calm water, first, areas of ripples appear and then disappear as they become waves. As the wind continues to gust, the waves become larger. How large the waves become depends on (1) the wind speed, (2) the length of time the wind blows, and (3) the **fetch,** the distance the wind blows over the water.

If the winds are strong, they might topple the wave, making foam. These foam-topped waves are called whitecaps. If you see whitecaps, you know that the wind is blowing at 30 km an hour or more.

Wind-caused waves can travel thousands of kilometers from where they were formed. The waves that travel such a distance are long, wide waves that are not very high. These waves are called **swell.**

As a wave moves from open water and nears the shore, it starts to change. The wave slows as the lower part drags on the ocean bottom. The wave height starts to increase. Then the wave top bends forward and finally breaks, forming foamy **breakers.** The zone where the breakers occur is called **surf.** It is swell that forms breakers in the surf along the shore. The breakers are caused by the ocean bottom near the shore, the long gentle slope. Here the waves begin to build up far from the shore. They travel a long distance before breaking.

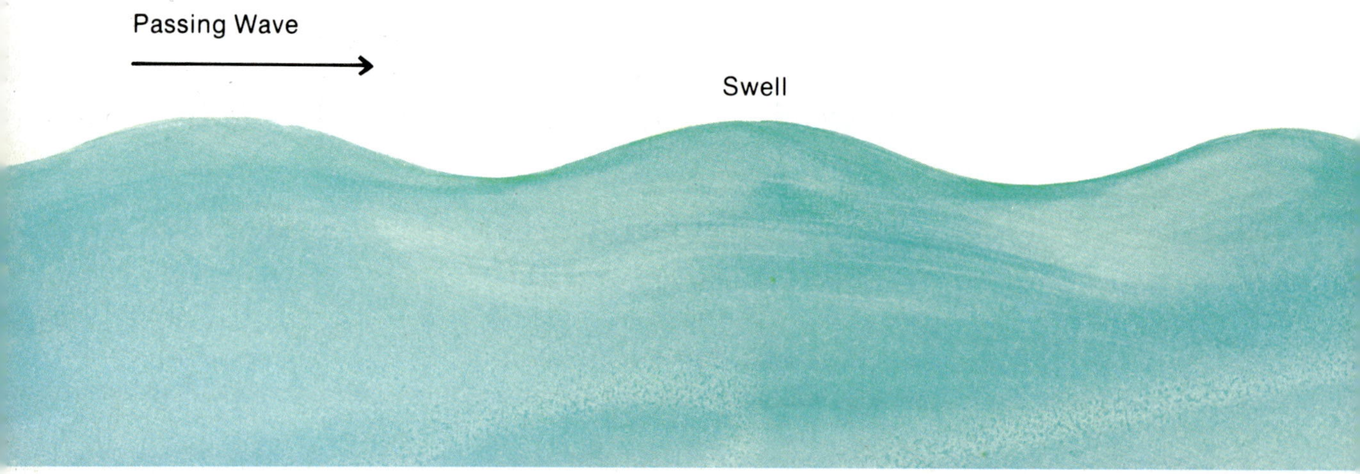

Wind-made waves are easy to see because they happen often. Wind-made waves have short wavelengths. Waves made by gravitational attraction or earthquakes have large wavelengths.

Tides can also be thought of as waves. They have a wavelength of nearly 22,000 km. It takes about 12 hours for the water to go up and down from such a wave. Tides are made by gravitational attraction of the sun and the moon. More information about tides can be found on page 68.

A wave made by an undersea earthquake is called a **tsunami** [tsoo-NAH-mee]. The tsunami is the largest of all ocean waves. It may have a 200-km wavelength and move across the ocean at hundreds of kilometers per hour. When this wave reaches the shore, a huge breaker forms. In some cases this breaker has been over 66 m tall, causing much damage as it moves on shore. The Hawaiian Islands average a tsunami every four years.

Almost all of the water in the ocean is always moving. Probably only the water in the bottom of some deep basins does not move. Waves move the water on the surface. Below the surface, the water is moving in a way you can't see. It is moving in streams called **currents.** A current is a moving stream of water in a body of water. Perhaps you have felt a current in a swimming pool. Water pumped through the filter returns to the pool through small pipes in the side. This makes currents in the pool. You can't see a current, but you can feel it.

Do You Know?
The world's greatest tide occurs in the Bay of Fundy and is over 16 m.

Table 10-1 The Beaufort Scale

Force	Wind Speed *	Description
0	0	Calm
1	2	
2	5	Light
3	9	
4	13	Moderate
5	18	
6	24	Strong
7	30	
8	37	Gale
9	44	
10	52	Storm
11	60	
12	68	Hurricane

* (knots)

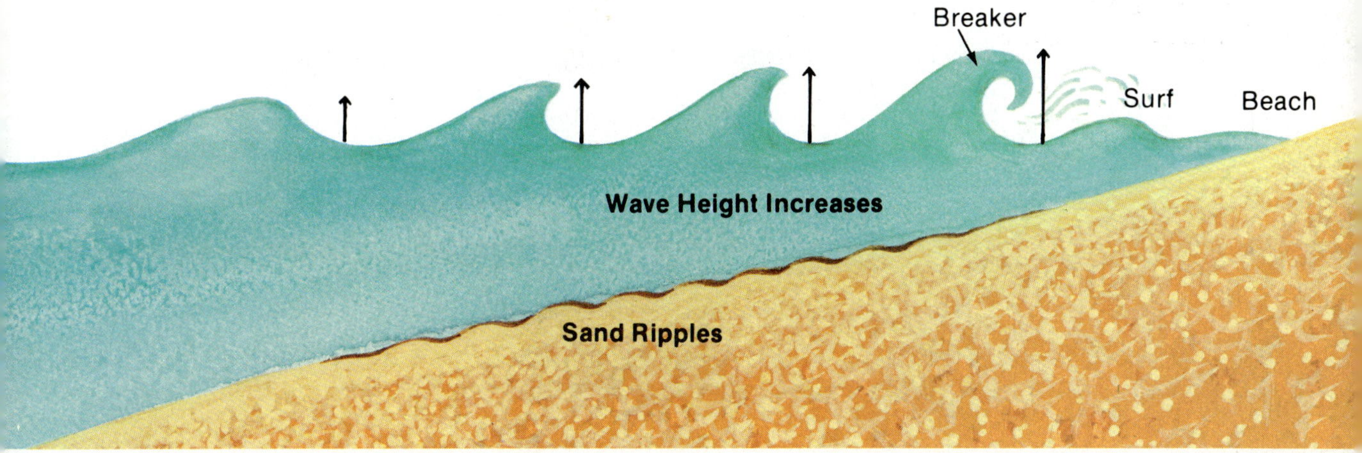

Figure 10.7 *top:* Sediments suspended in the water show the direction of longshore and rip currents.

Figure 10.8 *bottom:* The shape of the shore bottom may cause a longshore current to turn and become a rip current.

Ocean currents are made by (1) waves, (2) wind, and (3) differences in water density. Water density is changed by temperature and by how much salt is in the water. Evaporation, fresh water from rivers, and melting ice change the salt concentration and thus the density.

Currents made by waves occur in a small area near a shore. There is not much forward movement of water by waves in the ocean. It is different near the shore. Breakers pile up water near the shore. This water divides into two currents. One current is pushed along by waves, producing a **longshore current.** This current moves along the shore until it finds a place where it can head out to sea. When it does, it produces a second current called an **undertow** or a **rip current.** There is usually no surf where a rip current occurs. The current usually occupies a channel in this placc. A rip current continues for a kilometer or two out to sea. Rip currents are strong and can carry a swimmer far from the shore. You should swim across a rip current rather than try to swim against it.

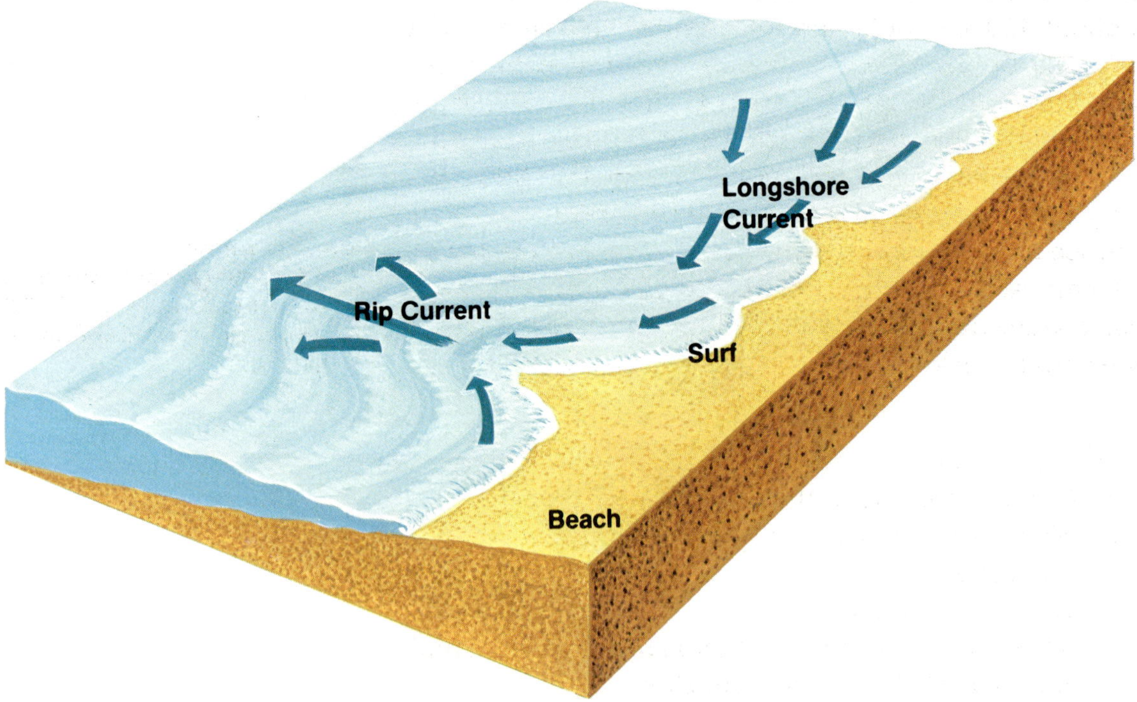

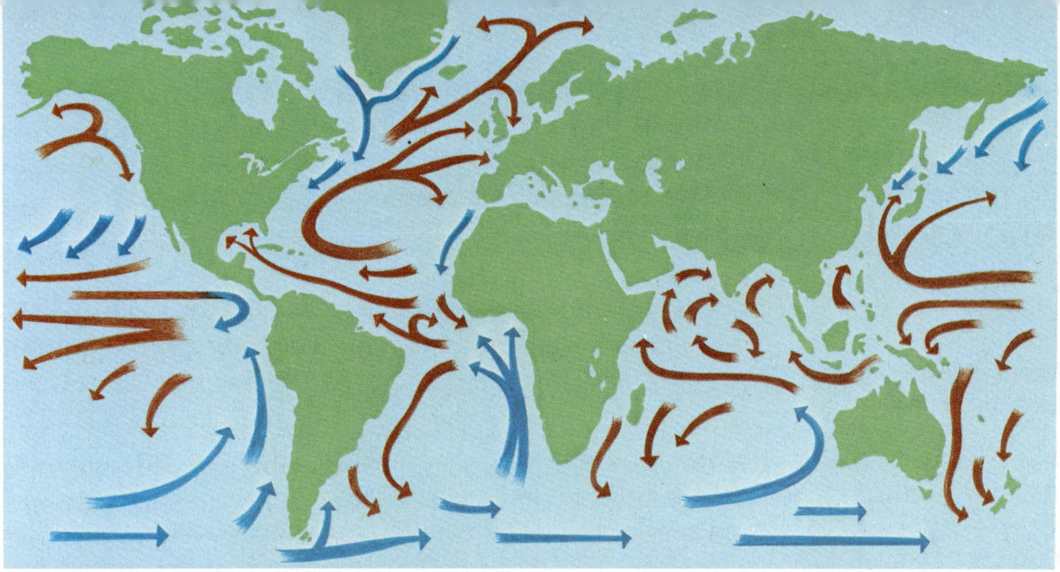

Figure 10.9 The major ocean currents of the world

There are also currents that move over long distances in the ocean. These are shown in Figure 10.9. These currents are started by density differences and then pushed by prevailing winds. These currents are also influenced by the Coriolis effect (see page 99) and the shape of the ocean bottom. They are like giant rivers of water in the ocean. The Gulf Stream, for example, is a stream of water that is moving up to 120 km per day. It is about 150 km wide and may extend to a depth of a thousand meters. Some of the currents are even larger and move more slowly. The California current is broader, but it is moving at only up to 7 km per day. Both the Gulf Stream and California current, however, carry huge volumes of water great distances.

Since all of the oceans are connected, all of the major currents are related. Just as the atmosphere has a worldwide circulation pattern of air, the ocean currents make up a worldwide circulation pattern of water.

Do You Know?
The volume of water moving in ocean currents is about 10,000 times greater than the flow of all large rivers.

Study Questions for 10.1

1. Name three things that make ocean waves.
2. What determines how big a wind-made wave becomes?
3. What is a current?
4. Name one ocean current that is near the shore and one current that is far from shore.

10.2 Beaches and Shorelines

You will find out

- how beaches are formed;
- what beach sand is made of;
- how sand-deposited shore features are formed.

Suppose you want to go to a place on the earth where you can see the most obvious and ongoing change in the land. Where would you go? The place where you should go is where the land and the ocean meet. The waves and currents never stop. They are very good at changing land, as you will see.

The place where the land and water meet is called the **shoreline.** The **shore** is the area where waves are active. Most of the shores of the earth are covered with beaches. A beach is a deposit of sand or gravel that covers the shore. Other shores are rock cliffs with no sand deposits at all. The constant pounding and grinding action of waves slowly wears these cliffs away. On the other hand, shores in sheltered bays have mud that has been deposited by tides and longshore currents. Eventually, a shallow bay could fill to form a broad, flat muddy area called a **tidal marsh.** Many tidal marshes are located on the eastern coast of the United States because of the low flat terrain and many rivers.

Figure 10.10 Examples of shorelines include a rocky shoreline (*top*), a sandy beach (*bottom left*), and a tidal marsh (*bottom right*).

Waves and currents not only grind up rock cliffs and deposit mud marshes but also move sand around. In fact, waves and currents moving sand around are what form beaches. Most all the sand making up a beach has been moved there from the ocean floor. Sand was brought to the ocean by rivers. Sand can also be small pieces of coastal rocks that have been ground up by waves. These sands were deposited on the ocean floor and then moved onto the beaches by waves.

Figure 10.11 A sandy beach can be made of quartz (*top left*), grains of shells and coral (*below left*), or volcanic rock (*below right*).

Most beach sand is made of many different minerals. The greatest amount is quartz, however, because quartz is very hard and weathers slowly. Beaches in tropical areas are sometimes made up of grains of shells or coral. Waves grind the shells and coral into grains and then deposit the grains on the shore. These beaches are usually white and softer than the hard, quartz-sand beaches. Along the east coast of Florida, you can see the beaches change from quartz sand to shell sand as you drive south. Some beaches in Hawaii are black. Black beaches are formed from grains of volcanic rock. You even have a choice in Hawaii, as some islands have a white beach on one side and a black beach on the other.

Figure 10.12 This photograph is a close-up of the grains of shells and coral composing some sandy beaches.

Figure 10.13 A sand bar

A beach is always changing. Wind, waves, and currents move the sand around. Some beaches grow larger during the summer and smaller during the winter. Small waves during the summer bring more sand to the beach. Winter storms with their high waves can carry much of the sand off the beach and out to sea. During the calmer summer months, waves and currents move the sand back onto the beach.

Moving sand is sometimes deposited by longshore current. This could happen if something deflects the current or if the current moves into open water. The deposited sand could be moved again by waves. But if enough sand is gradually deposited over a period of time, certain shore features are formed. A **sand bar** is a shore feature. It is a long pile of sand that is under water. Sometimes a sand bar can be seen at low tide. If the sand bar is connected to the shoreline, it forms another shore feature called a **spit.** A spit may grow as storm waves toss more sand on it. Eventually it may grow higher than sea level.

ACTIVITY

How Many Different Materials Are in Sand?

Materials

beach sand
magnifying lens
sheet of paper

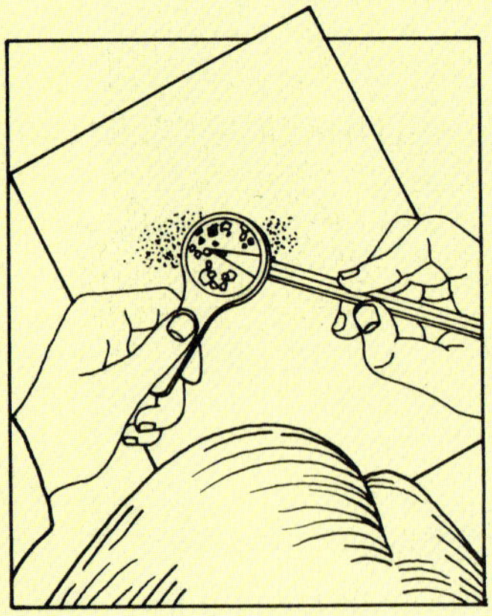

Procedure

1. Carefully pour about one-half teaspoon of beach sand in the center of a sheet of paper.

2. Record the overall color of the sand.

3. Use your pencil point to carefully spread the sand grains. Observe the grains with a magnifying lens.

4. Record your observations about the grains. How many different kinds of grains are there? Group the similar grains together.

5. Record other observations about the grains. For example, do they have corners or are they rounded?

Questions

1. How many different kinds of sand grains did you find?

2. Describe and sketch the different grains.

3. How many different minerals are in the grains?

Figure 10.14 *left:* A spit reaches out from the shore to the ocean.

Figure 10.15 *below:* A barrier bar and its long, straight beach

Another type of sand-deposited shore feature is found on the East Coast from Long Island to Florida, around the Gulf Coast, and in a few places along the West Coast. These features are long sand bars called **barrier bars.** Barrier bars are usually parallel to the coast and form a barrier between the coast and the ocean. They are separated from the coast by a relatively narrow body of water. Barrier bars accumulate sand to become **barrier islands.** Barrier islands have long, straight beaches on the seaward side. These beaches are popular recreational areas. Atlantic City, New Jersey, and Galveston, Texas, are both built on barrier islands.

Study Questions for 10.2

1. What is a beach?
2. How is a beach formed?
3. What are beach sands made of?
4. What shore features form from deposited sand?

10.3 Beneath the Ocean

You will find out

- what the three main divisions of the ocean's floor are called;
- what the landforms of the ocean basin are.

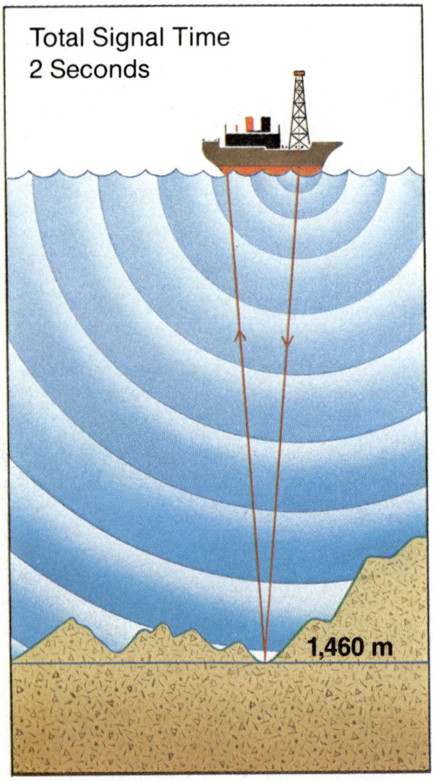

Figure 10.16 If a round trip for echo sounding signals takes two seconds, the one-way distance must be 1,460 m.

Rock cliffs, beaches, and tidal marshes are familiar parts of the ocean shore because you can easily see them. But what is underneath all the ocean water? How deep is the ocean? Early attempts to answer these questions were made from ships. A long length of wire with a heavy weight attached was lowered from a ship. The length of wire needed for the weight to reach the bottom was measured. Not many measurements were made in this way, but some data were collected. The limited data gave people the wrong idea. They thought the ocean floor was smooth and flat like a gigantic cereal bowl.

A new way to measure the ocean floor changed people's ideas about the ocean bottom. This newer way is called **echo sounding.** It uses a device that sends a sound signal through the water. The sound signal hits the bottom and bounces (echoes) back up to the ship. The echo sounder measures the time required for the sound signal to reach the bottom and to return. Sound travels through water at about 1,460 m per second. Suppose the ocean floor is exactly 1,460 m below the ship. How long

Figure 10.17 A scientist studies echo sounding data returning from the ocean floor.

would be required for the signal to reach the bottom and to return? One second would be required for the sound to travel to the floor and one second for it to return. Two seconds, then, would mean that the ocean floor is 1,460 m below the ship.

Some echo sounders draw a picture of the ocean floor. They make a continuous picture of the floor as the ship moves across the ocean. Much of what is known about the ocean floor comes from the use of such sounders.

Thousands of echo soundings have been used to gain an idea of what the ocean bottom is like. The bottom is no longer thought of as very smooth and flat. There are mountains underwater bigger than any on the land above water. There are canyons deeper than any canyon found on land. There are plains, hills, and valleys. Look at Figure 10.18. It shows what the earth might look like beneath the ocean. The ocean floor can be divided into three main areas: (1) the **continental shelf,** (2) the **continental slope,** and (3) the **ocean basin.** Let's take a closer look at each of these.

Do You Know?
If the tallest mountain on the earth (Mount Everest) were placed in the deepest trench (Mariana Trench), the mountaintop would still be about 2,850 m below the ocean's surface.

Figure 10.18 The features of the ocean floor are better understood because of echo sounding studies.

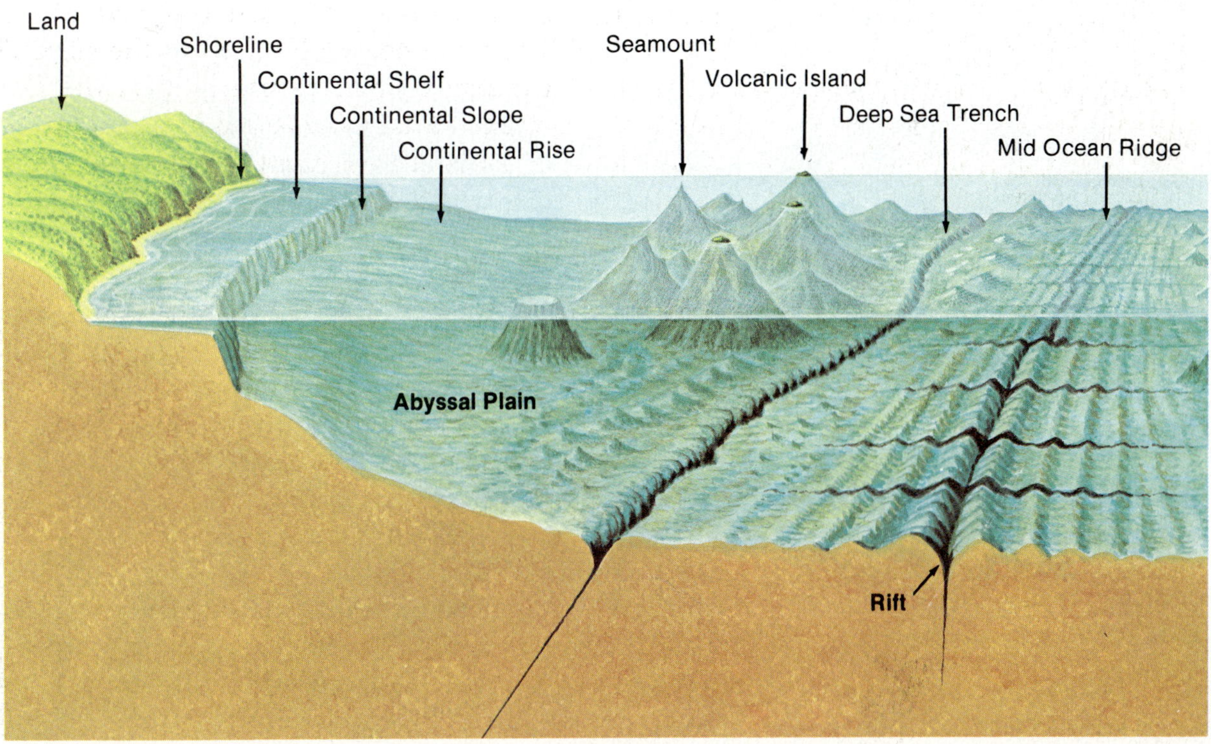

Figure 10.19 A physical map of the earth's ocean floor shows the locations of the larger ridges and trenches.

Each continent is surrounded by a somewhat shallow area called the continental shelf. The water gradually becomes deeper as you move from the shoreline and out over the shelf. Eventually, it will reach a depth of about 130 m. In Figure 10.19, you will notice that the continental shelf is wider in some places than in others. The average width of a continental shelf is 75 km.

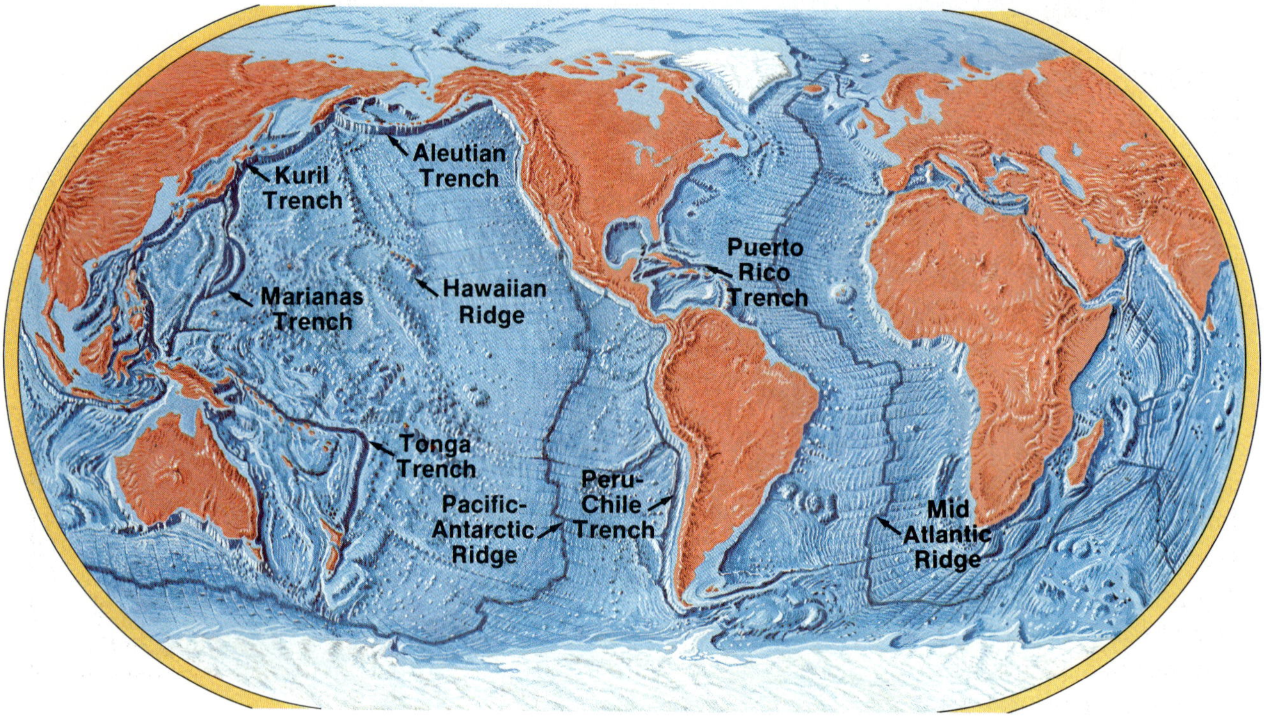

Beyond the continental shelf the depth starts to increase rapidly. This is the continental slope. The continental slope starts where the continental shelf ends and plunges rapidly down to depths of about 4,000 m. The slope is usually between 20 and 40 km wide. In many places the slope has gullies, valleys, and huge canyons. Some of these submarine canyons are larger than Arizona's Grand Canyon of the Colorado River. Submarine avalanches on the continental slope are thought to cause these valleys and canyons. It is not the shoreline but the outer edge of the continental slope that is the boundary of a continent.

Do You Know?
The *Glomar Challenger* in its cruises in 1970 and 1975 obtained conclusive evidence that the Mediterranean Sea has gone dry a number of times over the course of several million years.

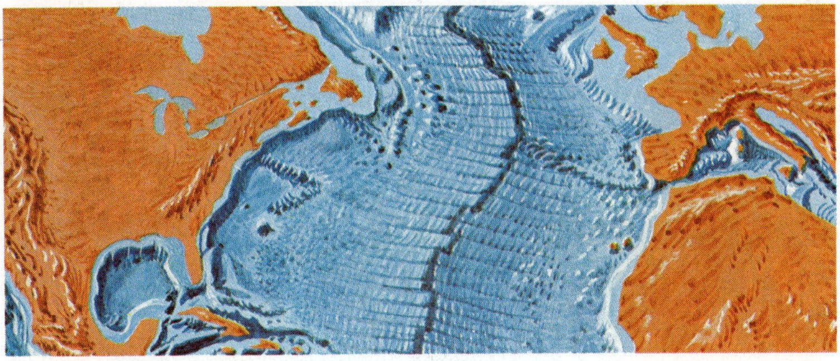

Figure 10.20 A closer look at the Mid Atlantic Ridge

Beyond the continental slope is the ocean basin, the bottom of the ocean floor. The basins are 4,000 to 6,000 m deep. The basin does have landforms as you can see in Figure 10.19. The basin is not like a gigantic cereal bowl. It is somewhat flat, but there are many hills, mountains, and mountain ranges. The great mountain ranges are called **ridges.** The mid Atlantic Ridge rises above sea level in some places, forming islands. Submarine volcanoes have also occurred across the ocean basin. If the volcano does not make it to the surface, it is called a **seamount.** Volcanoes that do make it to the surface make islands. The Hawaiian Islands are islands formed by volcanoes. The ocean basin also contains **trenches.** Trenches are long, narrow, and deep with steep sides. Trenches are often found next to a chain of islands or offshore, close to a coastal mountain range. Earthquakes and volcanoes often occur near these trenches. The names and depths of some of these trenches are shown in Table 10-2.

Table 10-2

Trench	Maximum Depth(Meters)
Aleutian	8,100
Kuril	10,400
Japan	9,700
Mariana	11,000
Mindanao	11,500
New Britain	8,300
New Hebrides	9,000
Tonga	10,800
Kermadec	10,800
Peru-Chile	8,000

More and more information about the ocean floor is being gathered by underwater cameras and deep-diving research vessels. Research vessels have been to the bottom of the trenches. Cameras have found strange new animals. Most of the ocean floor, however, is still unexplored, and there are still vast areas to be mapped.

Study Questions for 10.3

1. Describe the continental shelf.
2. Describe the continental slope.
3. Describe the ocean basin.
4. Describe the landforms found in the ocean basin.

What Did You Learn?

- A wave can be described by the crest, trough, wave height, and wavelength.
- Wind, earthquakes, and gravitational attraction make waves.
- How large wind-made waves become depends on the wind speed, how long the wind blows, and the fetch.
- Swell is long, low waves that travel thousands of kilometers.
- The wave bends forward and breaks as a wave approaches shore, making breakers.
- The zone where breakers occur is called surf.
- A wave made by an undersea earthquake is called a tsunami.
- Ocean currents are made by waves, wind, and differences in water density.
- Longshore and rip currents occur near the shore.
- A beach is a deposit of sand or pebbles on the shore.
- Shores are rock cliffs, sand or gravel beaches, or muddy tidal marshes.
- Sand is moved from the ocean floor to the beaches by waves and currents.
- Beach sand is made up of many different minerals. Most are tiny quartz grains.
- Beaches are always changing. Some grow larger in the summer and smaller in the winter.
- Sand-deposited shore features are sand bars, spits, barrier bars, and barrier islands.
- Echo soundings use sound signals to find the depth and shape of the ocean floor.
- The continental shelf, the continental slope, and the ocean basin are the three main divisions of the ocean's floor.
- The continental slope has valleys and canyons, some of which are larger than canyons on the surface.
- The bottom of the ocean is called the ocean basin. The basin has volcanoes and mountains that stick above the water, making islands.
- Mountain ranges called ridges occur in the ocean basin as do deep trenches.

Key Terms

crest
trough
wavelength
wave height
fetch
swell
breaker
surf
tsunami
current
longshore current
undertow
rip current
shoreline
shore
tidal marsh
sand bar
spit
barrier bar
barrier island
echo sounding
continental shelf
continental slope
ocean basin
ridge
seamount
trench

Biography

William Beebe (1877–1962)

William Beebe was born in Brooklyn, New York, in 1877 and grew up in New York City. He loved to read. When still a boy, he read the science fiction stories of Jules Verne and H. G. Wells. After graduating from college, he became a director of research for the New York Zoological Park (the Bronx Zoo).

William Beebe loved adventure, exploration, and writing. Following these interests, he took many trips to the jungles and across the ocean. He usually wrote books about his adventures and the animal life that he found. To him the ideal life would be to spend six months studying the oceans and jungles and then six months in New York reading and writing.

Beebe built an underwater laboratory in Bermuda and then wrote a book about life in the ocean. He became interested in deep-sea fish. Together with Otis Barton, he built a spherical steel diving chamber to explore the ocean depths. The steel chamber was attached to a cable and lowered from a ship. Beebe and Barton would ride down inside the chamber. After many earlier dives, they made a record dive to 923 m, deeper than anyone had ever been before. Beebe called his vessel a bathysphere. It was one of the first deep-sea diving vessels.

Beebe's book *Half Mile Down,* about the glowing fish and other things he found at the ocean bottom, makes great reading today. He later said he would never forget the memory of the magic of the ocean and all its life.

TO THINK ABOUT AND DO

Vocabulary

Number a sheet of paper from 1 to 8. Each statement below has a scrambled word or phrase. Write the unscrambled word or phrase next to the appropriate number on your paper.

1. Part of an ocean surrounded by land is a(an) ASE.
2. EARBERKS form when waves near the shore bend over and break.
3. ELSWL is long, low waves that travel far from storms.
4. Moving streams of water are called TURNRECS.
5. A deposit of sand or pebbles on the shore is a(an) HABEC.
6. Information about the ocean floor comes from HOEC GOSINSUND.
7. The CENTNOTLAIN LESFH is next to the land and less than 130 m deep.
8. The NOTLENTCAIN SOPEL goes from 130 to 4,000 m and is between the shelf and basin.

What Do You Remember?

Number a sheet of paper from 1 to 8. Write the letter of the correct choice next to the appropriate number.

1. The wave height is (a) the horizontal distance between two crests, (b) the horizontal distance between a crest and a trough, (c) the vertical distance between a crest and a trough, (d) the vertical distance between two crests.
2. An ocean wave in open water has (a) water moving with the wave, (b) the wave moving forward but not the water, (c) the water moving but not the wave, (d) neither the water nor the wave moving.

3. Long, low waves that travel many kilometers are called (a) surf, (b) tsunami, (c) swell, (d) whitecaps.
4. A wave produced by an undersea earthquake is called a (a) tsunami, (b) breaker, (c) swell, (d) whitecap.
5. A current that moves along the shore is called a(n) (a) rip current, (b) undertow, (c) longshore current, (d) breaker.
6. Beaches are made of deposits of grains of (a) rock fragments, (b) shells or coral, (c) quartz, (d) any of these.
7. Atlantic City, New Jersey, is built on a (a) beach, (b) sand bar, (c) spit, (d) barrier island.
8. Submarine canyons are found on the (a) continent, (b) continental shelf, (c) continental slope, (d) ocean basin.

Applying What You Have Learned

1. If the time interval indicated by an echo sounder is 10 seconds, what is the depth of the water?
2. There are similarities between the general pattern of ocean currents and the general circulation of the atmosphere. What would explain this?

Research and Investigation

1. Prepare a report on mineral resources that can be extracted from seawater or mined from the ocean floor.
2. Prepare a report on extracting fresh water from seawater.
3. Prepare a report on farming and ranching the ocean. Is this going on now?

Unit Four

The Earth's Structure

11 Minerals

Look at the unusual shape and form of the objects in this picture. These objects are just as they were when they were found in the earth. If you look in the right places, you can find many attractive and unusually shaped objects with beautiful colors. Why did these objects form in this shape? What material are they made of? Does that material always form in this interesting shape?

In trying to answer these questions scientists have made some interesting discoveries. The earth's crust is made from about 92 different building blocks. It is the combination in many different ways of these 92 building blocks that make up all the materials found in the earth's crust. In fact, there are about 2,500 different kinds of materials in the earth's crust. Each is unique in color, shape, and composition because of what it is made of and how it formed. In this chapter you will learn about these materials in the earth's crust and their importance to you.

Figure 11.1 *opposite:* These objects may look like an art sculpture, but they come from within the earth.

11.1 Atoms and Crystals

You will find out

- what the parts of an atom are;
- what model of atoms most scientists accept;
- why crystals have their shape.

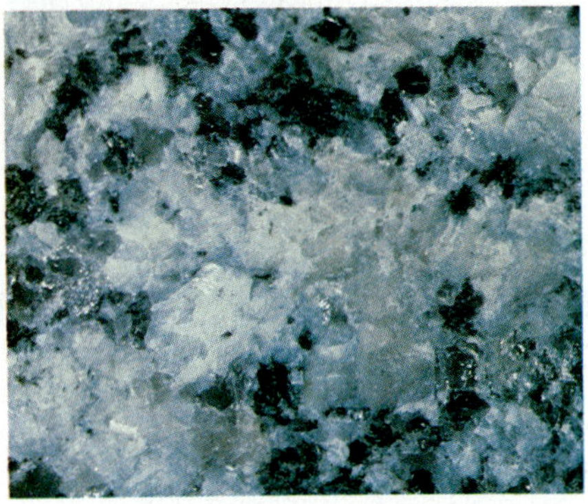

More than 2,000 years ago a Greek scholar named Democritus had a good idea. He thought that all solid substances were made of small indivisible particles. They were supposed to be something like very tiny balls. Somehow the particles held firmly together to form different substances.

Have you ever hit a stone with a hammer to break it into smaller pieces? If you break a stone into very, very small pieces and then look at the pieces through a microscope, you begin to see what the stone is made of.

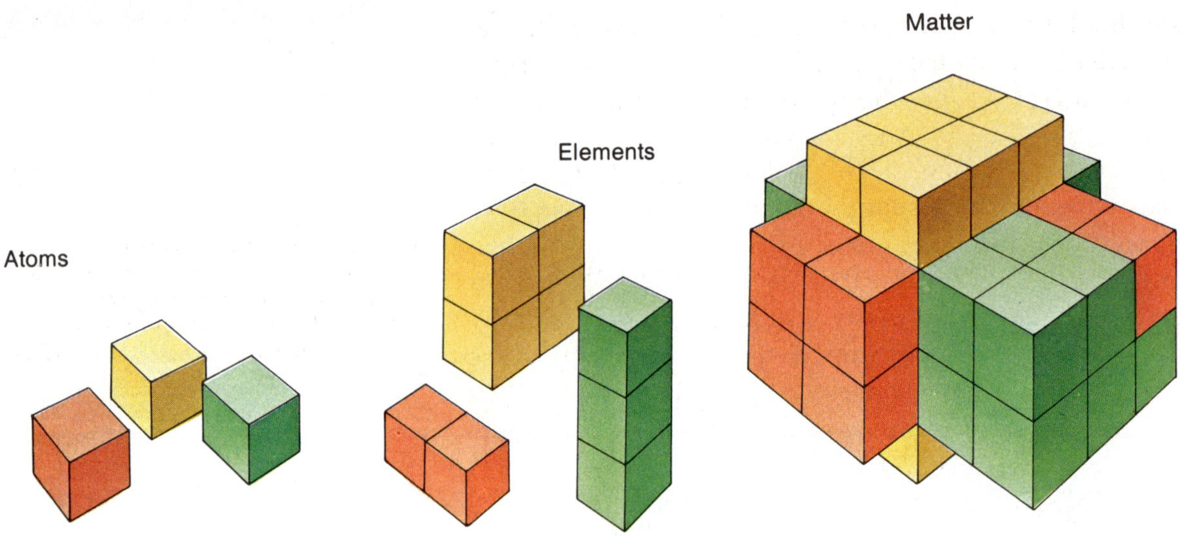

Figure 11.2 *top:* A close-up of granite shows small pieces that are meshed together.

Figure 11.3 *bottom:* The building blocks of matter

Do You Know?
Democritus was the first person to propose the word *atom*.

If you kept breaking a stone, you'd eventually find the building block of all material, an **atom.** There are 92 naturally occurring atoms. An atom is the smallest unit of an **element.** An element is made of only one kind of atom. Atoms of elements exist alone, or they can combine. When atoms combine, they form **matter.** Atoms are the building blocks of matter. Matter is anything that takes up space and has mass.

If atoms are the building blocks of all matter, what are atoms like? It turns out that Democritus was almost right, but not quite. What, then, is today's model of an atom? Since atoms are so small you can't see them, you must imagine what an atom looks like. Think of a tiny sphere in which, at the center, is a still smaller inner sphere, called the **nucleus** [NOO-klee-uhs]. The large sphere is very small, but even so, it is more than 600,000 times the size of the inner sphere. Yet, in the nucleus are still smaller particles! These smaller particles are of two kinds, called the **proton** [PROH-tahn] and the **neutron** [NOO-trahn]. They are defined by their electric charge.

Protons have a positive charge. Neutrons have no electrical charge. Each atom of an element has a certain number of protons. The simplest atom has one proton in the nucleus. The naturally occurring atom with the highest number of protons has 92. Outside the nucleus is another kind of particle that carries a negative charge. It is called the **electron** [ee-LEK-trahn]. The electrons really fly around. Even though the electrons are moving rapidly around the nucleus, they are believed to move in a kind of pattern. Figure 11.5 shows a model atom.

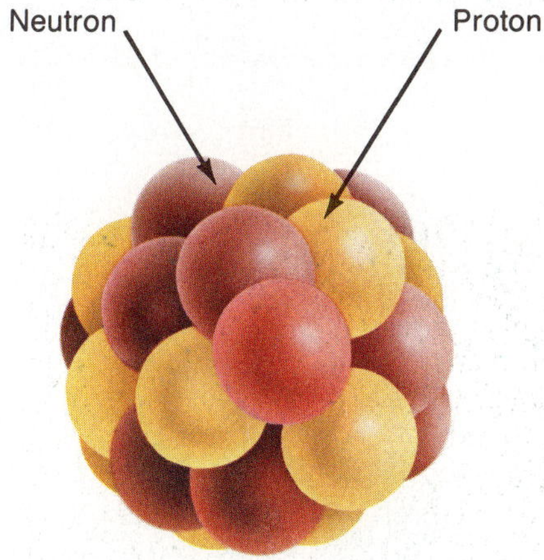

Figure 11.4 The nucleus of an atom is made of tightly packed protons and neutrons.

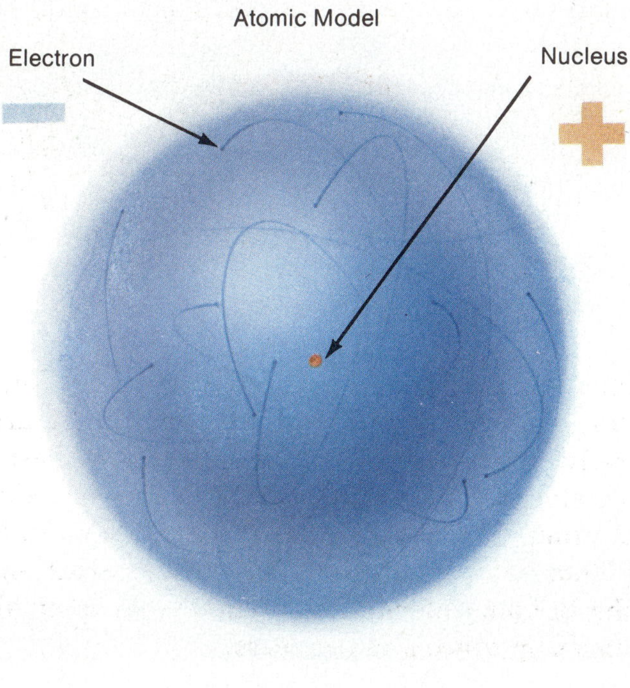

Figure 11.5 Scientists believe the modern model of an atom includes a tiny nucleus surrounded by rapidly moving electrons. The number of electrons equals the number of protons in any one atom.

The way an atom behaves is largely determined by the number of its electrons and protons. Some atoms easily gain electrons from neighboring atoms, and others easily lose electrons. When this occurs, the atoms become electrically charged and are called **ions** [EYE-uhns]. Unlike electrical charges attract one another. Thus an ion that has a positive electrical charge will stick tightly to an ion that has a negative charge.

Ions do not join together in just any old way. As you see in Figure 11.6, they pack together in a special arrangement. It is the way ions pack together that determines the form or special shape matter will take. Notice the geometric shape and smooth surfaces of the solid shown in Figure 11.7. The reason for the shape of this solid is that ions making up the solid are packed together in special repeating patterns. Solids with repeating patterns of packed ions are called **crystals** [KRIS-tuhls]. Each different kind of crystal has a special shape because of the way its ions are packed together in these repeating patterns.

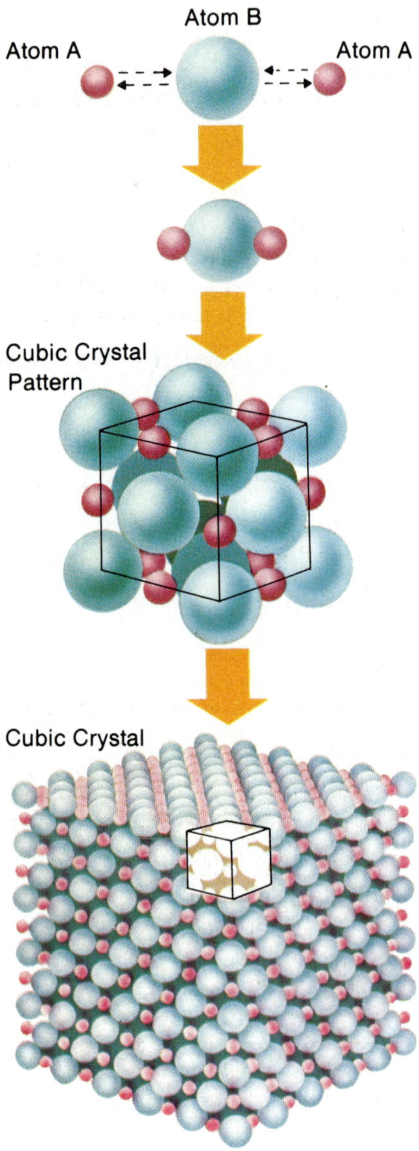

Figure 11.6 *above:* Ions often join in specific patterns that produce regularly shaped crystals.

Figure 11.7 *right:* The cubic shape of these halite crystals results from the way the ions join together.

Figure 11.8 Two examples of the many shapes that different crystals can have

Do You Know?
Since 1912, X-ray methods have been used to study the pattern arrangement of atoms in crystals.

ACTIVITY

Materials

safety glasses
ring stand
beakers
Bunsen burner
copper sulfate or
potassium alum
water
stirring rod
small glass jar with lid

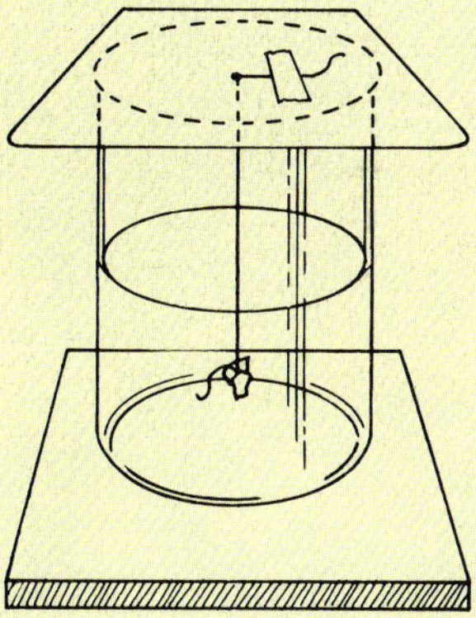

What Will the Crystal Look Like?

Procedure

1. Collect your materials. Put on safety glasses.

2. Fill the beaker about half full with water. Heat to boiling.

3. Remove the beaker and add the chemical slowly. Stir constantly. Try to dissolve all of the chemical.

4. Pour a little of the solution into the jar lid (about 4 mm deep).

5. Let the solution cool overnight.

6. Select a "seed" crystal from the jar lid.

7. Tie a human hair around the seed crystal.

8. Pour the solution from the beaker into the glass jar.

9. Suspend your crystal as shown.

10. Place your crystal where it will be safe for a week.

Questions

1. Draw a sketch of your crystal.

2. Did your crystal have the same shape as the crystals in the chemical you started with?

Study Questions for 11.1

1. Draw a sketch of a model of an atom.
2. Label the parts of the atom in your sketch.
3. How is a crystal formed?

11.2 The Physical Properties of Minerals

You will find out

- what a mineral is;
- what are some physical properties of minerals;
- how to test for some physical properties of minerals.

Do You Know?
A mineral is named by the person who discovers it. It can be called anything he or she wishes.

Do You Know?
Snow is a mineral.

Crystals have an important place in any study of the earth's crust. In Figure 11.9 you can see four different crystals in a piece of the earth's crust. These four different crystals are also called **minerals.** Generally you find minerals as crystals. In addition to their crystalline structure, minerals are also naturally occurring substances, and they have never been a living organism.

Figure 11.9 The large piece of granite at the top of the picture is made of the four minerals shown at the bottom. *left to right:* quartz, hornblende, feldspar, mica

The minerals making up the earth's crust are usually found as solids. Each mineral solid has its own set of physical properties. The differing properties make it possible to tell one mineral from another. There are about seven major properties that are useful in identifying minerals. They are color, streak, luster, hardness, cleavage, fracture, and density.

Figure 11.10 *left:* These are some examples of the many colors of quartz.

The **color** of a mineral is obvious, but it is not very useful for identification. Some minerals are always the same color. Other minerals, though, may vary in color. For example, quartz may be clear, milky, rose, or yellow.

A more accurate test of a mineral is to rub it across an unglazed tile called a streak plate. The color of the line of finely crushed mineral left on the plate is called a **streak.** The streak of the same minerals usually shows the same color even though the minerals themselves sometimes have a different color. Hematite may vary from steel gray to black in color, but it will always show a red-brown streak.

Luster [LUHS-tuhr] is the way light looks coming from a mineral's surface. Minerals looking like metal are called metallic. Others have a nonmetallic look, like silk, pearls, grease, glass, resin, and diamond. Luster is helpful for identification, but it must be used along with other properties.

Figure 11.11 *below:* The streak color of a specific mineral is always the same.

Figure 11.12 *left to right:* Metallic, silky, glassy, and greasy are examples of some mineral lusters.

Figure 11.13 Scratch tests help determine the hardness of minerals.

Do You Know?
An Austrian mineralogist named Friedrich Mohs (1773–1839) in 1822 proposed the hardness scale we use today.

Table 11-1 Mohs' Hardness Scale

Hardness	Mineral	Test Material
1	talc	fingernail
2	gypsum	fingernail
3	calcite	penny
4	fluorite	knife
5	apatite	glass
6	feldspar	file
7	quartz	will
8	topaz	scratch
9	corundum	glass
10	diamond	↓

Hardness is the resistance of a mineral to being scratched. It is a good test to use in identifying minerals. You will use a relative scale in making hardness tests. If your fingernail can scratch the mineral, it is softer than 2.5. Your fingernail has that hardness. A penny is about 3; a piece of glass, about 5 to 5.5; a knife, about 5.5; and a file, nearly 7. The test is made by trying to scratch the mineral or by using the mineral to try to scratch the test substance. If the mineral can scratch the test substance, the mineral is harder than the test substance. If the mineral is scratched by the test substance, the mineral is not as hard as the test substance. Table I is the scale used by mineralogists.

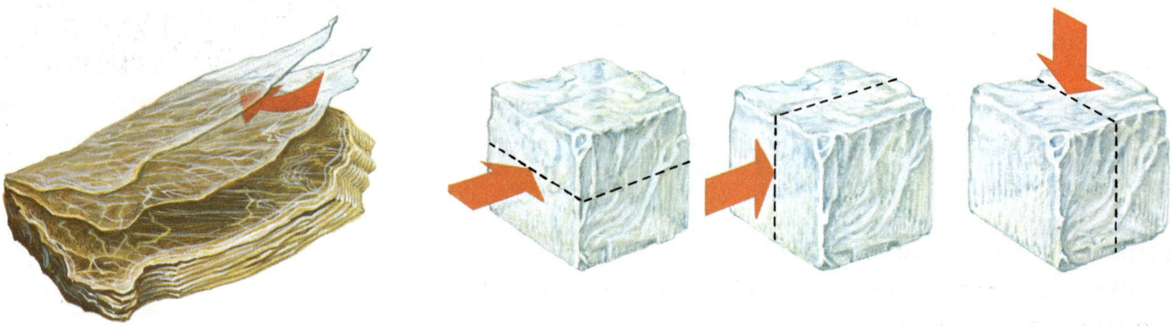

Figure 11.14 *above:* Parallel cleavage occurs in mica. *above right:* Minerals such as halite cleave in three directions.

Cleavage [KLEE-vuhj] is the ability of some minerals to break along smooth planes that are parallel to each other. Cleavage is determined by how the atoms are located in the minerals. Mica will split into very thin sheets. Calcite and halite will break in three directions.

If you strike a piece of either mineral, it will shatter into little pieces shaped like the big one. Cleavage is a good test to use in identification.

Fracture is the way the mineral breaks if it doesn't have cleavage. A fracture may make jagged edges, splinters, or irregular shapes. In obsidian, the fracture may look shell-like.

Figure 11.15 Two examples of fracture are obsidian (*far left*) and quartz (*left*).

Figure 11.16 *below:* The heft test is a simple way to compare densities.

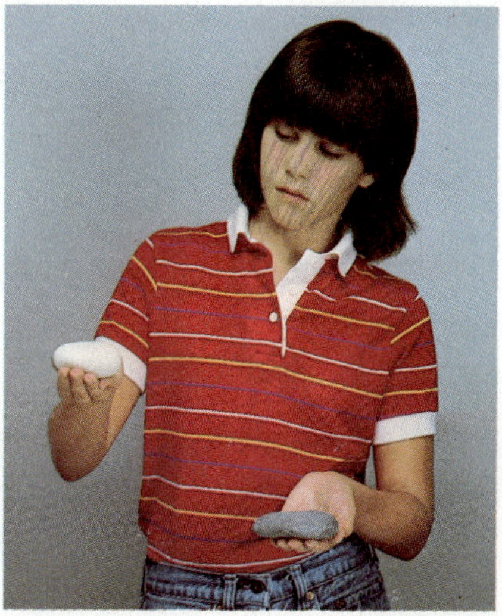

Density is the comparison of how heavy something is to the space it occupies. Imagine a piece of cotton and a piece of iron the same size. Which is denser? You know the iron is denser. Some minerals are denser than others. Take two minerals about the same size and put one in each hand. Which is heavier? The heavier one of the same size is also the denser one. This kind of hand test is useful only as a general comparison of the density of the minerals.

Some other properties and tests that can be used for a few minerals are salty taste, greasy feel, ease or difficulty of melting, and responses to a magnet.

Study Questions for 11.2

1. Name three characteristics of a mineral.
2. Define *luster, cleavage,* and *fracture.*
3. Describe how to test for streak, hardness, and density.

11.3 Identification of Minerals

You will find out

- that each of the many minerals found in the earth's surface has its own set of properties;
- what methods are used in identifying minerals.

There are more than 2,500 different minerals found in the earth's crust. How can you tell which mineral is which? Pick up a mineral. Look at it carefully. Even though each mineral has its own set of properties, choosing the correct mineral out of more than 2,500 does not seem easy. The mineralogist has a system to do this. Often what the mineralogist does is eliminate all the minerals it cannot be. The mineral that remains then is what it is. Sound confusing? Here's how it works.

Figure 11.17 Common minerals

You have a mineral you want to identify. Suppose you first make a hardness test and find it has a hardness of 7. That eliminates all minerals with a hardness other than 7. You then make a streak test and find it is white. That eliminates all minerals that do not have a white streak. By making two tests you have eliminated about 1,900 minerals that do not have both of these properties. You can eliminate many more with a few more tests. Then at last you have a good idea of what you are holding. In other words, you determine the mineral by showing what it cannot be.

Figure 11.18 "This must be . . . ?"

ACTIVITY

What Is the Name of These Minerals?

Materials

minerals
test equipment for hardness and streak
luster set

Procedure

1. Copy the physical properties chart.
2. Select one mineral.
3. Determine its properties.
4. Record the properties on the chart.
5. Determine its name from the list of minerals and their properties on pages 514-515.
6. Repeat steps 1 through 4 for the other minerals.

Questions

1. List the minerals you identified.
2. How did you identify each mineral?
3. List the minerals in your set in order of hardness from softest to hardest.

Mineral Number	Color	Streak	Luster	Hardness	Cleavage	Fracture	Estimate of Density	Mineral Name
1								
2								
3								

Study Questions for 11.3

1. Tell why you can identify each of the many minerals found in the earth's surface.
2. Describe the system used to identify a mineral.

What Did You Learn?

- Atoms are the building blocks of all matter.
- Atoms are made of protons, electrons, and neutrons, all of which are defined on the basis of their electric charge.
- The way an atom behaves is largely determined by its number of electrons and protons.
- Atoms that gain or lose electrons are called ions.
- Ions combine because unlike charges attract each other.
- The arrangement of ions determines the shape of matter.
- Crystals are solids that have repeating patterns of ions.
- Minerals are naturally occurring materials in crystal form and were never living.
- Each kind of mineral has its own set of physical properties.
- The physical properties of minerals are useful in determining the kind of mineral.
- Color is not reliable to use for mineral identification.
- Streak is the color of the crushed mineral as seen on a streak plate.
- Luster is the way light looks coming from the mineral's surface.
- Hardness is a measure of the resistance of the mineral to being scratched.
- Cleavage is the ability of the mineral to break along smooth planes that are parallel to each other.
- Fracture is the way a mineral breaks if it doesn't have cleavage.
- Density is the comparison of how heavy something is to the space it occupies.
- There are more than 2,500 different minerals found in the earth's crust.
- One way of determining the mineral is by showing what it cannot be.

Key Terms

atom
element
matter
nucleus
proton
neutron
electron
ion
crystal
color
mineral
streak
luster
hardness
cleavage
fracture
density

Biography

René Just Haüy [ah-WEE] (1743–1821)

Haüy was born in the small town of Saint Just in northern France. His family lived in poverty. He would have followed his father in making cloth if he had not been helped by others. As a child, Haüy displayed a strong interest in religion and religious music. His constant attendance at church attracted the prior of the local abbey to become interested in young Haüy. Others aided in seeing that he received a good education, and he thus became a botanist.

Haüy was walking through the Royal Gardens in Paris and noticed a large group of people entering a lecture hall. A lecture on mineralogy was being given. He came away from that lecture impressed with the regularity and variety of crystals found in minerals. Later while visiting a friend who had a collection of minerals, Haüy dropped a fine group of calcite crystals. As he examined the broken pieces, he noticed that their shape was similar to another crystal in his collection. He returned home and selecting from his collection a number of calcite crystals of different shapes, he broke them all. In every case the broken pieces were the same shape. Haüy did not hesitate to break up all the crystals in his collection. Furthermore, he broke up all the crystals he could obtain from friends. In all minerals that displayed cleavage, he believed that he could find some nuclear form about which their crystals had been built. This concept, expanded upon by others, led to the science of mineralogy. René Just Haüy is called the father of mineralogy. And all because he was willing to break things apart in order to see them better. Haüy became a professor of mineralogy under Napoleon's reign. When Napoleon was ousted in 1814, Haüy was discharged. He died a very poor man.

TO THINK ABOUT AND DO

Vocabulary

Make a sentence using each group of the following words. The sentence must be a logical, complete statement. You may change the form of the word if you want.

1. atom, element, matter
2. atom, proton, electron, neutron
3. crystal, mineral
4. luster, hardness, streak, cleavage, fracture
5. density

What Do You Remember?

1. Democritus had a good idea. What was it?
2. Define *element*.
3. What are the parts of the nucleus of an atom?
4. Define *ion*.
5. What is a naturally occurring substance in crystalline form?
6. How do you make a streak test?
7. What determines the luster of minerals?
8. Describe how to make a hardness test of a mineral.
9. Arrange the following in order from hardest to softest: penny, file, fingernail, glass, knife.
10. What would you need to see to indicate that a mineral has excellent cleavage?
11. How can you compare the density of two minerals the same size?

Applying What You Have Learned

1. Name something in your classroom that was once a mineral. Describe three of its physical properties.
2. Suppose two minerals have the same physical properties, but look different; what would you do to identify the minerals?
3. What are the properties of some minerals that make them desirable for jewelry?

Research and Investigation

1. Assemble, mount, and label samples of some local minerals.
2. Set up a display of minerals, crystals, and gems in the school office.
3. Arrange for the library to display their books on rocks, minerals, and gems.
4. Visit a jeweler and ask to see some of their unmounted colored gems.
5. Attend a meeting of a local gem and mineral club.
6. Obtain a tumbler and tumble some minerals. Show how tumbling changed the appearance of the minerals.

Agate quartz from Mexico

12
Rocks and the Rock Cycle

How did this gigantic mass of rock get to be in this location? Did some beings from another planet build it here? Notice the long lines that divide the rock mass into columns. Even the Indians once had a legend about this rock. They said a huge bear made these lines by trying to climb the rock.

What you are seeing was at one time liquid rock in the center of a volcanic mountain cone. The surrounding cone is gone, carried away by wind and water leaving only the harder rock core. The lines were formed when the liquid rock cooled. The lines are cracks that formed as the rock shrank during cooling, creating the huge columns in the rock. The name of this huge mass of rock is Devil's Tower, and it is located in Wyoming.

The rock you now see in Devil's Tower formed from cooling liquid rock. But not all rocks look like this, nor are they all formed from hot liquid rock. What are some other types of rocks you can find and how did they form? This chapter focuses on how to identify a rock and read its history.

Figure 12.1 *opposite:* Devil's Tower in Wyoming is an example of a volcanic neck.

12.1 How Rocks Are Made

You will find out
- what processes form rocks;
- what the three types of rocks are;
- what is meant by the rock cycle.

What is a rock? Although you may think this is a simple question, it is not so simple as it seems. Rocks are identified by the minerals in the rock and how these minerals came together. But the task of learning about rocks isn't very difficult, as there are only three basic types of rocks. Each type is named by the way it was formed.

Figure 12.2 Half Dome, Yosemite National Park, is made of granite, an igneous rock.

Figure 12.3 *right:* The eruption of Heimaey volcano in Iceland

Have you ever seen a rock being formed? Most people say no. But if you've seen a stream of liquid rock coming from a volcano, your answer should have been yes. Some rocks are formed by the process of cooling liquid rock. These rocks are called **igneous** [IHG-nee-uhs] **rocks.** The word *igneous* comes from the Latin word meaning "fire."

Do You Know?
Geologists believe igneous rocks are the ancestral material from which all other rocks are made.

Figure 12.4 *left:* Sedimentary rocks are exposed by the wearing action of a river.

Figure 12.5 *below:* This muddy river is carrying sediments that will eventually settle on the ocean bottom.

Igneous rocks exposed on the surface of the earth slowly break apart. When exposed to air, water, and changing temperature, some of the minerals in the rocks begin to crumble. This loose rock is pulled downhill whenever possible by gravity. At times wind, water, and ice move some of the pieces along. Eventually these tiny pieces of igneous rock are carried to the ocean by rivers and glaciers. Over millions of years the bits and pieces settle to the bottom of the ocean. Layer upon layer builds up. These layers that settle to the ocean bottom are called **sediments** [SEHD-uh-muhnts]. In time the sediments are pressed and cemented together. Rocks that formed from these settled bits and pieces of other rocks are called **sedimentary** [sehd-uh-MEHN-tuh-ree] **rocks.** *Sedimentary* comes from a Latin word meaning "to settle." Most sedimentary rocks are formed on the shallow sea floor near the shore.

Do You Know?
Some chalk, or limestone, is sedimentary rock. It contains microscopic shells of sea animals that accumulated on the ocean bottom millions of years ago.

A third type of rock is called **metamorphic** [meht-uh-MAWR-fihk] **rock.** *Metamorphic* comes from Latin and French words meaning "change." And that is just what has happened to these rocks. They are rocks that were once igneous or sedimentary rocks, but they have been changed. These changes are caused by heat and pressure. The ideal conditions for these metamorphic changes are generally deep in the ground. At depths of 12 to 16 km beneath the surface, temperatures may range from 200°C to 800°C. Under these conditions the existing rock is changed, often without melting. It may undergo a change of the mineral grains, a change in or an enlargement of crystals, or a change in the chemical composition.

Figure 12.6 *above:* One rock resulting from metamorphic change is quartzite which is made from sandstone.

Figure 12.7 *right:* The altered patterns of layers in this metamorphic rock result from extreme forces of heat and pressure.

The distinction between rock types is often shady, not clear or sharp at all. Look at Figure 12.8. This is a model of the changes that rocks undergo through time. No one has ever seen one rock change into another. Geologists believe that changes do occur, and also that they occur over and over. There is no such thing as the eternal hills or the unchanging mountain. In time—a long, long time—the rocks change. Over and over they change from one kind of rock to another. These endless changes are part of a cycle that is called the **rock cycle.** The sedimentary rocks you see were once igneous or metamorphic or other sedimentary rocks. Someday they will again be forced deep into the earth where they will melt. The rock cycle is endless.

Do You Know?
Geologists estimate that the time it takes for a rock to turn to sediment and back to rock again is a few hundred million years or less. The earth is at least 4 billion years old. It is unlikely that any large block of the earth's first rock could exist.

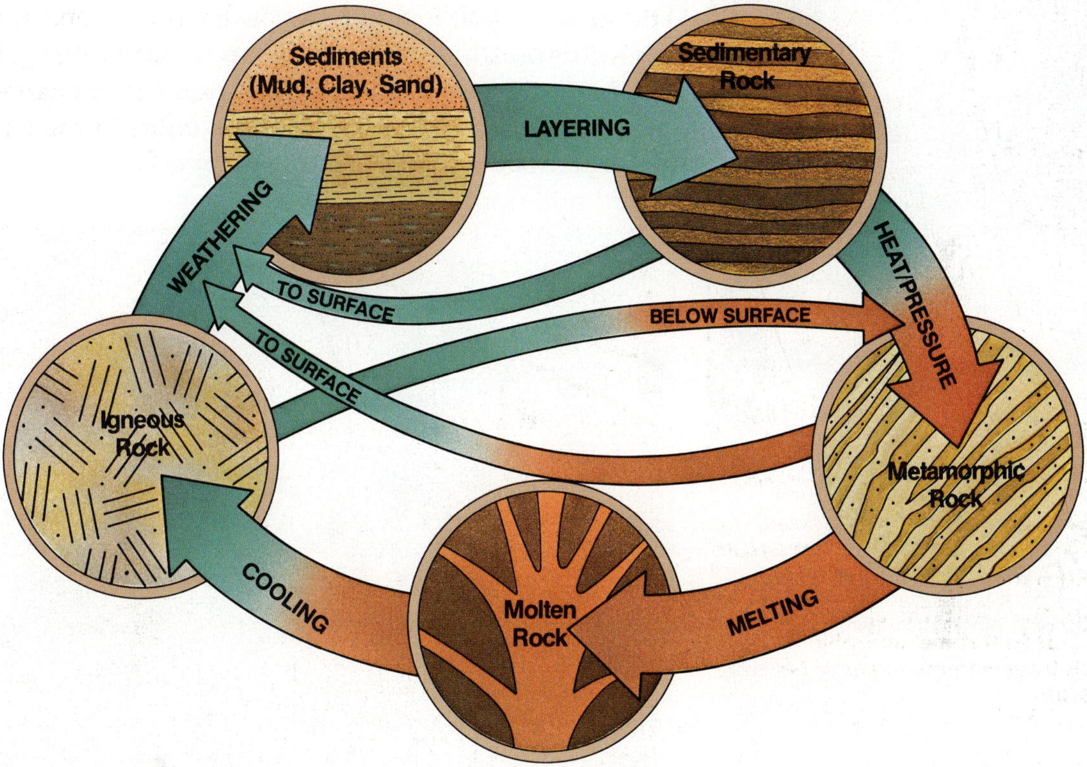

Figure 12.8 The rock cycle

Study Questions for 12.1

1. What are the major types of rocks?
2. List the processes that form the three types of rocks. Are the names given each type appropriate?
3. Describe what is meant by the rock cycle.

12.2 Igneous Rocks

Do You Know?
Most of the rocks returned from the moon were igneous rocks.

Figure 12.9 *above:* Three colors of igneous rock

Figure 12.10 *right:* Large and small crystals in igneous rocks

Figure 12.15 *below:* Porphyry is an igneous rock with large crystals surrounded by small ones.

You will find out

- how to read the cooling history of igneous rocks;
- what properties of igneous rocks can be used to identify them;
- what uses are made of igneous rocks.

How can you tell which rocks are igneous? Generally igneous rock can be classified by two characteristics. The first is their color pattern. Igneous rocks tend to separate into three color patterns. Some are formed of light-colored minerals; some contain only dark minerals; and some igneous rocks share both light and dark minerals. The colored materials appear to be generally mixed and not layered.

The second characteristic is the crystalline form of the minerals in the rocks. The crystals can be of various sizes depending on how the minerals cooled. An igneous rock with large, visible crystals probably cooled deep in the earth at a slow rate. Smaller crystals, in general, mean more rapid cooling.

What about the cooling history of a spotted igneous rock which has both large and small crystals? It is believed that at first this rock cooled slowly, forming large crystals. Then its cooling rate changed, and small crystals were formed. An igneous rock with large crystals surrounded by small ones is called a **porphyry** [PAWR-fuh-ree]. Crystal size is an important property in rock identification.

Textures or properties of igneous rocks may vary. Glassy textures are formed when the rock cools very fast. Fragmented textures occur when volcanic ash and dust are cemented together. Bubbly textures develop when gas bubbles are trapped in a rock as it cools. Sometimes this kind of igneous rock, called pumice, is so filled with gas bubbles it will float on water.

Figure 12.12 *left:* Examples of textures of igneous rock are (1) glassy obsidian, (2) fragmented basalt, and (3) bubbly tuff.

Figure 12.13 *below:* Granite is often used to construct buildings.

What are some of the uses of igneous rocks? Look around your school or town. You'll probably find a number of places where granite was used in building construction. Igneous rocks are also used to make statues, fountains, and tombstones. They also supply copper, iron ore, diamonds, silver, gold, nickel, chromium, and platinum.

How are igneous rocks identified? The process is similar to how you identified minerals. By examining the properties of a rock you can decide what it isn't. By using the eliminating process you finally determine what it is. Each property examined will eliminate whole groups of rocks. By elimination you come to the only rocks possessing the properties of the rock in your hand.

Do You Know?
Igneous rocks comprise 93 percent of the outer 10 km of the earth's crust.

ACTIVITY

What Are the Names and Cooling Histories of These Igneous Rocks?

Materials
igneous rock set
magnifying glass

Procedure

1. Select a rock from your set.

2. Look at your rock chart. Ask the first question. If the question is appropriate, record the names of the rocks with that physical characteristic.

3. Ask the second question. Record the answer.

4. Determine the amount of quartz or dark mineral in your rock.

5. Continue until you know what kind of rock you are holding.

6. Repeat this procedure with the other rocks in the rock set.

Questions

1. What are the names of your coarse grained rocks? What is their cooling history?

2. What are the names of your fine grained rocks? What is their cooling history?

Igneous Rock Classification Chart

	Obsidian	Granite	Rhyolite	Diorite	Andesite	Gabbro	Basalt	Pumice	Porphyry	Tuff
Is the rock fine grained?			●		●		●	●		
Is it coarse grained?		●		●		●				
Is it glassy?	●									
Does it contain quartz?		●	●	○	○					
How much quartz?		> 10%	> 10%	0 – 10%	0 – 10%					
Is it mostly dark mineral?						●	●			
Is it mostly light mineral?				●	●					
Does it have bubbles?			●		●		●			
Will it float?								●		
Is it spotted?									●	
Is it made of broken pieces?										●

Key:
● Always present
○ Sometimes present

1. Granite 2. Andesite 3. Basalt 4. Tuff 5. Diorite
6. Gabbro 7. Porphyry 8. Obsidian 9. Rhyolite 10. Pumice

Figure 12.14 Common igneous rocks

Study Questions for 12.2

1. How is crystal size used in interpreting the cooling history of an igneous rock?
2. Explain how the properties of igneous rocks are used in identification.
3. List three uses of igneous rocks.

12.3 Sedimentary Rocks

You will learn

- how sedimentary rocks form;
- how to identify rocks that form from sediment;
- what uses are made of sedimentary rocks.

Figure 12.15 *above:* Sedimentary rock

Figure 12.16 *below:* The formation of sedimentary rock

Three processes are involved in making sedimentary rock. The first process in making sedimentary rock is called **precipitation** [pree-sihp-uh-TAY-shuhn]. Ocean water has many substances dissolved in it. Precipitation occurs when chemical reactions form a solid that settles out of solution. This is followed by a second process called **compaction** [kuhm-PAK-shuhn]. In compaction, materials are pressed together by the weight of the layers above. The final process is called **cementation** [see-mehn-TAY-shuhn]. Water carrying mineral matter in solution passes through the sedimentary layers. This mineral matter crystallizes and holds the layers together like glue.

Slowly but surely the bits and pieces that form sedimentary rocks are moved toward the ocean. As they become smaller and smaller, other things also happen. Minerals continue to break up, and some of the mineral matter goes into solution in the water. All exposed igneous and metamorphic rocks continue this process of breaking up to the finest possible pieces. Finally you have the three basic materials that make up sedimentary rocks. These are quartz-sand, clay, and calcium carbonate in solution.

Suppose the pieces of rock do not break up completely. What is the sediment like? Near the place where the rock is breaking apart, sediments begin as a mixture. You'll find sharp-edged pieces, mineral pieces, clay, and mineral matter in solution. Most of the bits and pieces will look like the rock from which they came. As the bits and pieces move along, they get worn down. They become rounded, and the smaller pieces move farther from the source rock. Smaller pieces wash farther downstream or downhill. Larger pieces fall out upstream or uphill. The history of a sedimentary rock is revealed by what happens to the bits and pieces.

In general, rocks containing large sharp-edged pieces were formed near the source rock. Rocks containing small pieces that have been rounded formed farther from the source rock. A sedimentary rock such as some limestones and shale with no visible fragments is very likely to have formed near shore in the ocean far from the source rock.

What uses are made of sedimentary rocks? You could make a long list. Here are a few things obtained from these rocks: coal, petroleum, natural gas, aluminum, iron, and copper. Sedimentary rocks are also used as sand, gravel, road base, and building stone. Limestone and sandstone have been widely used in buildings. Sedimentary rocks have great economic importance for people all over the world.

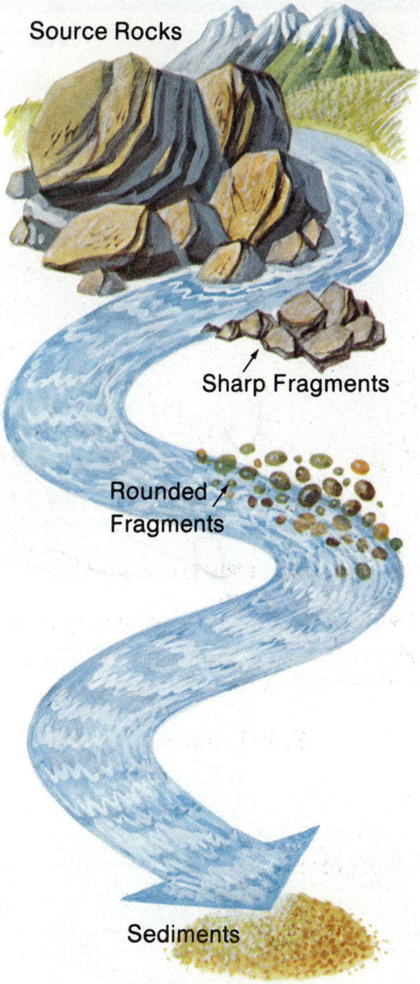

Figure 12.17 *above:* As rock travels farther from its source, the shape of the fragments change.

Figure 12.18 *left:* One of the many uses of sedimentary rock

Do You Know?
One kind of sedimentary rock stands out as the most valuable: oil shale. Found in the western United States, oil shale contains enormous quantities of oil. It has been estimated that more than 12 trillion barrels of oil are contained in this oil shale.

ACTIVITY

What Are the Names and Histories of These Sedimentary Rocks?

Materials

sedimentary rock set
magnifying glass

Procedure

1. Select a rock from your set.

2. Look at the chart. Ask the first question. If the question and answer are appropriate, record the rock names with that physical characteristic.

3. Ask the second question. Record the answer.

4. Repeat this procedure until you know what kind of rock you are holding.

5. Repeat this procedure with the other rocks in the rock set.

Questions

1. What are the names of your sedimentary rocks made of fragments? What are their histories?

2. What are the names of your sedimentary rocks made from precipitation from sea water? What is their history?

Sedimentary Rock Classification Chart

	Conglomerate	Breccia	Sandstone	Shale	Limestone	Dolomite	Chert	Coquina	Coal
Contains fragments?	●	●	●	●	●			●	
Round fragments?	●								
Angular fragments?		●							
Bubbles when acid applied?					●			●	
Bubbles only when acid applied to powder?						●			
Harder than a knife blade?							●		
Black color?									●
Scratches limestone?						●	●		
Contains plant material?									●

Key:
● Always present

How are these characteristics used in identifying sedimentary rocks? Look carefully at the rock and at the pieces making up the rock and again use the process of elimination. Each identified characteristic will eliminate other rocks without that characteristic. This will help you select its name and describe its history.

Figure 12.19 Common sedimentary rocks

1. Conglomerate 2. Breccia 3. Sandstone 4. Siltstone 5. Shale 6. Limestone 7. Dolomite 8. Gypsum 9. Coquina 10. Coal

Study Questions for 12.3

1. Describe how sedimentary rocks form.
2. Explain how the properties of a sedimentary rock are used in identification.
3. List three uses of sedimentary rock.

12.4 Metamorphic Rocks

You will learn

- how to interpret the degree of change in rocks;
- how to identify rocks that have been changed from pre-existing rocks;
- what uses are made of metamorphic rocks.

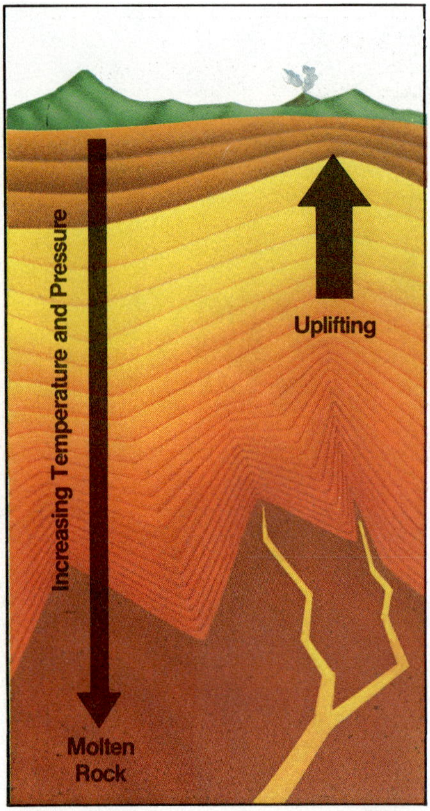

All metamorphic rocks start out as something else and end up as metamorphic rocks. Most processes bringing about this kind of change take place deep inside the earth's crust. In Figure 12.20 you can see layer upon layer of sedimentary rock. This continuous piling up of layer upon layer over a large area increases pressure on the bottom layers. As the layers increase, tremendous pressure and heat develop. Changes are occurring at all levels. But in the lower levels the changes are greatest. Heat, pressure, and contact with mineral-rich liquids act to change one kind of rock into another. Geologists can read the changes that have taken place by looking at the arrangement of minerals in the rock.

Figure 12.21 shows a picture of a metamorphic rock. Try to read its history. This rock formed on the surface. In time it was buried five to ten km under rocks of the earth's crust. There it was squeezed and heated. Pressure appears to have flattened the fragments, and some material recrystallized around the fragments. Over much time the rock was forced upward to the surface. This cycle may be repeated more than once for the same rock.

Figure 12.20 *above:* The formation of metamorphic rock

Figure 12.21 *right:* Flattened fragments are clearly visible in this example of metamorphic rock.

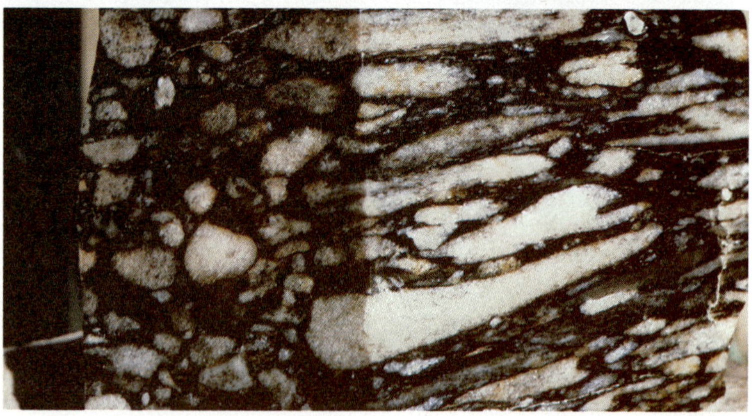

Figure 12.22 is a model of metamorphic change. In general, the degree of change in a rock depends upon its depth in the earth and how long it stays at that depth. The model of metamorphic change can also be applied to the way minerals arrange themselves during metamorphism. In a low-grade change the minerals recrystallize in flat or long slender shapes that line up in the rock. In high-grade change the minerals recrystallize to form bands of light- and dark-colored minerals. The way the minerals are arranged determines whether or not the rock will break in layers.

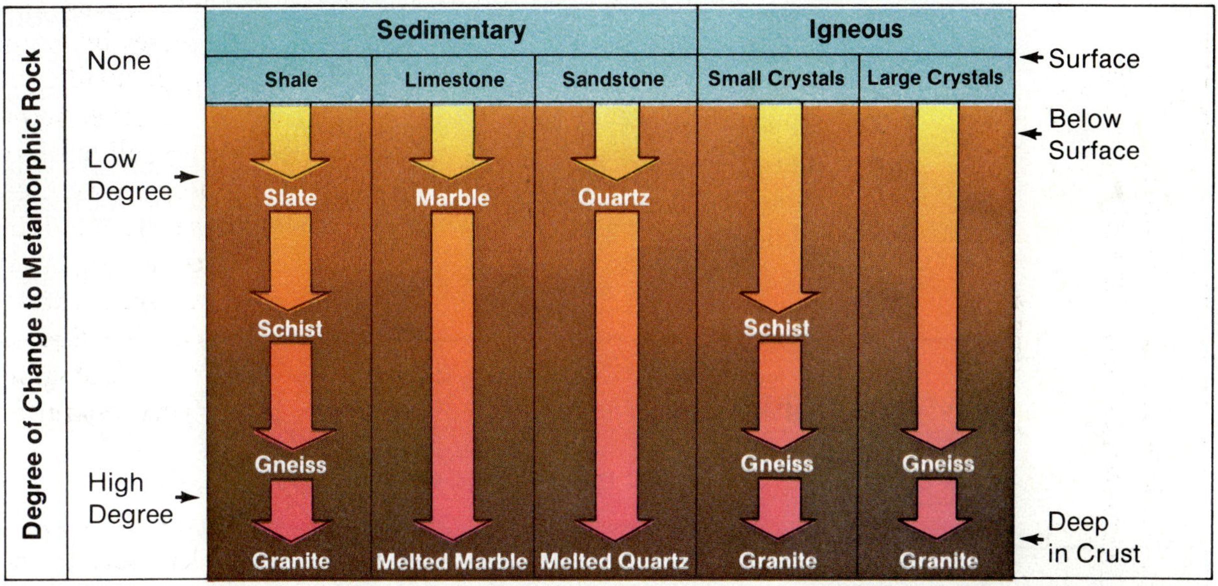

Figure 12.22 The change to metamorphic rock varies with the type of rock, depth, and time.

Metamorphic rocks are also identified by their physical properties. Each property, such as layered or no visible grains, combined with another property, eliminates many rocks and leaves only a few kinds for your final identification. If you now find the rock also breaks in layers, you are likely to have a piece of slate.

Metamorphic rocks are widely used for building stone. Marble frequently is slabbed and used to make decorative walls and columns. The graphite lead in your pencil may have come from a metamorphic rock. Other useful materials from this rock type are talc, asbestos, soapstone, and garnet to make abrasive stones.

Figure 12.23 Marble was used for the statue of Abraham Lincoln at the Lincoln Memorial in Washington, D.C.

ACTIVITY

What Are the Names of These Metamorphic Rocks?

Materials

metamorphic rock set
magnifying glass

Procedure

1. Select a rock from your set.
2. Look at your rock chart. Ask the first question. If the question is appropriate, record the names of the rocks with that physical characteristic.
3. Ask the second question. Record your answer.
4. Repeat this procedure until you know what kind of rock you are holding.
5. Repeat this procedure with the other rocks in the rock set.

Questions

1. What are the names of your layered metamorphic rocks? What is their degree of metamorphism?
2. What are the names of your nonlayered metamorphic rocks?

Metamorphic Rock Classification Chart

	Slate	Schist	Gneiss	Quartzite	Marble
Is the rock layered?	●	●	●		
Can you see large mineral grains?			●		
Can you see medium sized grains?		●			
Is the rock fine grained?	●				
Does it break into layers?	●	●			
Is it banded?			●		
Does it sparkle like sugar?					●
Does it bubble with acid?					●
Will it scratch a knife blade?				●	
Does it contain quartz?				●	

Key:
● Always present

How many different minerals are found in metamorphic rocks? Table 12.1 shows that most metamorphic rocks are made from seven minerals. With the exception of garnet the same minerals are also abundant in igneous rocks. Sedimentary rocks are made primarily of only four minerals. The great variety of rocks you find is due to different arrangements and different amounts of these minerals.

Table 12.1

Mineral Type	Igneous	Metamorphic	Sedimentary
Olivine	Abundant	Rare	Rare
Garnet	Rare	Abundant	Rare
Augite	Abundant	Common	Rare
Hornblende	Abundant	Abundant	Common
Muscovite	Abundant	Abundant	Common
Biotite	Abundant	Abundant	Common
Kaolinite	Rare	Rare	Abundant
Potassium Feldspar	Abundant	Abundant	Common
Sodium Feldspar	Abundant	Abundant	Common
Quartz	Abundant	Abundant	Abundant
Calcite	Rare	Common	Abundant
Dolomite	Rare	Common	Abundant

Key: Rare (green) Common (blue) Abundant (red)

1. Schist 2. Gneiss 3. Slate 4. Quartzite 5. Marble

Figure 12.24 *left:* Common metamorphic rocks

Study Questions for 12.4

1. Describe the model of metamorphic change.
2. Explain how the properties of metamorphic rocks are used in identification.
3. List three uses of metamorphic rocks.

What Did You Learn?

- Igneous rocks are formed by the process of cooling liquid rock.
- Igneous rocks exposed on the surface of the earth slowly break apart.
- The bits and pieces of igneous rock eventually reach the ocean and settle to the bottom to form sedimentary rock.
- As the layers of sediments accumulate, the lower layers gradually become rock by the processes of compaction and cementation.
- Rocks buried deep in the earth's crust are changed by heat and pressure into metamorphic rocks. The heat, pressure, and mineral-rich fluids deep in the earth form new minerals out of old ones.
- The rock cycle is a model showing the changes rocks undergo through time.
- Reading the cooling history of igneous rocks is based on a knowledge of cooling rates and crystal size.
- An igneous rock with large crystals surrounded by small ones is called a porphyry.
- Igneous rocks are identified by studying the properties of the rocks. By eliminating what the rock isn't, you finally come to find out what it is.
- As bits and pieces of rock move toward the ocean, three kinds of sediments are formed—quartz-sand, clay, and calcium carbonate in solution.
- Reading the history of a sedimentary rock is possible by carefully looking at the pieces forming the rock.
- All metamorphic rocks start out as something else and end up as metamorphic rock.
- The history of a metamorphic rock can be read by looking at the arrangement of minerals in the rock.
- Rocks deepest in the earth are changed the most.
- Only a few different minerals make up igneous, sedimentary, and metamorphic rocks.
- The great variety of rocks is due to different arrangements and different amounts of only a few minerals.

Key Terms

igneous rock
sediment
sedimentary rock
metamorphic rock
rock cycle
porphyry
precipitation
compaction
cementation

Career

Geochemist

Geochemists study the chemical composition of and the actual or possible chemical changes in the crust of the earth. To do this work requires advanced study in the fields of mathematics, chemistry, and geology.

Dr. Rosemary Vidale, a geochemist, works at Los Alamos National Laboratory and lives in Los Alamos, New Mexico. She earned a B.A. degree in chemistry from Oberlin University in Ohio and a Ph.D. from Yale University. When she's not working on her field of special interest, the transport and concentration of chemical elements in the earth's crust, Dr. Vidale enjoys several hobbies. These include photography, hiking, reading, camping, and music. She regularly gives geology travelogues and talks to school groups and the public on such topics as hot springs, volcanoes, and mineral and rock identification. She leads local geology field trips for interested groups.

Her research findings are useful in determining a safe disposal method for chemically dangerous wastes.

TO THINK ABOUT AND DO

Vocabulary

Unscramble the following words and find the word spelled in the squares.

t a m o [r] m e p h i c
s e [o] n i g u
t o m [c] a p c i o n
[k] o r c
a t t e n [c] i o n e m
r y p h [y] o r p
[c] r e p i t a p i o n i t
h e a [l] s
n o d [e] s i m t e n a t i

What Have You Learned?

1. Rocks that are formed by the process of cooling liquid rock are called what?
2. What causes the different sizes of minerals making up igneous rocks?
3. Where do most sedimentary rocks form?
4. List the three processes that form sedimentary rock.
5. What happens to sedimentary rocks that are forced deep into the earth's crust?
6. Under what conditions are new minerals formed from old ones in making metamorphic rocks?
7. How would you describe the cooling history of an igneous rock with small crystals?
8. What causes glassy textures in igneous rocks? Fragmented textures? Bubbly textures?
9. Low-grade metamorphic change causes the minerals to recrystallize in what shapes? What do high-grade changes cause?
10. What is the name for the model of how rocks change through time?

Applying What You Have Learned

1. Suppose you find an igneous rock with dark-colored minerals and large crystals. Where was it likely to form in the earth's crust? Explain your answer.
2. Develop a model that describes the process of changing a sandstone—a sedimentary rock—to quartzite—a metamorphic rock.
3. Look at the rock cycle, Figure 12.8. (a) What has to happen to a sedimentary rock to become another sedimentary rock? (b) What has to happen to change a sedimentary rock into a granite? (c) What has to happen to change an igneous rock into a metamorphic rock?

Research and Investigation

1. Make a list of the buildings in your community in which different kinds of rocks can be seen. List the kinds of rocks in each building.
2. Design a field-trip guide to buildings in your community that are made of different kinds of rocks.
3. Visit a sand and gravel operation.
4. Find out how much sand and gravel is used to build one block of sidewalk.

A granite quarry

13

Internal Structure of Earth

A carved gully or a deep canyon caused by rushing water is something you can easily see. You can also see soil and sand being carried away by muddy rivers. But deep inside the earth there are some things you cannot see. You can see the evidence that they happened, though, in earthquakes and volcanoes. In this picture you see the terrible destruction caused by the earthquake in Anchorage, Alaska, on March 27, 1964.

What caused the earth's surface to move with such destruction? Possibly something happened to squeeze or stretch the earth's outer layer. Some people in the past thought that perhaps the earth was growing larger. This growth would stretch and crack the rocks in the earth's outer layer. Others have suggested that the earth was a cooling molten planet with a shrinking and crackling surface. Now scientists have constructed a model of the internal structure of the earth to answer some of these questions. How they obtained the information they used to make their model is developed in this chapter.

Figure 13.1 *opposite:* Sudden destruction comes from forces within the earth.

13.1 A Model of the Earth's Interior

You will find out

- what the three main layers of the earth are;
- what the characteristics of these layers are believed to be.

What do you think of when you hear that something is "as big as a mountain"? The size of a mountain is hard to understand. When you see a mountain from a distance, it looks big. Yet when you are on the mountain, you see only small boulders and rocks. It is difficult to look at all the small rocks and understand the size of the mountain. Scientists had this same problem when they tried to understand the whole earth. They couldn't see the whole earth. They could only see bits and pieces of information about the earth. But over a long period of time these bits and pieces of information helped the scientists build a model of the earth. According to this model, shown in Figure 13.2, the earth is made of three main layers.

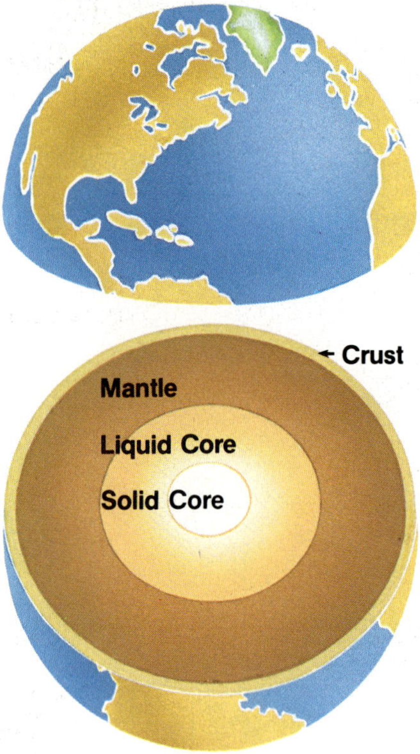

Figure 13.2 The main layers of the earth

The outside layer of earth is called the **crust.** It is a relatively thin layer of solid rock made of sedimentary, metamorphic, and volcanic rocks. You cannot see the crust because it is mostly covered with small particles of rocks, soil, and sand. The seas and oceans also cover the crust.

The crust is not the same thickness everywhere. Its thickness varies from 20 to 50 km across continental United States. The average thickness is 32 km. The crust is thinner beneath the oceans, averaging less than 8 km thick. The crust makes up 1 percent of the earth's volume but only 0.4 percent of its mass.

Figure 13.3 The thickness of the earth's crust varies at different locations.

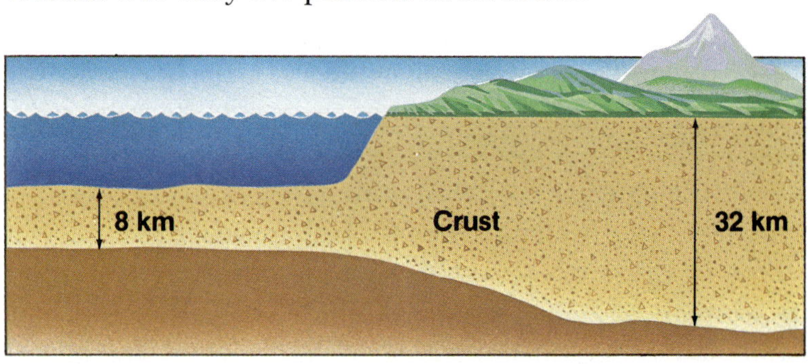

The layer below the crust is called the **mantle.** It is a very thick layer that goes from the bottom of the crust down to about 2,900 km. The mantle is made of solid rock. But the rocks in the mantle are different from the rocks in the crust. The mantle itself consists of layers. It has rigid upper and lower parts surrounding a middle layer that is able to move slowly like very thick syrup. The mantle makes up 84 percent of the earth's volume and about two thirds of the earth's mass.

The center part of the earth is a large, dense **core.** The core begins below the mantle and continues the remaining 3,400 km to the center of the earth. The core is believed to be made of a solid inner part surrounded by a liquid outer layer. Both are thought to be made of iron and nickel. The core makes up 15 percent of the earth's total volume and about one third of its mass.

As you can see, the model of the earth's interior has three basic parts. It is like a huge soft-boiled egg. The model has a center core (the yolk), a middle part (the white), and a thin outer layer (the shell).

How do scientists know so much about the earth's interior? No one has ever traveled more than a few kilometers beneath the earth's surface. Even the deepest wells have not been drilled much more than 8 km deep. Yet the center of the earth is over 6,000 km from the surface. How can scientists make a model of something they cannot see? To make a model of the earth's interior, it has been necessary for scientists to collect and fit many bits and pieces of evidence together.

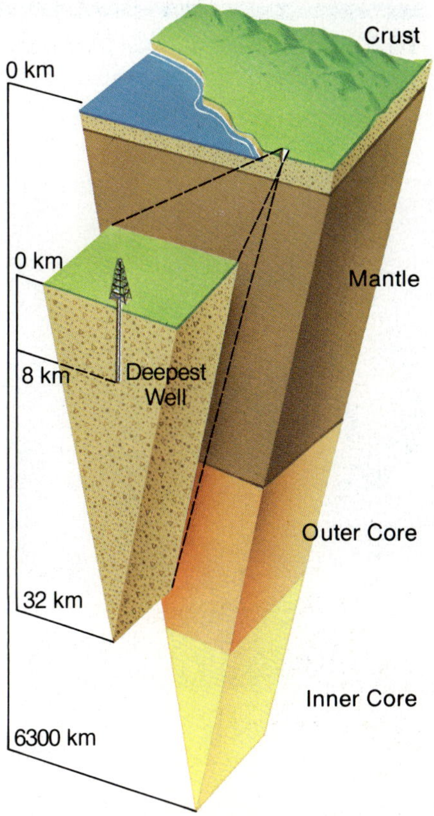

Figure 13.4 The deepest wells of 8 km barely pierce the earth's surface.

Study Questions for 13.1

1. Name the three main parts of the earth.
2. Describe each of the main parts of the earth.
3. Explain how the model of the earth is like a soft-boiled egg.

13.2 Indirect Evidence from Space

You will find out

- what evidence the moon provides about the earth's interior;
- how meteorites provide evidence about earth.

Figure 13.5 Studying the earth by studying the moon helped scientists determine the mass of the Earth.

Nearly all of the information scientists have used to make their model of the earth's interior has been learned indirectly. The first evidence about the earth's interior came from a study of the earth's moon.

The moon's path around the earth depends on several factors, one of which is the mass of the earth. Remember that gravitational attraction is related to mass. Using information they knew about the moon's motion and distance from the earth, scientists were able to determine the mass of the entire earth.

Knowing the earth's mass alone did not tell scientists much about the earth's interior. But now the scientists were able to determine the average density of the earth by comparing its mass with its volume. Surprisingly, the earth's density was found to be more than 5 grams per cubic centimeter. This was a surprise because scientists knew that rocks on the earth's surface had an average density of less than 3 grams per cubic centimeter. The whole earth's average density, then, is almost twice the density of the rocks on the surface. This must mean that the earth's interior is much more dense than surface rocks.

Table 13-1. Volume, Mass, and Density of the Earth

	Average Radius (km)	Volume (millions of km^3)	Mass (g X 10^{24})	Average Density (g/cm^3)
Total Earth	6,371	1,083,230	5,976	5.52
Oceans	3.8	1,370	1.41	1.03
Glaciers	1.6	25	0.023	0.9
Continental crust	35	6,210	17.39	2.8
Oceanic crust	8	2,660	7.71	2.9
Mantle	2,883	899,000	4,068	4.5
Core	3,471	175,500	1,881	10.71

Figure 13.6 *left:* Earth is believed to have the same composition as meteorites like these.

Figure 13.7 *below:* Earth's core matter weighs more than an equal volume of crustal matter because the core is more dense.

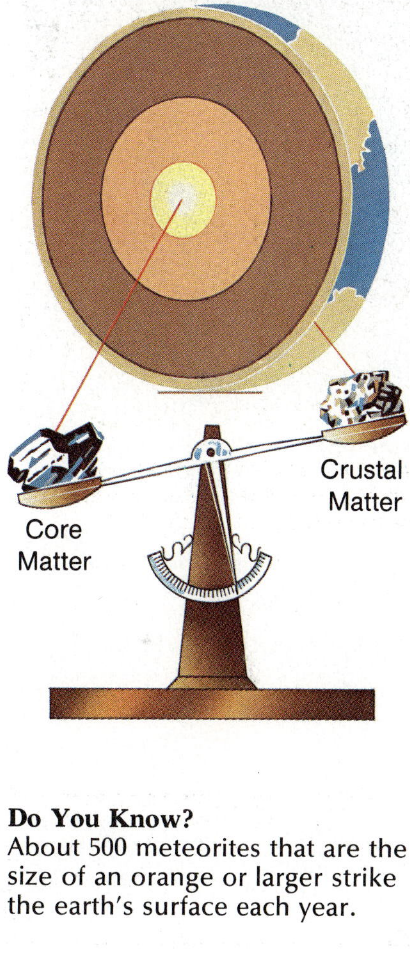

More evidence about the earth's interior came from a study of meteorites, which sometimes fall to the earth's surface from space. Meteorites that have been recovered are mainly of two types, rock and metal. The rock meteorites are made of silicate minerals. The metal meteorites are mostly made of iron and nickel. Meteorites are believed to be mostly odds and ends which were left over from the time when the solar system was formed. It is reasonable, therefore, to assume that meteorites and the earth are largely made of the same materials.

As the earth was formed, scientists believe that the earth passed through a molten stage. During this molten period heavier metal materials would have settled to the earth's center, or core. Lighter silicate materials would have risen to the surface and eventually formed a crust. Since iron and nickel are found in the metal meteorites, it appears these metals were present when the earth was formed. An iron and nickel core would greatly increase the average density of the earth. This hypothesis fits nicely with what has also been learned about the earth's density from the moon. Thus, meteorites indirectly provide a second piece of evidence about the earth's interior.

Do You Know?
About 500 meteorites that are the size of an orange or larger strike the earth's surface each year.

Study Questions for 13.2

1. How did the moon provide indirect evidence about the earth's interior?
2. Why do scientists believe that the earth has a center part made of iron and nickel?

13.3 Earthquakes and Seismic Waves

You will find out
- what causes most earthquakes;
- how earthquake waves are measured;
- what the three kinds of seismic waves are.

Figure 13.8 *above:* A disturbance causes vibrations to move through matter in waves.

Figure 13.9 *right:* An explosion of dynamite starts vibrations moving through the earth.

You have learned that meteorites and the moon provide indirect evidence about the earth's interior. Evidence comes from within the earth too. Have you ever felt the earth vibrate? A passing truck or train can cause vibrations to move through the ground. These vibrations move through the ground much like the ripples you see move across a bowl of water if you tap the bowl with a pencil. Sometimes there are much larger vibrations or waves in the earth's surface. These vibrations can cause the ground in some places to roll like waves on the ocean. These sudden waves in the earth are called **earthquakes.**

Some earthquakes are small and just rattle windows. Other earthquakes are large and can cause large buildings to crumble. Small or large, an earthquake is the result of some disturbance in the earth.

What causes a disturbance in the earth? Relatively small disturbances can be produced by a passing train, an explosion, or a landslide. Larger disturbances, though, are produced by a movement deep in the earth's crust or mantle. Various pressures cause part of the crust or mantle to be pushed together or pulled apart. A section breaks along a huge crack and moves. The place where the movement happens is called a **fault.** At a fault, rocks can break and move past one another, causing a disturbance that results in an earthquake.

Do You Know?
Each year, on the average, there are 10 earthquakes with widespread destruction, 100 with local destruction, 1,000 that do some damage, 100,000 that can be felt, and about 1,000,000 that are detected only by sensitive instruments.

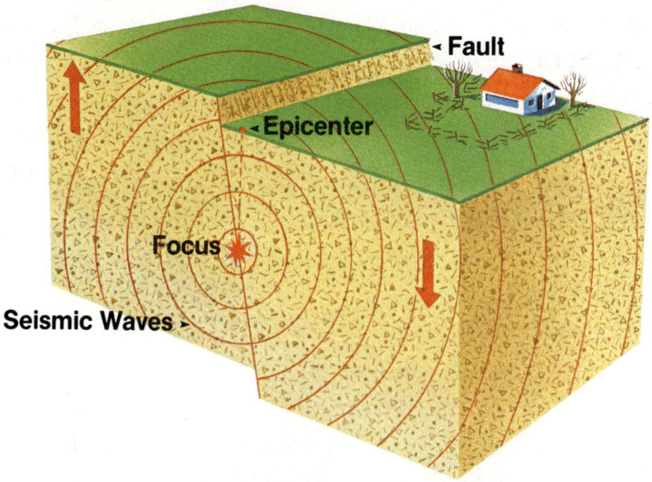

Figure 13.10 *above left:* Rocks break and move along a fault. The Richter scale is a numerical system that measures the amount of energy released during an earthquake.

Figure 13.11 *above:* The San Andreas Fault in California

The place along a fault where an earthquake begins is called the **focus** [FOH-kuhs] of the earthquake. Vibrations move outward from the focus. These vibrations are called **seismic** [SIZE-mihk] **waves.** Measurements of seismic waves have shown that the focus of most earthquakes is in the crust and top part of the mantle down to about 75 km. But the focus of some earthquakes has been measured as deep as 700 km.

Seismic waves move outward in all directions from the focus of an earthquake. Traveling downward, some of the seismic waves can travel all the way through the earth. Traveling upward, some of the waves reach the surface. The place on the surface directly above the focus is called the **epicenter** [EHP-uh-sehn-tuhr]. News reports usually identify the epicenter of an earthquake.

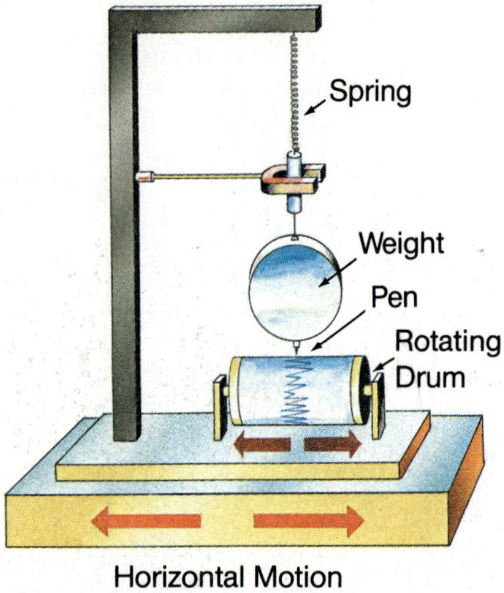

Figure 13.12 This seismograph measures horizontal movement of the earth's crust during an earthquake.

Scientists measure seismic waves with an instrument called a **seismograph** [SIZE-muh-graf]. One type of seismograph has a massive object suspended from a spring. Since the object is not attached directly to the earth, it tends not to vibrate much even as the earth vibrates under it. A pen is attached to the relatively stationary massive object so that it will mark on a sheet of paper on a slowly turning drum. Since the pen does not move but the drum vibrates with the movement of the earth, a wavy line is made on the paper. Thus, a record is made of how much of the ground is vibrating.

Scientists have found that there are three kinds of seismic waves. These are the **primary waves,** the **secondary waves,** and the **surface waves.** The primary waves, or *P* waves, are vibrations that cause a material to vibrate back and forth. A clap of thunder is a sound wave in the air that is like *P* waves in the earth. They move in the direction the waves are traveling as shown in Figure 13.13. *P* waves are the fastest seismic waves and can travel through any material. In addition, *P* waves speed up as the density of the material they pass through increases. Denser materials have closer molecules, which let the waves pass through more rapidly. The changing density of materials also causes the waves to curve or change direction just as light waves are refracted when passing from air to water.

Do You Know?
Some seismic waves travel at 24 times the speed of sound, or 30,000 km per hour, through the earth.

Figure 13.13 This is a model of how seismic waves move through the earth.

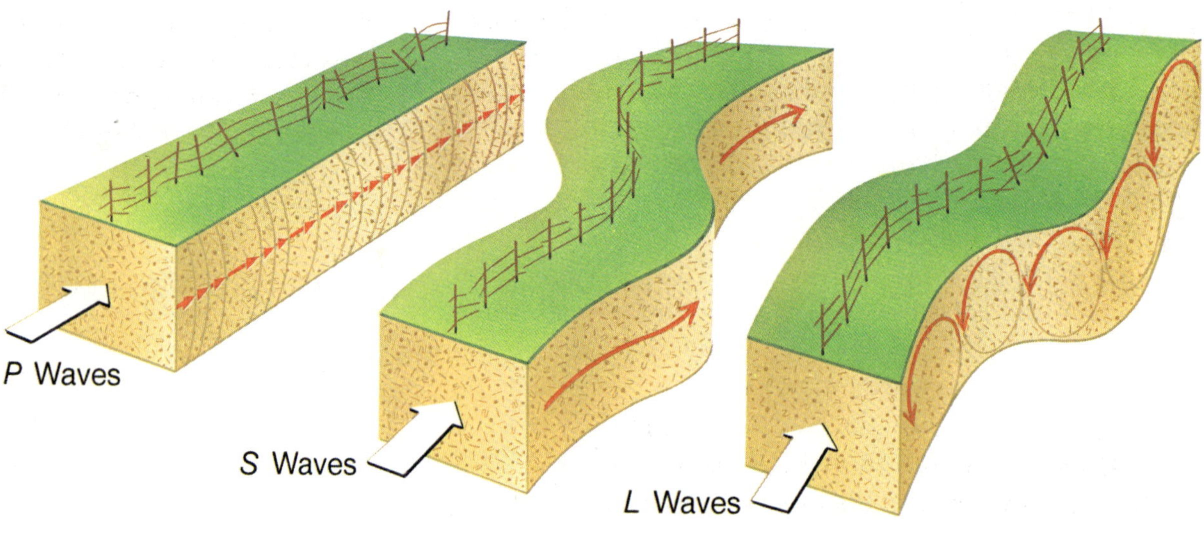

The secondary waves, or *S* waves, are vibrations that cause a material to move from side to side. They move at right angles to the direction the waves are traveling. *S* waves are slower than *P* waves, and they cannot pass through liquid. The molecules in a liquid can move by each other easily, so the *S* waves just cause some molecules to slip back and forth. Since the molecules just slip back and forth, the *S* waves do not pass through the liquid.

The surface waves, or *L* waves, are the slowest of the seismic waves. *L* waves move along the earth's surface much like waves on the ocean. As the surface of water rises and falls with each passing wave, the earth's surface moves up and down as each *L* wave passes. The *L* waves cause most of the damage from an earthquake.

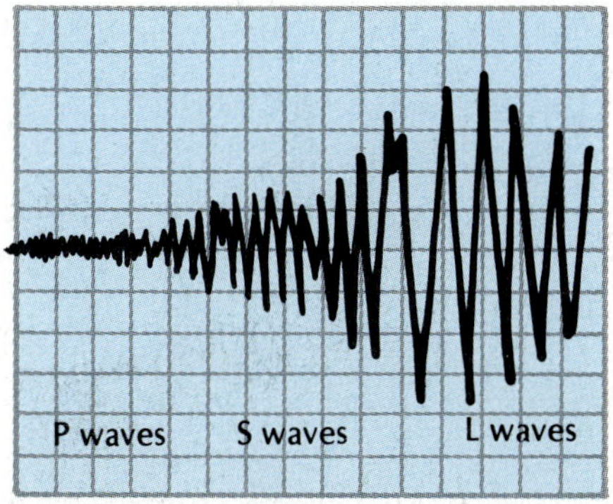

Figure 13.14 Seismic waves as they would look on a seismographic record

ACTIVITY

How Do Earthquake Waves Move?

Material

Slinky toy

Procedure

1. Place the Slinky on the floor. Hold one end while your partner stretches the other end about three meters across the floor.

2. Pull about 15 of the coils together and then let them go.

3. When the Slinky is still, move one end of the Slinky rapidly from side to side.

Questions

1. What did you observe when you pulled the coils together and then released them?

2. What type of earthquake waves move like this?

3. What did you observe when you moved one end of the Slinky from side to side?

4. What type of earthquake waves move like this?

5. Which wave form moved faster through the Slinky?

Study Questions for 13.3

1. What is the cause of most earthquakes?

2. How are earthquake waves measured?

3. What are three types of seismic waves?

13.4 Evidence from Within the Earth

You will find out

- how the crust was discovered;
- what the structure of the mantle is;
- why scientists believe the core has a solid center surrounded by a liquid outer layer.

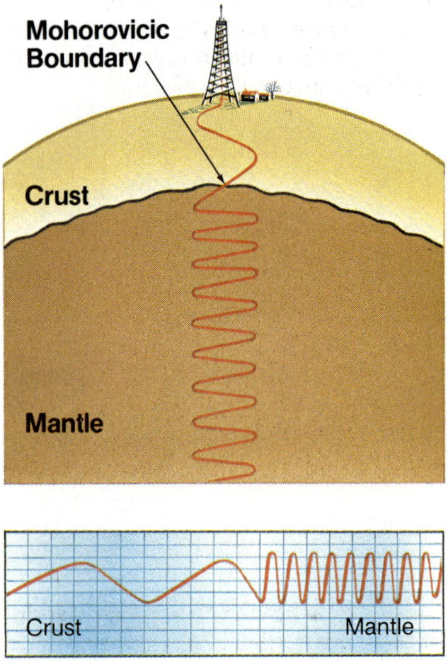

Figure 13.15 By studying seismic waves, scientists discovered the Moho boundary between the crust and the mantle.

When you open a soft-boiled egg, you know what to expect. But no one has ever "opened" the earth or visited the interior. Yet scientists know a great deal about the earth's interior. They have used indirect evidence from the seismic waves to develop their information.

The earth's crust, for example, was discovered early in the history of seismic-wave studies. Andrija Mohorovicic [MOH-hoh-ROH-vuh-sihk] found that seismic waves suddenly increase in speed below a certain depth. This increase could be accounted for by a change in the density of the rocks. There seems to be a boundary with a different kind of rock below it. The boundary where the seismic waves increase in speed is now called the **Moho** [MOH-hoh]. The Moho boundary marks the bottom of the crust and the top of the mantle.

Recent evidence indicates that part of the mantle acts like a thick syrup. During the Ice Age, huge amounts of ice accumulated in some places. The weight of this ice caused the crust and upper mantle to sink. The ice has now melted, of course, and the crust and upper mantle are moving back up. Some of these places are still rising today at about one centimeter per year. It is as if the weight of the ice squeezed part of the mantle away like toothpaste in a tube, causing the land to sink. After the ice melted, the mantle part started slowly moving back. With increasing or decreasing weight the crust and upper mantle sink or rise as if they are floating on a thick syrup layer.

Figure 13.16 Some places on the earth are rising. This fact led scientists to wonder about the nature of the mantle.

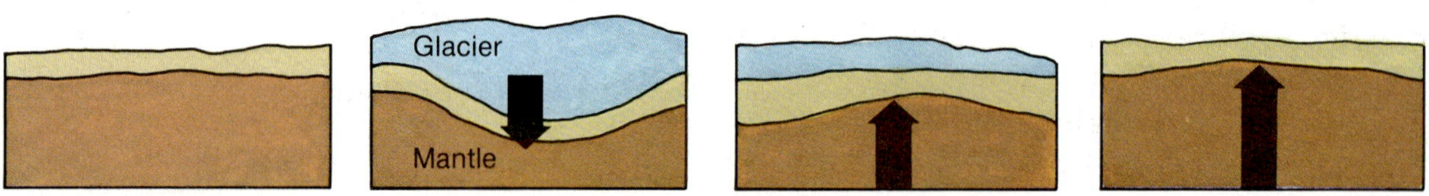

There is seismic evidence, however, that the mantle is a solid, not a liquid. *S* and *P* waves move all the way through it. Since *S* waves do not move through liquid, it must mean that the mantle is solid.

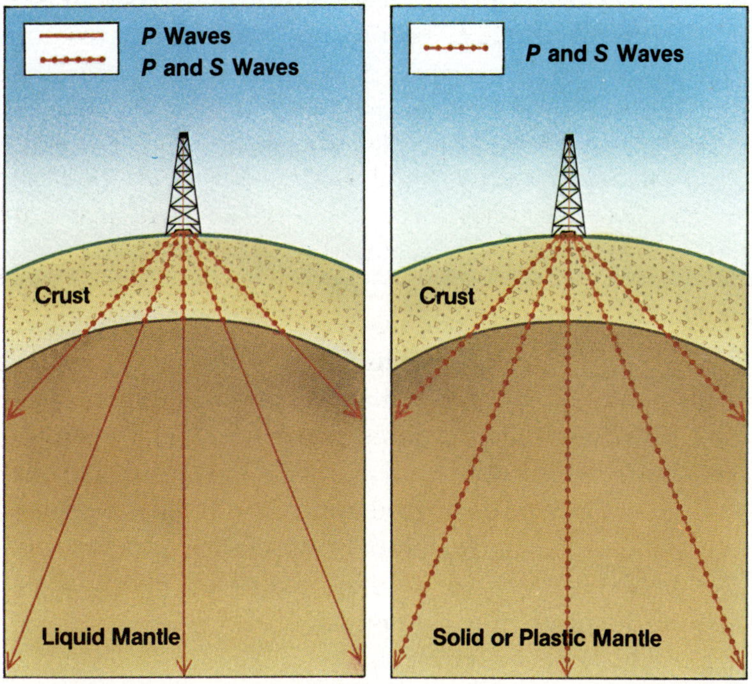

Figure 13.17 *left:* Evidence from seismic waves implies that the mantle is not a liquid but rather has the solid and liquid properties of a plastic.

Figure 13.18 *below:* The mantle seems to have properties like some plastics. *top:* Solid and brittle; *bottom:* Soft and stretchable

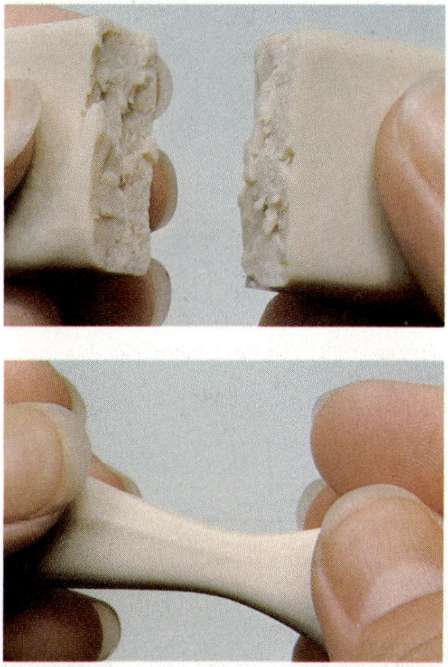

How can the mantle act like both a solid and a liquid? Perhaps it is a **plastic** material. Plastic here does not mean a material such as a plastic bag. A plastic material is one that is a solid but can flow over a long period of time.

Since rocks in the mantle are subjected to high pressure and high temperature, that fact could account for their having plastic-like properties. The mantle would then be able to flow slowly, adjusting to stresses such as the weight of ice. *S* waves could also move through a mantle that responds like a plastic.

More studies of seismic waves suggest that not all of the mantle is plastic. Below the solid crust is a zone of plastic, easily-flowing material. The plastic layer is from 75 to 250 km below the surface, and it goes all around the earth. Below this plastic layer, the rest of the mantle is a dense solid.

Seismic-wave evidence also points to something unexpected in the earth's core. When scientists study records of seismic waves that have passed through the earth, they find a pattern. The pattern forms three zones on the earth's surface. See Figure 13.19. No matter where the earthquake occurs, the same pattern is observed. One zone receives *S* waves and *P* waves that have curved and increased in speed. Another zone receives no waves and is called a shadow zone. The third zone receives *P* waves only.

Do You Know?
The earth has a magnetic field that causes compasses to point north. The magnetic field is believed to originate from the liquid outer core.

ACTIVITY

How Does a Plastic Substance Act?

Materials

dish or container
paper cup
plastic spoon
cornstarch
water
metal, wood, and plastic objects

Procedure

1. Place one tablespoon of cornstarch in a dish. Add water a few drops at a time and stir the mixture. Stir until a thick paste is formed.

2. Pour the mixture onto a dish. Give it time to pour. Remember that a plastic flows over a long period of time. If it will not pour at all after 5 minutes, add a few drops of water and stir. Repeat until the mixture will slowly pour. Record your observations.

3. Place metal, wood, and plastic objects on the mixture. Record your observations.

4. Remove a large piece of the mixture from the dish. Can you bend it? Can you break it? Record your observations.

Questions

1. In what ways did the mixture appear or act like a liquid?

2. In what ways did the mixture appear or act like a solid?

3. In what ways did the mixture appear or act like a plastic material?

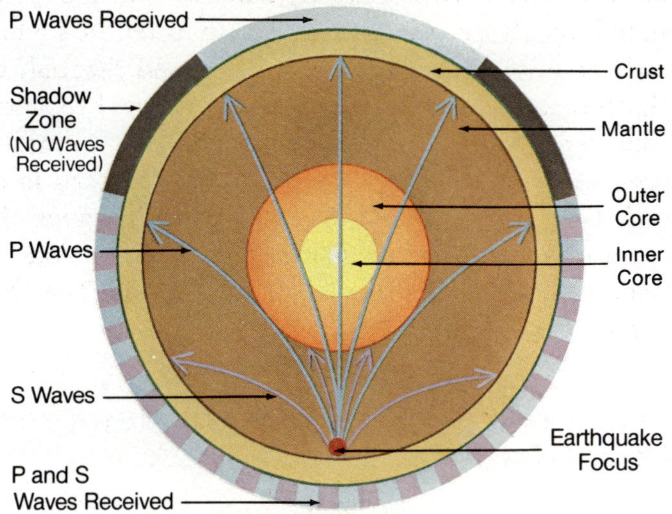

Figure 13.19 *left:* The patterns followed by seismic waves support the model of solid, plastic, and liquid layers within the earth.

Figure 13.20 *below:* This profile of the earth's interior is based on the latest information from seismic studies.

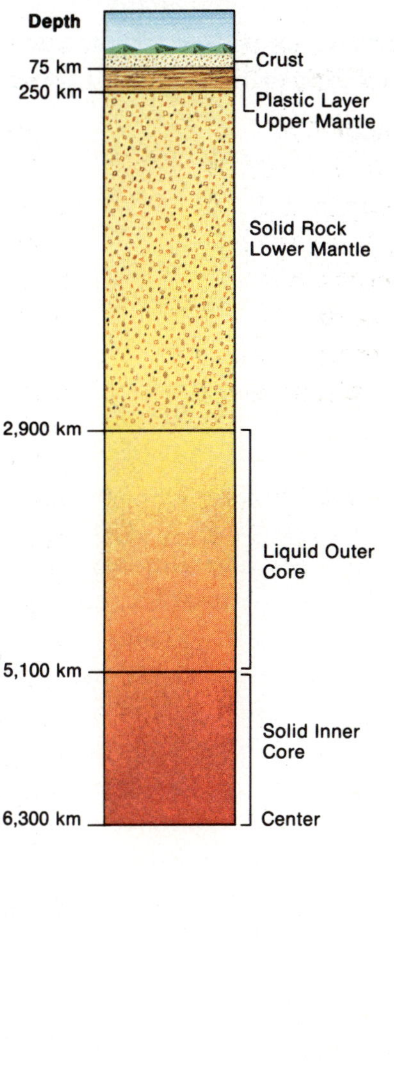

How can this pattern be explained? First, the direction of travel of *P* waves is changed when they reach a material that is less rigid, such as a liquid. This occurs at a depth of 2,900 km, bending the waves to produce a shadow zone on the surface. Second, no *S* waves appear in or beyond the shadow zone. Since *S* waves are stopped by liquid, it must mean that at least part of the core is liquid. Finally, the *P* waves that pass directly through the earth gain speed and arrive on the opposite side sooner than they would if the core were all liquid. This leads to the possibility that there is a solid inner core surrounded by a liquid outer core. A solid inner core would cause *P* waves to increase their speed. Thus, the evidence suggests that the earth has a core with a solid inner part and a liquid outer part.

Study Questions for 13.4

1. How was the presence of the earth's crust discovered?
2. What is a plastic substance?
3. What evidence points to a layer in the mantle that acts like a plastic substance?
4. What is the evidence that points to a liquid and solid inner core?

What Did You Learn?

- The earth is made up of three main layers: the crust, the mantle, and the core.
- The crust is the thin outside layer of the earth.
- The crust makes up 1 percent of the earth's volume and 0.4 percent of its mass.
- The mantle extends from an average of 32 km below the surface down to the core at 2,900 km.
- The mantle makes up 84 percent of the earth's volume and two thirds of its mass.
- The core makes up 15 percent of the earth's volume and one third of its mass.
- Information about the earth's interior had to be learned indirectly.
- Indirect evidence from the moon indicates that the earth's interior is very dense.
- Indirect evidence from meteorites indicates that part of the earth's interior is iron and nickel.
- When pressure builds up in the earth's crust, the rocks may break and slide past each other, forming a fault.
- Most earthquakes are caused by rocks breaking and sliding by each other on a fault.
- Vibrations that occur with an earthquake are called seismic waves.
- The focus of an earthquake is the place on a fault where an earthquake begins.
- The epicenter of an earthquake is the place on the surface directly above the focus.
- Seismic waves are measured with a seismograph.
- There are three major types of seismic waves: primary, secondary, and surface.
- Indirect evidence from seismic wave patterns supports a plastic layer in the mantle and a liquid outer core surrounding a solid inner core.
- A plastic material is one that is solid but can flow over a period of time.

Key Terms

crust
mantle
core
earthquake
fault
focus
seismic wave
epicenter
seismograph
primary wave (*P* wave)
secondary wave (*S* wave)
surface wave (*L* wave)
Moho
plastic

Career

Seismologist

An earthquake happens. Very soon the newspapers and TV are describing what happened, where the earthquake was, what time it happened, and how strong it was. Where do they get their information? The information comes from a seismologist, a scientist who studies earthquakes and the earth.

Kate Hutton is a seismologist at the California Institute of Technology. She collects information on earthquakes. The information is studied by students. It is also used by reporters. Sometimes she appears on TV to answer questions about a big earthquake.

Kate Hutton felt her first earthquake when she was 14 when her family was living in Taiwan. It woke her up, and when she went back to bed, she took a book to find out about earthquakes. Later, she took science and mathematics courses in college. After college, she had a job at the Goddard Space Flight Center. She used two radio telescopes focused on the same quasar to determine if the earth's crust was moving. After that project, she moved to the California Institute of Technology to work on recording earthquakes in southern California. Today she supervises the collection of data from over 200 seismology recording stations. She likes her job because it is interesting and because it has enough surprises not to become boring. She likes the idea that what she does may someday lead to the ability to predict earthquakes. That may help save someone's life when a big earthquake hits. She says that her job keeps her on her toes.

TO THINK ABOUT AND DO

Vocabulary

Number a sheet of paper from 1 to 10. Each statement below has one or more scrambled words. Write the unscrambled word(s) beside the appropriate number on your paper.

1. Most earthquakes occur along a *ULTFA.*
2. *MISSICE SVEWA* move out from the focus of an earthquake.
3. Seismic waves are measured with a *MISSROPEHAG.*
4. The *RECO* is at the center of the earth.
5. The *LENTAM* is between the *TURCS* and the core.
6. The place directly above the focus of an earthquake is the *CENTERPIE.*
7. *DEARSONYC SEAVW* cannot pass through a liquid.
8. A *LISTPAC* material is one that is a solid but can flow.
9. The *HOOM* marks the bottom of the crust.
10. The *COFUS* is the place along a fault where an earthquake begins.

What Do You Remember?

Number your paper from 1 to 10 and write *true* or *false* for each statement beside the appropriate number. If the statement is false, change the underlined word or words to make it correct.

1. <u>P waves</u> cannot travel through a liquid.
2. Some meteorites are made of stony material while others are made of <u>nickel and iron</u>.
3. The average density of the earth is <u>the same as</u> the density of rocks on the surface.
4. The core has a <u>liquid</u> outer part and a <u>solid</u> inner part.
5. The core is <u>thicker</u> than the mantle.
6. The <u>epicenter</u> is the place where an earthquake begins.
7. The <u>mantle</u> of the earth has a plastic layer with a rigid layer above and below.

8. There is a plastic layer in the crust.
9. The crust is thicker beneath the oceans.
10. The core is made of sedimentary rocks.

Applying What You Have Learned

1. One seismograph recorder shows a time lapse of seven minutes between the recording of *P* and *S* waves. Another recorder shows a difference of nine minutes. What does this tell you about the location of the recorders?
2. Make a scale model of the earth's interior. Use Styrofoam or some other suitable material to construct the model. Figure out what scale you will use—for example, 1 millimeter to represent 100 kilometers.

Research and Investigation

1. There have been several attempts to drill through the earth's crust into the mantle. This would provide *direct* evidence about the earth's interior. Investigate in the library to find out about some of these projects. Investigate Project Mohole and prepare a report for your class.
2. Attempts are being made to predict earthquakes. Use your library to research the problems related to earthquake prediction.

Do You Know?
What to Do in an Earthquake
1. Duck and cover.
2. Wait until the earth stops shaking.
3. Look at surroundings and move to safety. If there is open space, move outside.
4. Don't move back inside, except for very quick trips for food, supplies, and a radio.

14 Plate Tectonics

These once flat layers of rock have undergone a remarkable change since they were formed. What caused these rock layers to be folded is part of an incredible detective story you will find in this chapter. This story is so incredible that some people refuse to believe it. Let the information in this chapter provide the evidence. As the plot unfolds, you will find clues to help you interpret the evidence. In the end you'll have the final say—"I'm a believer"; "I'm a doubter"; or "I'm going to wait and see." You'll be the judge.

The plot is simple enough. All the continents were once joined together in a supercontinent. The supercontinent then broke up into big pieces. Some pieces became ocean floor. Some pieces became continents and ocean floor combined. Even today pieces are still moving. In 50 million years our world will look very different.

Can solid rock drift across the earth's surface like a raft on water? Many people have said "No." What does the evidence show? You will have to study the evidence and decide for yourself.

Figure 14.1 *opposite:* Folded limestone rock is an eerie sight that challenges the observer to explain how the folds were made.

14.1 The Setting: The World Today and Continental Drift

You will find out

- that the earth's crust is made up of plates that drift or move about;
- the types of plate boundaries;
- where different types of plate boundaries are found.

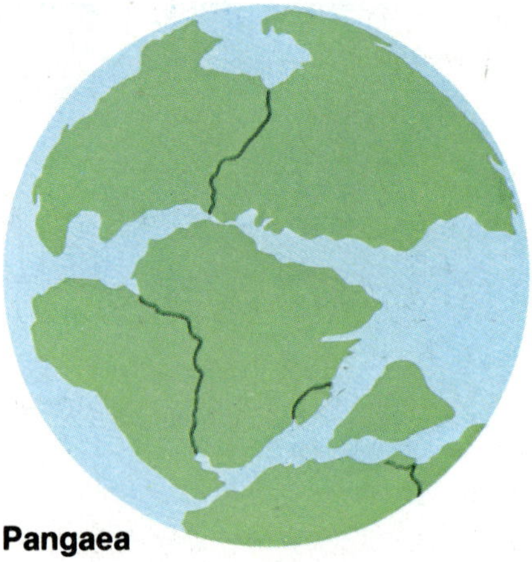

Figure 14.2 *above:* Pangaea is believed to have been a massive combination of today's separate continents.

Figure 14.3 *below:* These are the positions of the continents as they appear today. The red line outlines the Pacific Plate.

The world map you see below is familiar. That map will look the same for the rest of your life. But did the world always look this way? What did the world look like 300 million years ago?

Imagine the continents were at one time all joined together. Some of the continents look like they would fit together. For example, North America, South America, and Africa would fit together. If all the continents were attached there would have been a supercontinent. The world would have one large landmass and one large ocean. This supercontinent is called **Pangaea** [Pan-JEE-ah].

Pangaea broke into huge pieces. Each piece drifted across the earth's surface to its present location. The movement of these pieces is called **continental drift.** The way the pieces fit together now makes our current picture of the world.

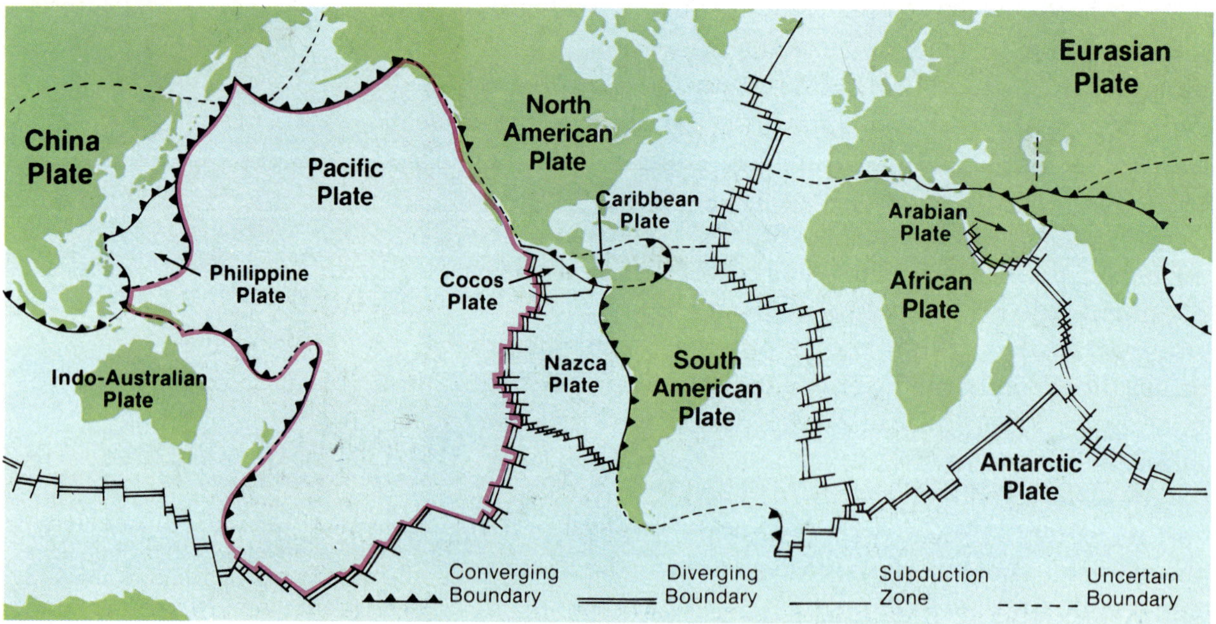

How fast do continents drift? On the average they drift about 5 cm a year. In your lifetime—or about 72 years—some of the continents will have drifted about 360 cm or 3.6 m. The world changes so slowly that most people think of it as unchanging.

ACTIVITY

Did These Two Continents Fit Together?

Materials

scissors
plastic tape
colored pencils

Procedure

1. Copy the two map sections. Carefully label and color each section as it appears in the original.

2. Carefully cut out the two continents you have copied.

3. Try to fit the two continents together.

Questions

1. Do the two continents fit together?

2. Is there a best fit? How do you know?

3. What new information did you use?

4. The age of
rock G is __?__ years old.
rock H is __?__ years old.
rock X is __?__ years old.

5. When did the continents come together?

6. How long did the continents stay together?

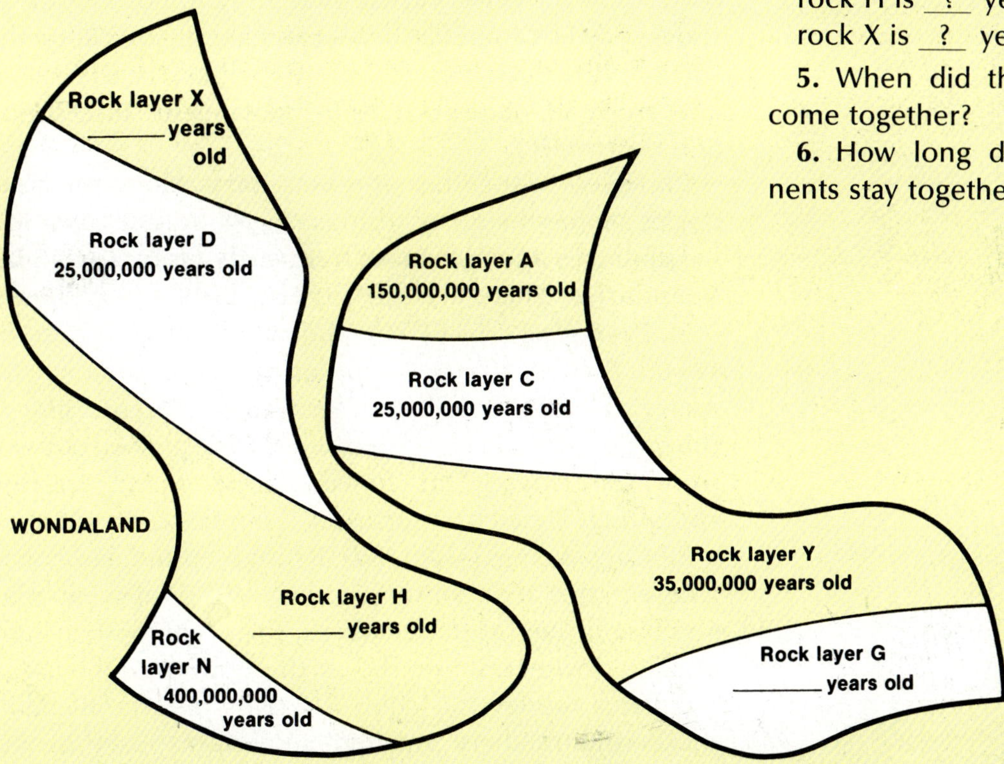

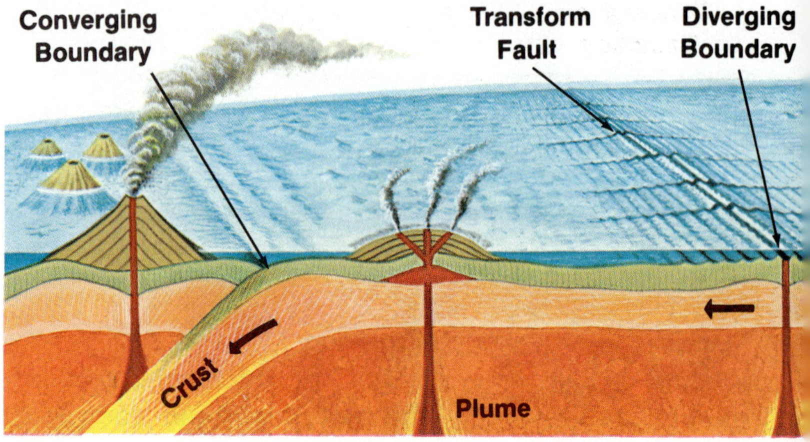

Figure 14.4 *right:* The interaction of moving crustal plates produces a variety of effects at the plate boundaries.

Figure 14.5 *below:* Diverging boundaries lead to the appearance of new crustal material.

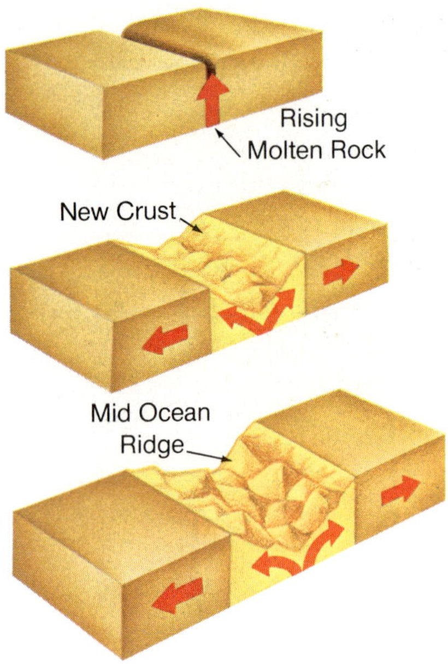

Take another look at the world map. Run your finger along the red line. Notice that your finger traced around a big area—the Pacific Ocean. It is one of the larger pieces of earth's crust that move across the earth's surface. Such big blocks are called **plates.** Plates are rigid sections of the earth's crust. Plates are hundreds or even thousands of kilometers across and about 100 kilometers thick. **Plate tectonics** [tehk-TAHN-ihks] is the theory and study of plate formation, movement, interaction, and destruction.

Plates come together at a **boundary.** Along the boundaries some exciting things happen. For example, earthquakes and volcanoes frequently occur along plate boundaries. There are three types of plate boundaries.

1. **Diverging** [dy-VUHR-jing] **boundaries,** occur between plates that are separating. New molten rock material rising from below forms a mid ocean ridge on the ocean floor. The rock hardens and is pushed outward from the ridge. This movement is called **sea-floor spreading.** New crust forms at the mid ocean ridges.

2. **Converging** [kuhn-VUHR-jing] **boundaries** occur where plates are coming together. When an ocean plate and a continental plate collide, the ocean plate usually buckles downward into the earth. This type of boundary is called a **subduction** [suhb-DUHKT-shuhn] **zone.** Along the boundary there are high mountains, earthquakes, volcanoes, and deep ocean trenches. Deep trenches are found in the oceans at these places where one plate

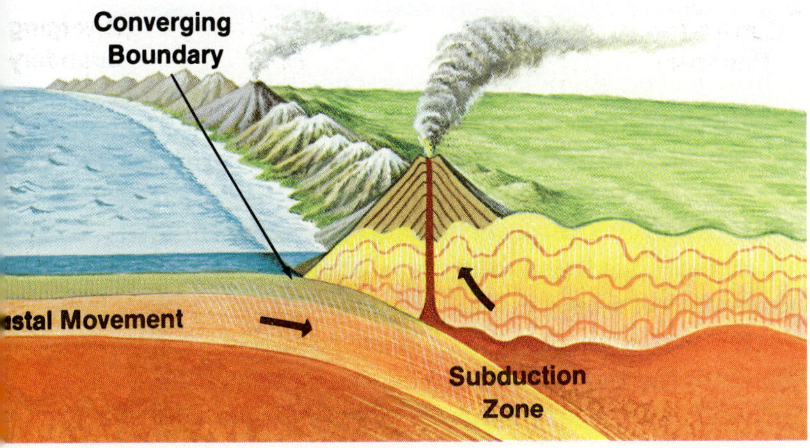

Figure 14.6 *below:* Changes at a transform fault

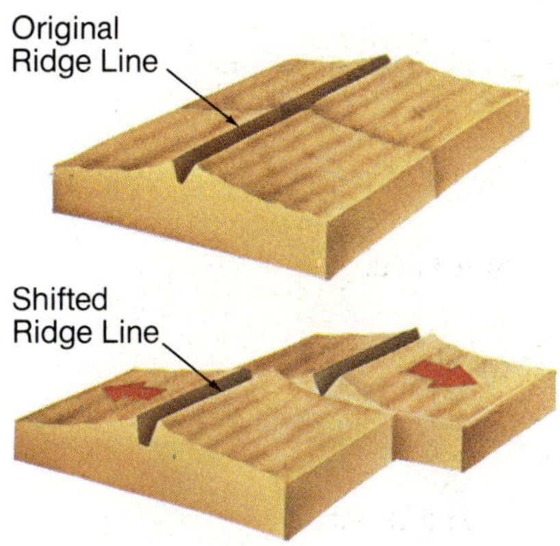

is being pushed under another. Crustal material is returned to the earth's interior at boundaries of this type. At other places plates of continental crust may collide with each other. High mountains are formed at this boundary.

3. **Transform fault** boundaries occur where plates slide past each other. Notice how the ocean ridge is offset in Figure 14.6. Many such offsets occur along mid ocean ridges. These offsets are called transform faults.

Can you find each type of plate boundary on the world map? What type of boundary is found along the mid Atlantic Ridge? Along a trench? Along the coast of California? Along the border of India and Asia?

Before learning more about the plot of this incredible detective story, take one more look at the world map. There are six major plates. They are the African Plate, the Eurasian Plate, the American Plates, the Pacific Plate, the Indo–Australian Plate, and the Antarctic Plate. Some of the minor plates are the Philippine Plate, the Cocos Plate, the Nazca Plate, and Caribbean Plate.

Study Questions for 14.1

1. Describe what is meant by a plate in the term *plate tectonics.*
2. Name the types of plate boundaries.
3. Describe the kinds of plate movements along each plate boundary. Give at least one location for each type.

14.2 What the Rock Record Tells About Plate Movements

You will find out

- how fossils support the theory of plate movements;
- why the age of the ocean floor is important;
- what is meant by magnetic stripes.

Do You Know?
The fossil record shows that India drifted alone about 50 million years.

No one has ever directly measured continental drift. We infer the speed and direction of movement based on information stored in the rocks. Events that have happened to the earth in the past are recorded in the rocks. By studying the record you can infer what has happened. No one can see what the earth looked like 100 million years ago. But scientists can get an idea of what it looked like from the evidence they collect. Although the evidence may be clear, people can interpret the data differently. That is why scientists often disagree with each other.

Figure 14.7 The locations of fossil remains of *Lystrosaurus (top left)* and *Glossopteris (bottom left)* are shown on the world map *(right)*.

Lystrosaurus ●

Glossopteris ○

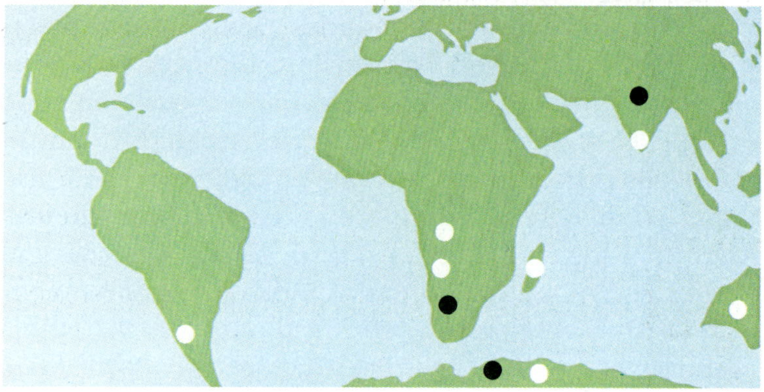

Both plant and animal **fossils** can be used as supporting evidence for plate tectonics. A fossil is any trace of a plant or animal preserved in rock. Two fossils ***Glossopteris*** [glahs-SAHP-tur-us] and ***Lystrosaurus*** [li-struh-SAWR-us], provide strong supporting evidence for plate tectonics. Fossil remains of *Glossopteris,* a plant, are found in rocks that are the same age on five continents and some islands in between. The continents are South America, Africa, Antarctica, India, and Australia. Notice that *Lystrosaurus* is found on the same continents as *Glossopteris* with the exception of Australia and South America.

Do You Know?
Lystrosaurus was a heavily built reptile. It lived on plants and was amphibious. Its size was about 1.2 m.

Isn't that strange? Part of the evidence would require the continents to be together. Another part of the evidence seems to indicate that the continents were not together. You don't have all the evidence you need to prove the incredible story. The fossil evidence is strong, but not conclusive. What can you believe? Perhaps not all the rocks have been looked at.

Another bit of evidence from the rock record supporting plate tectonic theory is found on the floor of the ocean. Scientists have made measurements of the age of the ocean floor. One way they do this is to collect rocks from the ocean floor by drilling. The age of the rocks can then be measured. This is the procedure carried out by the ship *Glomar Challenger*. This ship has a complete research laboratory and drilling rig for collecting rock samples from the ocean floor. Some of the results of the research are shown below in the graph.

Do You Know?
The *Glomar Challenger* is 120 m long with 10,000 tons displacement. A 43 m high drilling derrick is amidships.

Figure 14.8 *below left:* The farther rocks are from the mid Atlantic Ridge, the older they are. Such evidence supports the theory of plate movement.

Figure 14.9 *below:* Out at sea aboard the *Glomar Challenger*

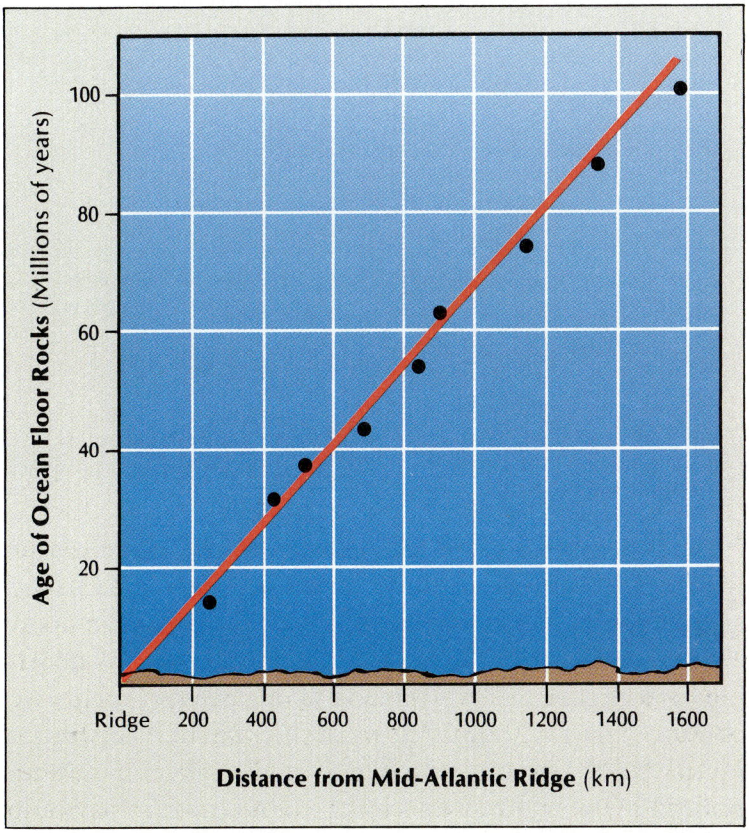

Figure 14.10 *above:* Examining rock samples taken from the ocean floor aboard the *Glomar Challenger*

Figure 14.11 *right:* Rocks of the same ages are found at increasing distances on opposite sides of the mid Atlantic Ridge.

Be sure you understand the evidence shown in the graph. Samples of rocks taken from the ocean floor can be dated. Dates of these show that the further away from the ridges one goes, the older the rocks. This is exactly the pattern to be expected if the plates are gaining new material at the ridges. Check the map that shows the age zones on either side of the mid Atlantic Ridge. Notice the striped pattern. This is very strong evidence supporting plate tectonics. So far the rock record supports the plate tectonic theory.

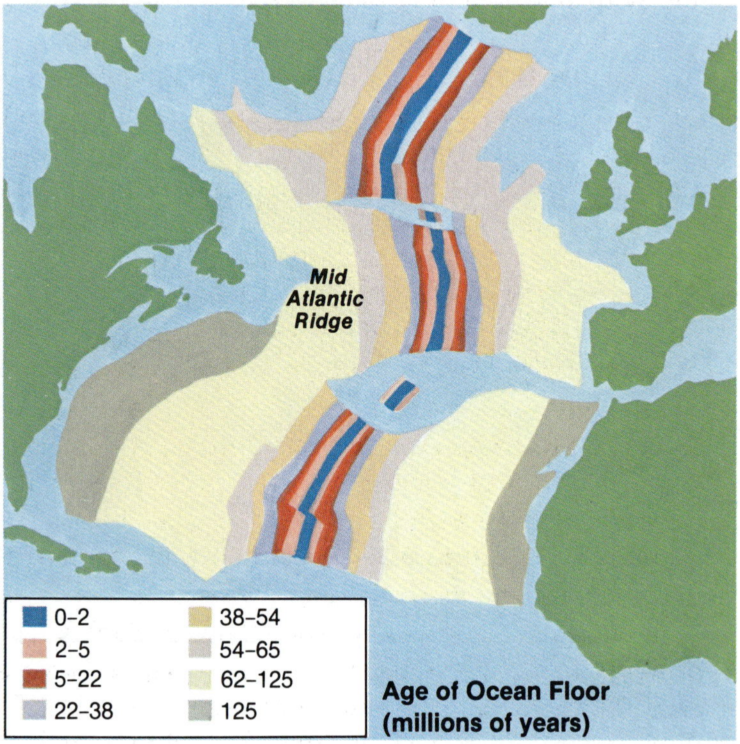

But detectives always like more evidence, especially if it comes from a new source. Scientists have also found striped magnetic patterns in the rocks of the ocean floor. This is an important idea. To understand it, think about how a compass works. One end of our present magnetic compass needle points toward the north. The other points south. A compass that turns freely will line up pointing nearly north and south to the earth's magnetic poles.

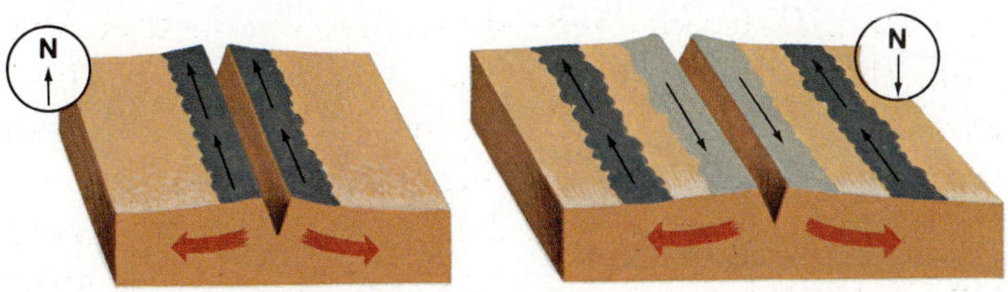

Figure 14.12 *left:* Magnetic-seeking minerals in newly-forming rocks align with the magnetic field present at the time of their formation.

Figure 14.13 *below:* Magnetic striping patterns and dating techniques help map crustal spreading in Iceland.

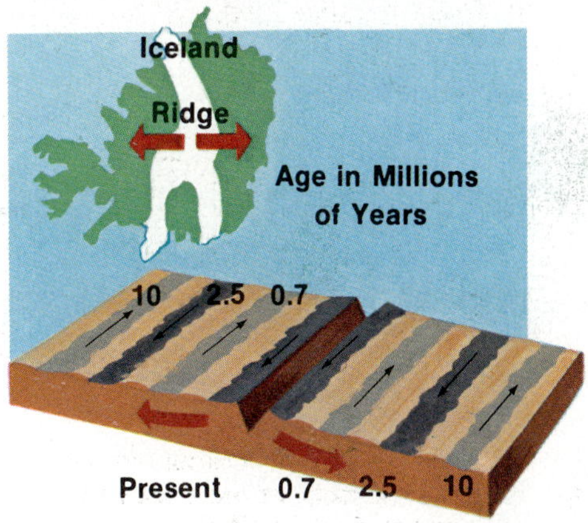

Now imagine a magnetic reversal occurring in the earth. The end of the needle that pointed north would actually turn around and point the other way. It would be reversed. This has happened many times, and it is called **magnetic reversal.** The crystals in molten rock align with the magnetic field of the earth. As the molten rocks cool and harden, this information about the magnetic field is locked into the rocks. The crystals in the rocks become a permanent record of the magnetic field of the earth at the time they cooled and hardened.

A record of the magnetic changes locked in the rocks on the ocean floor looks like a series of stripes. If the ages of the stripes were known, it would be possible to calculate how fast sea-floor spreading takes place. This would also be strong evidence for plate tectonics.

Since the rocks on the ocean floor can be dated, magnetic reversals can also be dated. It is possible to tell how many years ago a reversal occurred and how long it lasted. Look at the magnetic record in Figure 14.13. It shows the ages of some magnetic changes. Each stripe represents a period of time when the magnetism reversed. There have been many reversals in the last 76 million years. The map shows that the stripes match in age on either side of an oceanic ridge.

Study Questions for 14.2

1. How does the fossil record support plate tectonic theory?
2. How does information about the age of the ocean floor support plate tectonic theory?
3. How does the formation of magnetic stripes give additional support to the plate tectonic theory?

14.3 Evidence for Plate Movements

You will find out

- the location of volcanoes and earthquakes is related to plate movements;
- mountain ranges are the result of plate movements;
- how glaciers indicate that plates have drifted apart.

Earthquakes and volcanoes provide additional evidence that directly supports plate tectonic theory. Although volcanoes and earthquakes are two different things, they are sometimes related. For example, the movement of hot, liquid rock underground can cause earthquakes. But most earthquakes have no direct association with volcanoes.

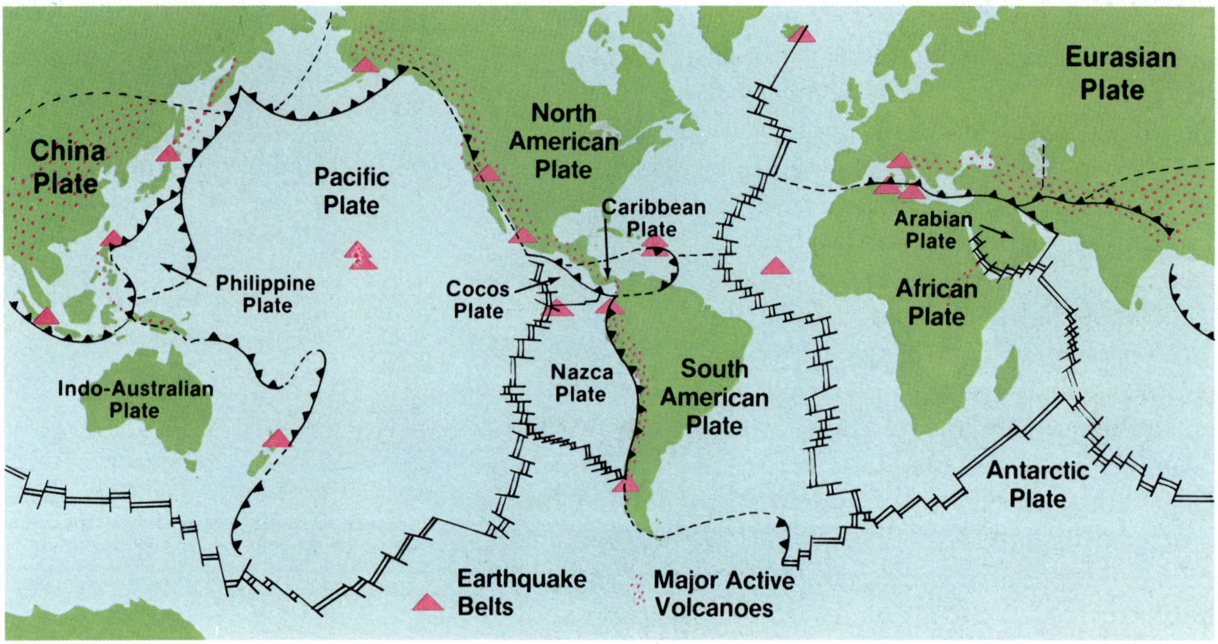

Figure 14.14 Compare areas of frequent earthquake and volcanic activity with the location of major plates.

Volcanoes frequently occur along subduction zones. Since earthquakes also occur with great frequency along plate boundaries, both may be found in nearly the same place at times. But earthquakes along plate boundaries are usually not directly associated with volcanoes. A careful study of the map shows that many earthquakes are found in areas unrelated to volcanic action, such as ocean ridges and transform faults.

In general, volcanoes and earthquakes are located where something is happening in the earth's crust. The most activity in the crust occurs along boundaries. A careful study of volcanic and earthquake activity then tells where plate boundaries are found. It appears that worldwide geologic activity supports plate tectonic theory.

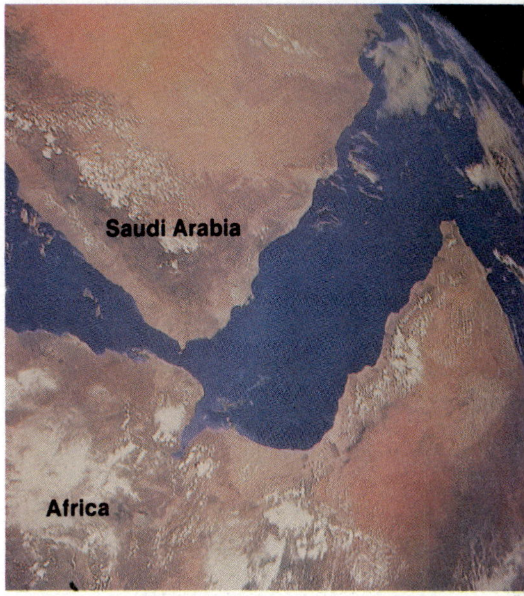

Figure 14.15 *left:* This man is measuring a widening rift at the boundary of the North American and Eurasian plates in Iceland.

Figure 14.16 *above:* A satellite picture of the Arabian peninsula may support the theory of plate movement.

There is still more worldwide geologic activity supporting plate tectonic theory. Think of the big mountain ranges—the Himalayas, the Rockies, the Alps, the Andes, and the Appalachians. All of these mountain ranges show evidence of being squeezed and pushed into their present shapes. Some parts of the Appalachians now occupy a width of 96 km. If you stretched out these same folded rocks, they would extend to 288 km. What could squeeze the rocks from 288 km down to 96 km wide? Plate collisions are one explanation. The origin of the Appalachian Mountains is explained by the diagrams in Figure 14.18.

Figure 14.17 *above:* The Blue Ridge Mountains of the Appalachians

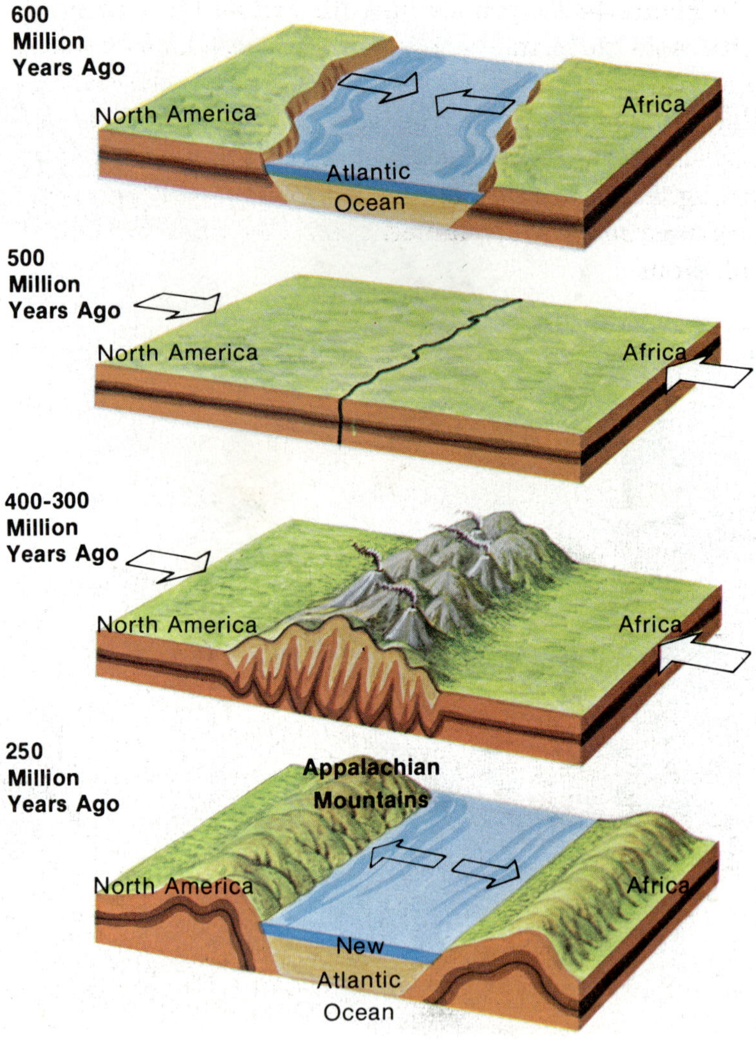

Figure 14.18 *right:* The story of the Appalachians is an example of massive folds in the earth's crust being produced by plate collisions.

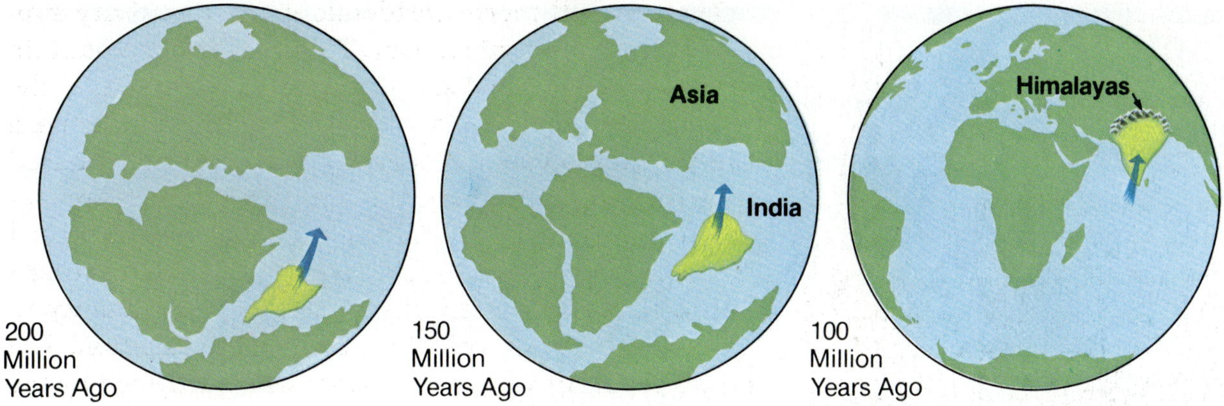

In Figure 14.19, you see how the Indian Plate collided with Asia to form the Himalayas. Collisions between plates are thought to be the major cause of the Appalachians, the Himalayas, the Alps, and the Andes. In the case of the Rockies, however, the evidence is not so strong. The Rocky Mountain Range is too complex. Its origin cannot be explained solely by plate boundary collisions.

Figure 14.19 *above:* India is believed to have moved northward, colliding with Asia to produce the Himalayas.

Figure 14.20 The Himalayas are still being forced upward as India continues to push against the Asian Plate.

ACTIVITY

What Kind of Plate Boundary Is This?

Material

pencil

Earthquake Number	Distance East (km)	Depth (km)
1	400	230
2	80	50
3	450	320
4	220	120
5	10	15
6	480	400
7	250	150
8	500	500
9	150	60
10	300	175
11	600	550

Procedure

The chart shows the location and depth of 11 earthquakes that occurred near South America. These earthquakes were selected from thousands in that area. They appear to be on a plate boundary.

1. Make a copy of the graph in your notebook or on a sheet of graph paper.

2. Plot the earthquakes on the graph.

3. Draw a line on the graph that represents the data. This would be a smooth line passing through or near the data points.

Questions

1. Suppose that you had been given 1,000 earthquakes for this area. Where on the graph would they likely plot?

2. These earthquakes lie on a plate boundary. What type of boundary is this one?

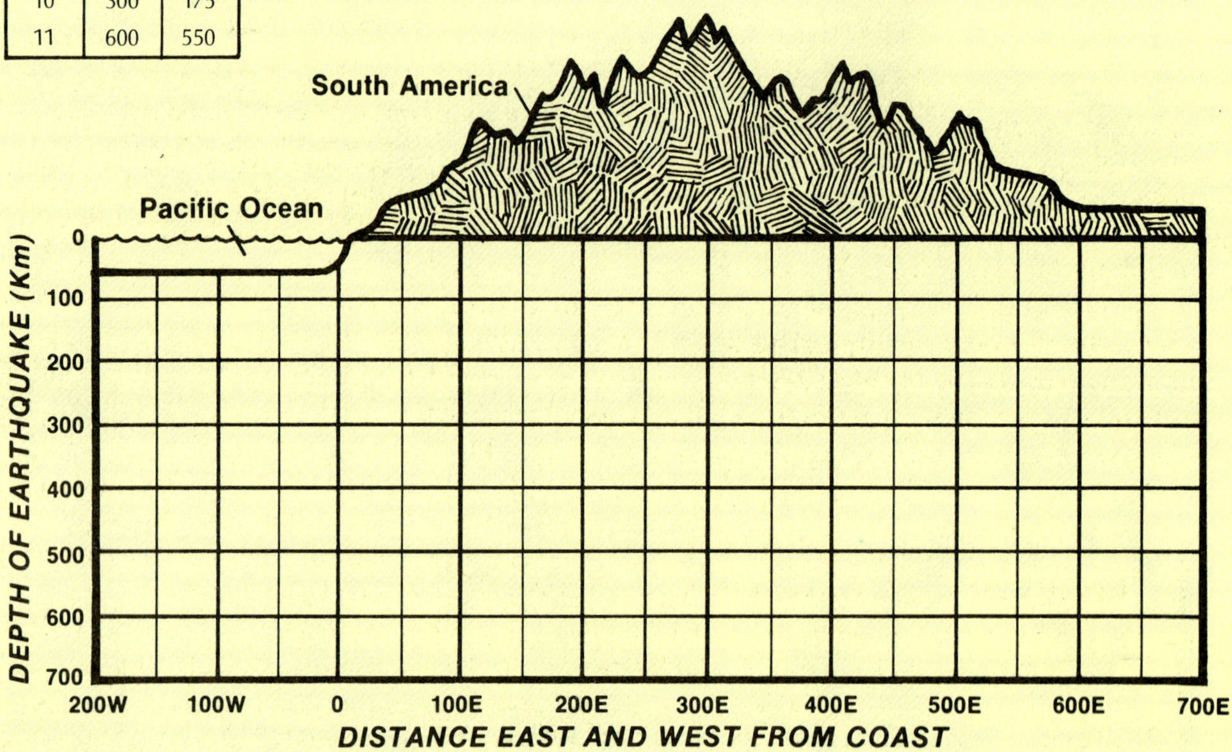

There is another bit of evidence supporting continental drift. It is found in the study of the large masses of ice called **glaciers.** In the past, glaciers covered enormous amounts of land area. The glaciers left a clear record of where they had been. A thick glacier scrapes across the rocks as it moves. This makes deep scratches in the rocks. Glaciers also carve deep valleys in the earth. Rocks carried by the glacier are deposited when the glacier melts.

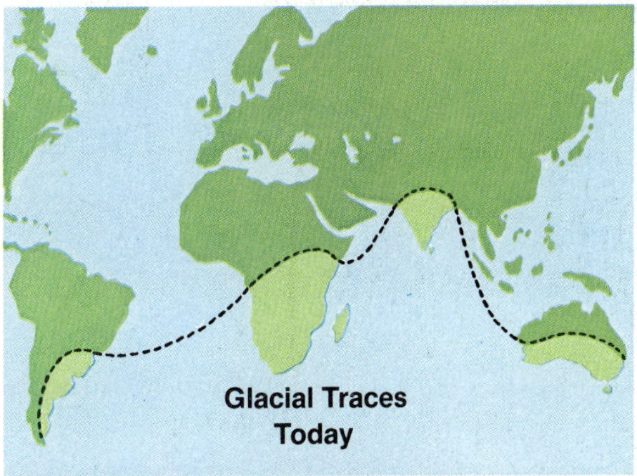

Figure 14.21 Glacial traces found on the continents today provide some evidence that the continents were connected in the past.

Traces of glaciers in South America, Africa, India, Australia, and Antarctica were found to have a related pattern. These traces were left on the continents about 300 million years ago. Glacial traces that appear to match up have been found on five continents separated by long distances today. Why? Look at the map that shows the glacial covering of Pangaea. Then look at the map that shows the same glaciated regions, but as they are today. Are there any other explanations for such worldwide glaciation than continental drift?

Study Questions for 14.3

1. Along which kind of plate boundaries are volcanoes and earthquakes generally found?
2. Name four mountain ranges whose origin may be related to plate movements.
3. A glacier covered part of Pangaea. Which continents did the glacier partially cover?

14.4 Weaknesses in the Arguments Supporting Plate Tectonic Theory

You will find out

- six reasons for questioning the plate tectonic theory.

Not everyone accepts the theory of plate tectonics. What are some of the observations people have made that cause them to doubt? First, there is no known mechanism within the earth to make the plates move. Many ideas have been proposed, but all seem to need additional supporting information. Look carefully at Figure 14.22 as you consider this problem.

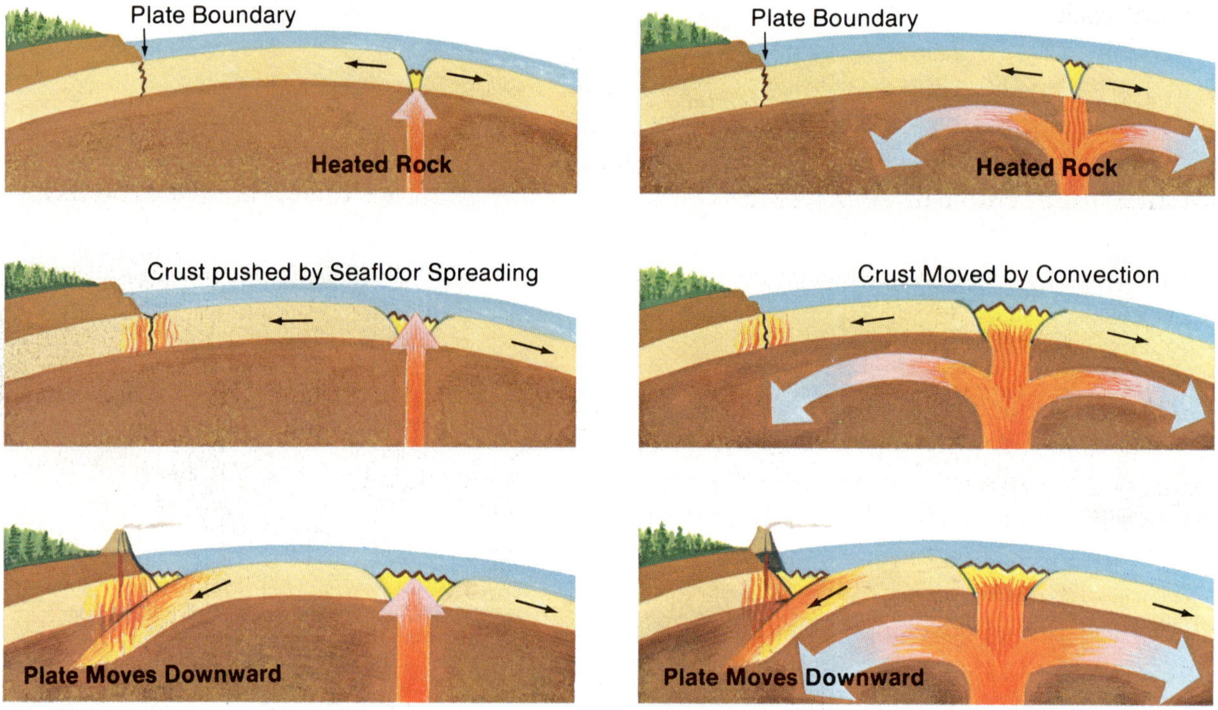

Figure 14.22 *left:* Are crustal plates pushed sideways by pressure from heated rocks that are rising to the surface and spreading? *right:* Do crustal plates move by floating on the convection currents of moving, heated rocks? Neither model explains why the plates move downward at their other boundaries.

Scientists think they have found evidence that heated rocks rise up in the sea-floor ridge. This heated rock may push sideways on the crustal plates. But that still leaves the question of why a plate would move downward under another plate. Another theory holds that the heated rock moves in a convection current. But even if the crustal plates floated along on the current, again, what moves them down? As yet there is no satisfactory explanation for this motion.

Second, what causes earthquakes to occur in the middle of a plate? The greatest earthquakes to hit the United States occurred near New Madrid, Missouri, in 1811 and 1812. New Madrid is near the center of a plate. If earthquakes occur because of moving plates, what causes earthquakes in the interior of a continent?

Third, South America, South Africa, India, and Australia seem to have drifted northward. In order for this to happen, there must have been a big trench in the north. This would be the trench into which crustal rock of the leading edge of the plate would disappear. There is no evidence of a trench that could swallow enough crustal rock.

Fourth, why doesn't the evidence agree? People who study fossil evidence say the continents broke away from Pangaea at a certain time. People studying the ages of the rocks say the continents broke apart at another time. Who can be believed?

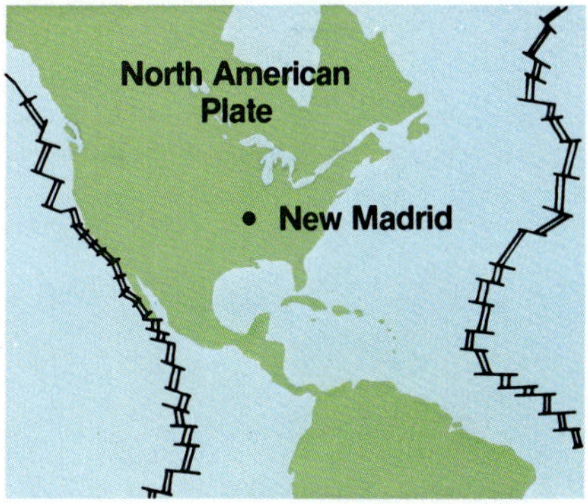

Figure 14.23 *above:* New Madrid is nowhere near a plate boundary.

Figure 14.24 *below:* Is the real story to be found in the fossils or in the rocks?

Fifth, some people believe that the Hawaiian Islands are formed over a hot **plume.** A plume is a localized, hot spot of material rising from the interior. Figure 14.25 shows how a plume might have formed the Hawaiian Islands. But is there another answer? For example, what happens if you have a stationary crust and a moving plume?

Figure 14.25 Scientists wonder if the crust moves over a plume (*left*) or if a plume moves beneath the crust (*right*).

Sixth, is it possible to look at the movement of the continents in another way? Suppose the earth is getting larger. The movements of the continents could be like the movement of spots that nearly cover a balloon. As the balloon increases in size, the spots would get farther and farther apart.

Figure 14.26 A ballooning earth?

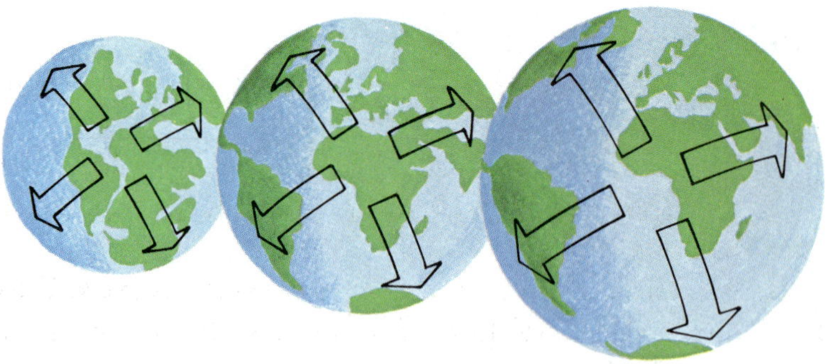

There are some good reasons to support plate tectonic theory. There are good reasons to question the theory. The weaknesses you have studied provide a stiff challenge to the "drifters." "Drifters" are the people who believe in continental drift. Think about the things that the theory doesn't explain. Could it be that the continents don't drift? Further discoveries and study will be needed to answer this question fully.

Do You Know?
Lunar reflectors set up by crews of *Apollo 14*, *15*, and *16* reflect laser beams that are used to measure continental drift as little as 5 cm.

Figure 14.27 What's your theory?

Study Questions for 14.4

1. List six reasons to question the plate tectonic theory.
2. Which of the six reasons do you think presents the strongest case against plate tectonics? Give your reasons.
3. Use the cycle of scientific processes on page 7 to develop a testable model of one of the reasons for questioning plate tectonic theory.

What Did You Learn?

- The earth's crust consists of 11 large plates that are relatively thin.
- There are three types of plate boundaries.
- Plates gain new material from the earth's interior along mid ocean ridges.
- Plate material is destroyed in subduction zones.
- The fossil record contains evidence supporting plate tectonic theory.
- The age of the ocean floor is older as you move away on each side from a mid ocean ridge.
- Magnetic reversals created matching magnetic stripes on the sides of mid ocean ridges.
- Scientific evidence can be interpreted in different ways.
- The location of most volcanoes and earthquakes is related to plate boundaries.
- Plate collisions cause some mountain ranges.
- Traces of glaciers on five continents now separated support plate tectonic theory.
- The cause of plate motion is not known.
- The northward drift of four continents has not been satisfactorily explained.
- The age of rocks and the fossil record do not always match.

Key Terms

Pangaea
continental drift
plates
plate tectonics
boundary
diverging boundary
sea-floor spreading
converging boundary
subduction zone
transform fault
fossil
magnetic reversal
glacier
plume

Career

Geophysicist

Geophysicists use the science of physics to study the earth and its surroundings. They use instruments to measure gravitation, magnetics, and electric and radioactive characteristics of the earth. Data from seismographs are used by geophysicists to learn about the interior of the earth.

To carry out this kind of work, you must have a strong academic background in geology, physics, mathematics, and computer science.

J. Tuzo Wilson is a geophysicist. He began college at the University of Toronto. In his early geological career, Wilson considered himself a non-drifter. Yet in a few years he became the supporter and architect for many of the ideas you have been studying. What brought about this change? He reports that the work with magnetic data helped change his mind. This work led him to study the islands in the Atlantic. If the islands had drifted, they must have left a clue. He found that the farther an island is from an ocean ridge, the older the island is. That was the clue he was looking for.

Wilson proposed the idea of hot spots—plumes. He was one of the first to explain how transform faults work. He became a super-salesperson for the drifters.

J. Tuzo Wilson is a person with a vision. This vision is not like seeing something that isn't there but a vision of reality—the world as it really is. Wilson has a knack for explaining the complicated in a simpleway. His creative insights give him the ability to see the grand design within a mass of confusing details.

TO THINK ABOUT AND DO

Vocabulary

Unscramble the words. Select the letter indicated in the bold boxes. What word is spelled out? ________

Words	Missing Word
ertpis	
railgce	
ainld	
rsiotlGsospe	
ofntrrasm	
tkarehaque	
covnlao	

What Do You Remember?

1. What continents touch the Nazca Plate? The Cocos Plate? The Philippine Plate?
2. Along what type of plate boundaries do you find a mid ocean ridge? A transform fault? Continental crust against continental crust?
3. Name the continent on which you'll find the Alps? The Andes? The Himalayas? The Rockies?
4. Define *fossil* and why fossils are important to drifters.
5. Where are the oldest rocks in the Atlantic Ocean found?
6. What is the pattern of earthquakes along a subduction zone?
7. What is a magnetic reversal?
8. What is a mid plate earthquake? Why is it important?

Applying What You Have Learned

1. On an average the continents drift apart about 5 cm a year. How far apart will the continents drift in 50 years? Is this number greater than the height of your teacher?
2. A rock sample taken from the ocean floor is magnetically dated and found to be 60 million years old. How far has the rock moved from the ridge? (Use the graph on page 273.)
3. Describe one mechanism that might cause the plates to move.
4. It is believed that the oldest rock on the ocean floor is less than 280 million years old. Suppose someone finds a large mass of ocean-floor rock older than that. What does that do to the plate tectonic theory? Explain.

Research and Investigation

1. Design a demonstration that explains how a transform fault works.
2. Design a 10-question questionnaire about plate tectonics. Use this questionnaire to survey how much the students in your school know about this subject.
3. Write an article for the school paper on the theory of plate tectonics.
4. Select one of the following people and write a paper describing what contribution each made to the theory of plate tectonics: H. Benioff, Robert Dietz, H. Hess, D. Matthews, F. Vine, and Alfred Wegner.

"Smoke" from the earth's interior billows from an underwater trench.

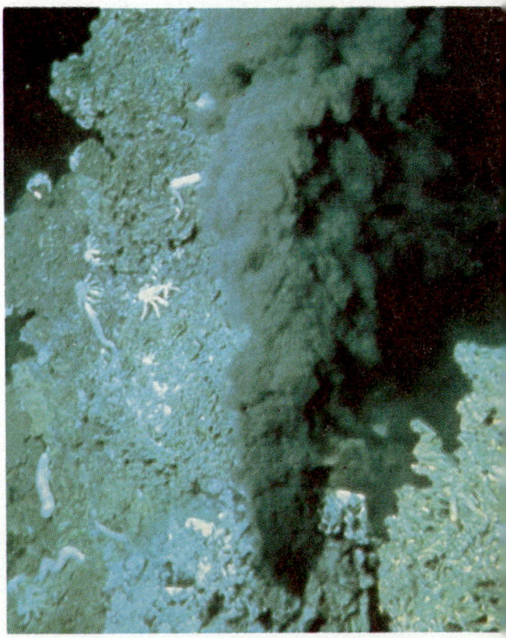

15 Mountains and Crustal Movement

You live on an earth that is constantly changing. This remarkable mountain scene may have been the bottom of a seafloor eons ago. It could have been a mountain twice as high as the one you see in the picture. How does the hard, rocky crust of the earth move up and down? Why are some mountains so high? Why are they where they are? What do geologists know about mountain building?

This chapter is about the types of mountains found on the earth and the processes that formed them. Mountains vary in size and shape. In Chapter 14 you learned that there is a relationship between mountains and movement in the earth's crust. But how are mountains formed and how long does it take to make them?

Answers to questions like these are the focus of this chapter. You'll find that learning to recognize clues to mountain building is an exciting challenge.

Figure 15.1 *opposite:* A part of the Canadian Rockies in Banff National Park, Alberta

15.1 Folding and Faulting

You will find out
- what a mountain is;
- how to identify two kinds of folds;
- what a fault is;
- how to identify different types of faults.

Figure 15.2 Mount McKinley

A **mountain** is any part of the earth's surface that stands much higher than its surroundings. The highest mountain in the United States, Mount McKinley in Alaska, is more than 6,100 m high. Pikes Peak is one of 14 peaks in Colorado that are more than 4,300 m high. Mount Whitney in California, the tallest mountain in the continental United States, is 4,418 m high.

Geologists find that most of the highest places on the earth are igneous or metamorphic rocks. You may have read in Chapter 12 that granite is a rock that cools slowly deep in the earth. If that is so, then these mountaintops must have once been deep in the earth's crust. Even if the highest rocks are sedimentary, they too started far below the place where they are now found. Sedimentary rocks form in the bottom of a sea. What could cause rocks from deep inside the earth or below the sea to be moved to such high places? **Folding** is one way mountains are produced. The easiest folds to recognize are those found in sedimentary rocks. Since these rocks are usually laid down as flat layers, any bending appears as a fold. Two easily spotted kinds of folds are **anticlines** [AN-tih-klyns] and **synclines** [SIN-klyns]. An anticline forms when the layers of rocks are bent upward into an arch. A syncline occurs when the layers of rocks are bent downward to form a trough. Both of these kinds of folds are shown in Figure 15.3. Whenever rocks are folded, the earth's crust is squeezed together or shortened.

Figure 15.3 Types of folding found in sedimentary rock

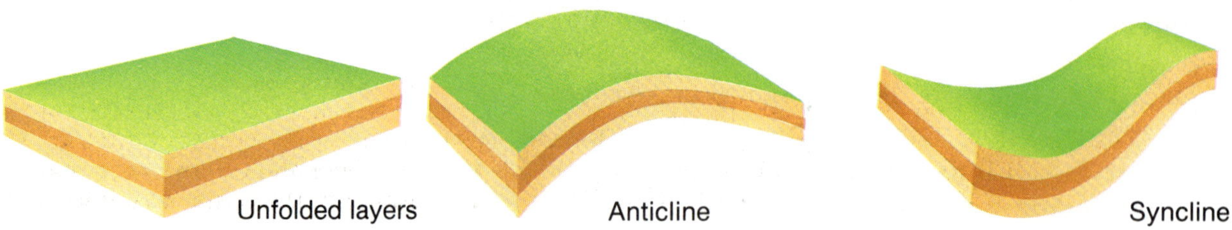

You can't fold a hard, brittle rock. But if the rock was flexible, somewhat like warm wax, it could be bent. Where would you find such rocks? The answer is deep in the earth's crust. Here high pressure and temperature turn rock into a flexible state. This occurs over a long period of time. If the rock is squeezed, it folds and wrinkles like a piece of cloth.

Folds are easily seen from a cross-sectional view. A cross-sectional view is like cutting a cake and looking at the cut side to count the layers. If you looked down on a fold, you might see a striped pattern of rock layers as shown in Figure 15.5B. If the fold is tilted and dips into the earth, you might see a curved pattern of rock layers on the surface as shown in Figure 15.5C. The zigzag ridges found in Pennsylvania are caused by folds that have been tilted like this.

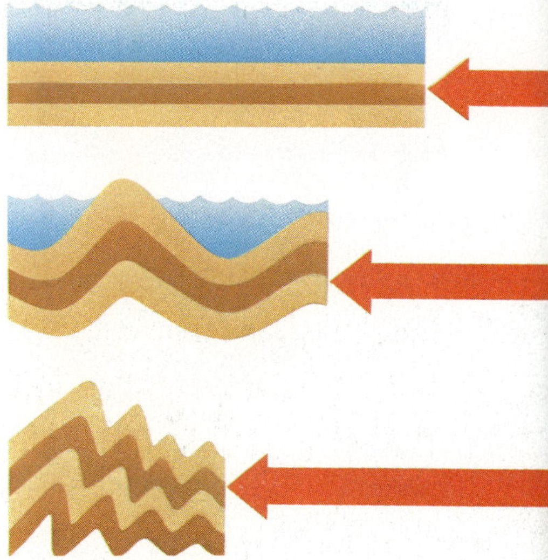

Figure 15.4 *above:* The arrows show how pressure may force rocks to fold.

Figure 15.5 *top left:* This cross section of a fold shows an anticline and a syncline. *bottom left:* The exposed layers of a tilted fold. *bottom right:* An aerial photo shows the erosion of tilted layers in Pennsylvania. The dotted lines show zigzag ridges caused by this erosion.

Do You Know?
In anticlines the oldest rocks are found in the center of the fold. In synclines the youngest rocks are found in the center of the fold.

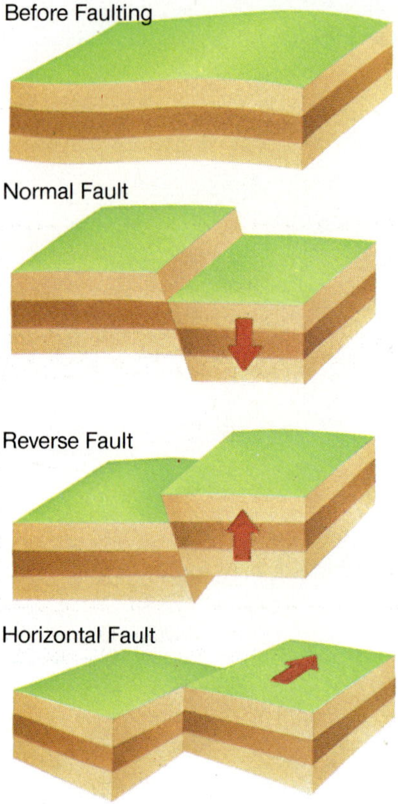

Figure 15.6 *above:* The type of fault produced depends on the direction in which the rock slips.

Figure 15.7 *right:* Examples of three types of faults

Do You Know?
You can see many faults in the sidewalks and gutters on the way to school. Remember that faults indicate movement and can vary a great deal in size. Look for mini faults around your school.

Faulting is another way mountains are produced. A **fault** is a plane along which the rocks break and slip. A fault can often be easily recognized because the rock layers don't match in places. This mismatch happens when rocks are squeezed but aren't flexible enough to fold. Sometimes they break and one section of rock slides by another. When the rocks have moved in this way, geologists say the rocks are faulted.

Figure 15.7 shows three types of faults. Figure 15.7A looks like the block on the right has slid over the block on the left. This is a **reverse fault.** A reverse fault happens when the faulting causes the crust to be shortened. A **normal fault** is shown in Figure 15.7B. A normal fault happens when the blocks slide to extend the crust. Figure 15.6C shows a **horizontal fault.** In a horizontal fault the blocks move sideways with respect to each other rather than over or under each other.

ACTIVITY

What Kind of Fault Is It?

Materials

scissors
pencil
ruler
one piece of cardboard

Procedure

1. Mark your piece of cardboard as shown and color the layers of rock.

2. Cut along the fault.

3. Hold the left side and slide the right side up. Observe what happens.

4. Hold the left side and move the right side down.

Questions

1. Which type of fault does step 3 show?

2. Which type of fault does step 4 show?

3. Diagram and label a reverse fault and a normal fault.

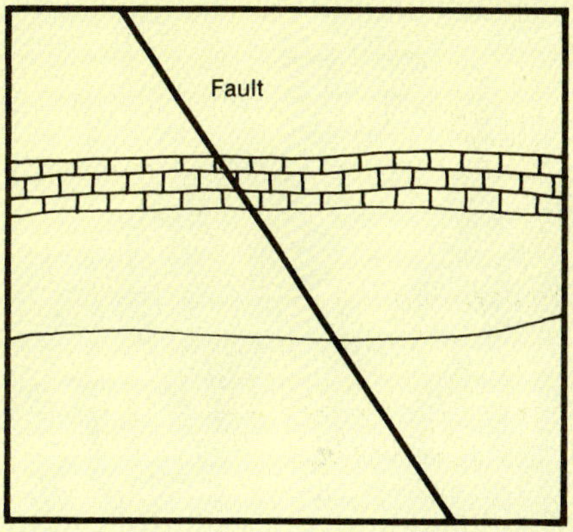

How are faults on the surface recognized? In Figure 15.8 you can see blocks of rock tilted at odd angles. Faults exist along sides of these blocks. Faults appear as straight lines on the surface as seen from the air. Sometimes a stream flows across a fault line. This may cause an abrupt change in the stream's path or offset as you can see in Figure 15.9.

Figure 15.8 *above:* Tilted surface rocks in Red Rock Park, Colorado

Figure 15.9 *left:* A stream's path is offset by a horizontal fault.

Study Questions for 15.1

1. What is a mountain?
2. Describe an anticline and a syncline.
3. Define what geologists mean by a fault.
4. Name three things that indicate a fault.

15.2 Volcanic Mountains

You will find out

- how volcanic mountains are made;
- what features can be found in volcanic areas.

Mount St. Helens in the state of Washington erupted in 1800 after 150 years of quiet. The eruptions lasted about 55 years. On May 18, 1980, after 125 years of quiet, another eruption took place. About 4 cubic km of rock and ash were hurled as high as 2,215 km into the air. Trees were flattened within a fan-shaped area 12.8 km long and 24 km wide. This was followed in March, 1982, with a great avalanche of mud.

Mount St. Helens is a **volcano.** Volcanoes are mountains formed from liquid rock. Liquid rock from deep within the earth sometimes forces its way through the crust. As this liquid rock cools, it piles up higher and higher around the opening of the crust.

What causes volcanoes? Since most volcanoes are found near plate boundaries, geologists believe they are caused by what happens deep in the earth's crust along

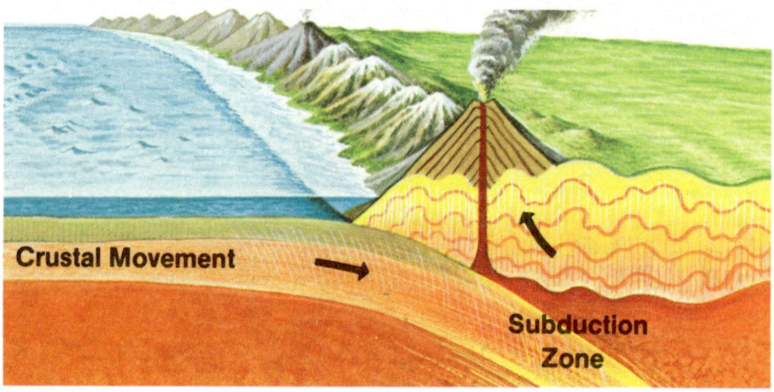

Figure 15.10 *above:* Mount St. Helens

Figure 15.11 *right:* The formation of a volcano at a subduction zone

these boundaries. Volcanoes are most likely to occur along plate boundaries where one plate is descending into the earth. See Figure 14.4. In subduction zones the rock becomes sufficiently heated in place to become liquid rock. Liquid rock deep in the earth is called **magma.** Magma may force its way up to the surface, making a volcano. Magma, once it reaches the surface, is called **lava.**

Mount St. Helens is one of a series of volcanoes extending from California to Canada that are located over a subduction zone. The volcanoes in this region have a record of long calm periods followed by short periods of eruptions.

Figure 15.12 shows three basic types of volcanoes. **Composite** or **stratovolcanoes** are built up from a combination of lava, solid rock, and ash. Mount St. Helens is a stratovolcano. **Cinder cones** are built of rock and ash. Parícutin in Mexico is a cinder cone. **Shield volcanoes** are built by a series of lava flows which pile one on top of another. The best known shield volcanoes are the Hawaiian Islands.

Figure 15.12 Photographs of (*A*) Mount St. Helens, a stratovolcano; (*B*) Parícutin, a cinder cone; and (*C*) Mauna Kea, a shield volcano. Matching diagrams show the structure of each type of volcano.

There is still another kind of volcanic activity. In eastern Washington, Oregon, and southern Idaho, geologists have found evidence of multiple lava flows. The accumulated lava is as much as 2 km thick, but there is no volcanic mountain. These flows came from huge cracks in the earth's crust. The flows cover an enormous area of at least 51,000 square km in size.

Figure 15.13 *left:* The formation of a lava flow; *right:* The lava field of the west wall of Grand Coulee Valley, Washington

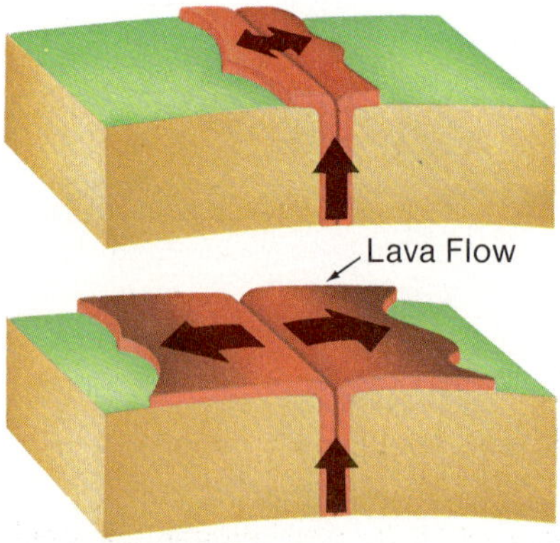

ACTIVITY

What Volcanic Eruption Was the Greatest?

Materials
pencil
graph paper

Procedure

1. Make a bar graph of the data in the table.

2. Label the vertical axis *Amount of Ejected Material.*

3. Label the horizontal axis *Date and Volcano.*

4. Enter data on the graph.

Questions

1. Which volcano ejected the greatest amount of material?

2. What is the total amount of material ejected from Mount St. Helens?

3. How does the total amount of ejected material from Mount St. Helens compare with that from Tambora?

Table

Volcano	Dates of Eruptions	Amount of Material Ejected (cubic kilometers)
Mazama	4600BC	40
St. Helens	1900BC	3.3
Vesuvius	79AD	2.7
Hekla	1104AD	2.0
Fuji	1500AD	1.3
St. Helens	1707AD	0.9
Tambora	1815AD	80.0
Krakatoa	1883AD	20.0
Pele	1902AD	0.3
Katami	1912AD	30.0
St. Helens	1980AD	4.0

What of the magma inside the earth that doesn't come to the surface? Look at Figure 15.14 and follow the movement of magma from deep in the earth.

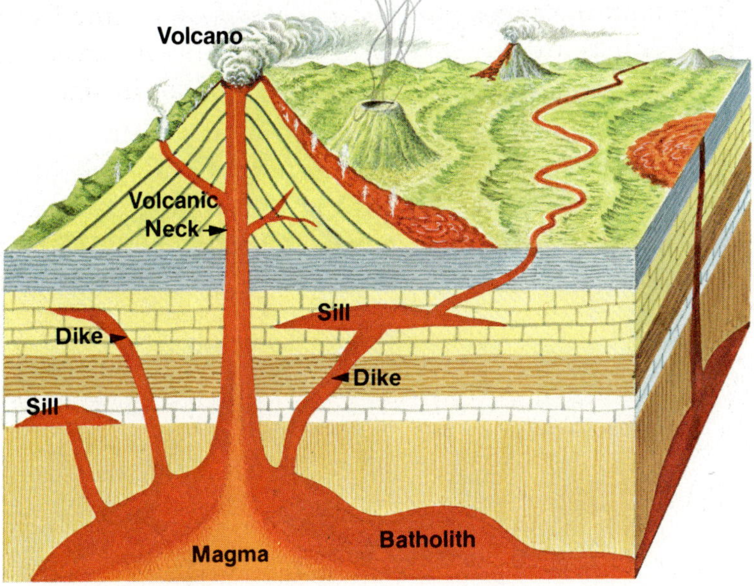

Figure 15.14 *left:* Magma formations beneath the earth's surface

Figure 15.15 *below:* The volcanic neck of Ship Rock is all that remains of a onetime volcano.

Deep in the earth the magma may lie in a huge pocket. These huge pockets of magma may cool and harden, forming a huge mass of rock called a **batholith** [BATH-uh-lith]. The magma may also force its way upward cutting through overlying layers of rock. The magma that cools and hardens in these vertical cracks forms **dikes.** Magma that spreads out in horizontal cracks between rock layers and cools and hardens forms **sills.** These features cannot be seen until erosion later exposes them to view. Sometimes a volcanic cone is completely removed by erosion, leaving a mass of igneous rock that hardened inside the mountain. This feature is called a **volcanic neck.** Ship Rock in New Mexico is a near-perfect example of a volcanic neck.

Study Questions for 15.2

1. What is a volcano and name the three types of volcanic mountains.
2. What features may be produced by magma inside the earth?

15.3 Other Mountain-Forming Processes

You will find out
- what processes form fault-block mountains;
- how complex mountains are formed;
- how the Rocky Mountains were formed.

If the formation of all mountains is related one way or another to movements of plates in the earth's crust, what about mountains that are not located near plate boundaries? This is a puzzling question because the internal processes that make mountains are not well understood.

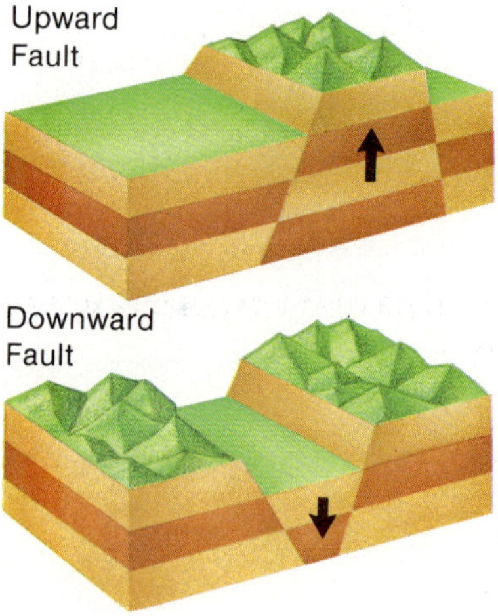

Figure 15.16 *left:* Fault-block mountains may result when land is uplifted or when land sinks. *right:* Frenchman Mountain in Nevada

An example of mountains that do not appear to be related to a plate boundary is the **fault-block mountains** in Nevada. A fault-block mountain is a landmass noticeably higher than its surroundings and bounded by one or more faults. You can see a photograph of a fault-block mountain in Figure 15.16. The faults aren't visible, but they're there alongside the blocks. Some movement of magma and hot gases from the earth's interior is also associated with the fault-block mountains in Nevada.

Geologists believe that this region of the plate, although not on a boundary, was once over an enormous plume, or hot spot. Heat coming from the interior of the earth warmed the crust under a large area. This caused

Do You Know?
The fault-block mountains in Nevada have no streams running through them.

the crust to bulge upward. As the uplifting continued, it cracked the crust. Volcanic rocks on the surface indicate there was some volcanic activity. Later the volcanic activity decreased. The area gradually cooled off when it moved from the hot spot or when the hot spot became inactive. Cooling rocks shrink in size. The uplifted land then slowly collapsed and broke into many blocks. The blocks settled at angles to each other, sliding along normal faults.

Figure 15.17 How fault-block mountains develop

ACTIVITY

How Are Fault-Block Mountains Formed?

Materials

a piece of file folder
ruler
plastic tape
colored pencils
scissors

Procedure

1. Copy the pattern of faults and rock layers shown onto the piece of file folder.

2. Cut along the faults.

3. Make a normal faulting of about 2 cm along fault number 1. Make normal faulting of 2 cm along the other faults.

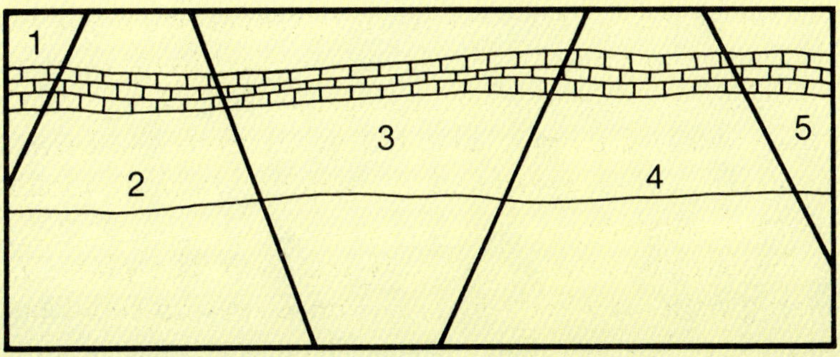

4. Now tape your model together.

Questions

1. Which of the blocks are mountains?

2. How would you describe a highway built across this area?

3. Why is *fault-block mountains* a good name for land that is shaped this way?

Geologists also recognize that a combination of building processes sometimes occurs. You can see evidence of this in the Appalachians, the Alps, and the Himalayas. For example, they each have long narrow zones of folded sedimentary rock alongside a region of metamorphic and igneous rocks. Mountain ranges that have this kind of structure were formed by a combination of mountain-building processes. These mountains are called **folded** or **complex mountains.**

What on earth would cause all this folding, faulting, and igneous activity? Remember that all mountains are related to crustal plate movements. Folded or complex mountains occur along boundaries where plates are moving toward each other. Figure 14.18 shows the model that explains the formation of folded or complex mountains. Notice that in this model the plates collide and press on each other. Slowly over time the crust folds and faults. In some cases such as the Appalachians there has been more than one collision.

The Rocky Mountains in western United States are different from any other mountain region of North America. The Rockies did not form along a plate boundary either. They occupy a region located between stable continental crust on the east and a less stable crust on the west. The Rockies are long, extending more than 4,800 km from Mexico through the United States into Canada. This region has a long history of mountain building. It is believed that several different processes, as shown in Figure 15.19, contributed to the formation of today's Rockies.

Figure 15.18 The Colorado Rockies

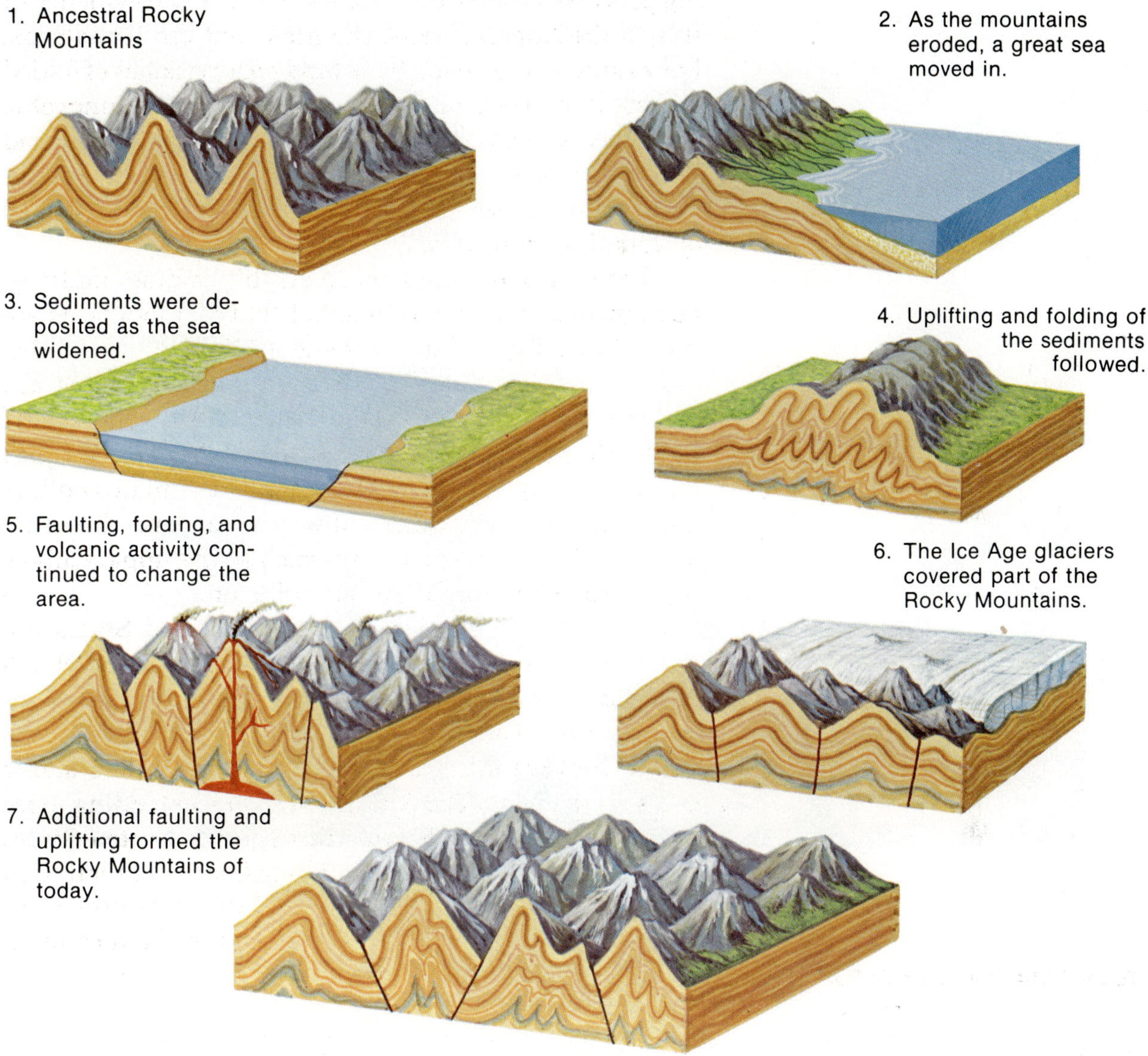

Figure 15.19 Repetitions of the geologic changes shown in this model produced the modern Rocky Mountains.

Study Questions for 15.3

1. Describe how fault-block mountains may have formed.
2. Describe the stages in forming folded or complex mountains.
3. On a time line indicate the major events in the history of the Rocky Mountains.

15.4 The Distribution of Mountain Ranges

You will find out

- where the major mountain ranges are located on land;
- where the major mountain ranges are located in the ocean;
- what is the geologic cycle.

A series of mountains makes a **mountain range.** The Sierra Nevada mountains in California are an example of a mountain range. These mountains extend from southern California to northern California and appear to belong to a single geologic unit. A **mountain system** is made of a series of mountain ranges. The Rocky Mountains extending from Mexico to Canada are a mountain system made of different ranges.

Figure 15.20 The Sierra Nevadas

Now look at the map on page 306 and notice that there appears to be one continuous system of mountain ranges after another from the tip of South America to Alaska. All these mountain ranges and systems connected together are called a **mountain chain.** Where are the mountains in this mountain chain located relative to a plate boundary? You were right if you answered that the mountains are located on or near the boundary. In fact, the cause of most all mountains can be traced to some kind of movement of the earth's crust.

For example, volcanic mountains such as the Pacific Cascades occur along subduction zones. Folded or complex mountains occur along converging plate boundaries. Here sediment is trapped between the colliding plates, forcing some sediment upward and some downward to make a mountain range like the Appalachians. Fault-block mountains are formed by movements and adjustments of the earth's crust even though they are not always located near a plate boundary. The Sierra Nevadas in California and the Grand Tetons in Wyoming are fault-block mountains.

Wherever mountains occur, geologists consider their presence as part of the **geologic cycle.** This cycle, or model, is built on the idea that the earth's crust is constantly changing. Refer to Figure 15.21 as you read the following description of the geologic cycle.

Do You Know?
A spectacular movie could be made if the earth's surface could be photographed with one frame for every 1,000 years. If this happened for 50 million years, it would make a movie one hour long. What you would see is the rise and erosion of several folded or complex mountain ranges.

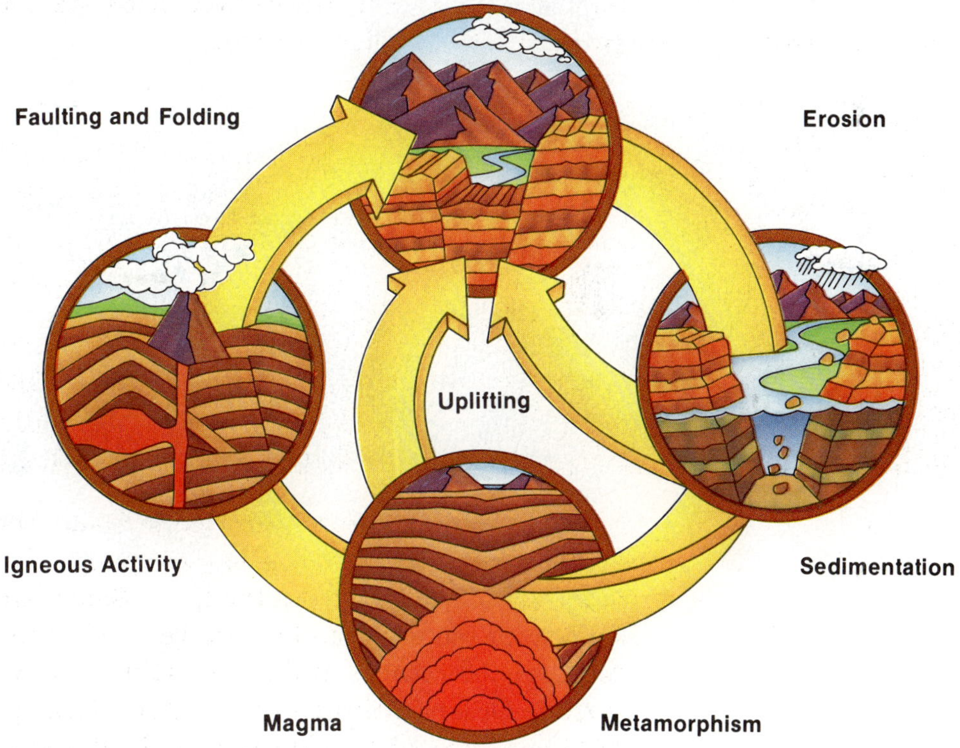

Figure 15.21 The geologic cycle

Mountains wear away to become sedimentary rocks. These rocks may be uplifted to form mountains or may be changed to metamorphic rocks. These metamorphic rocks may be uplifted to form mountains or in some cases melted to become igneous rocks later. Igneous rocks can be uplifted to form mountains. The processes continue through time.

Mountain Ranges
1. Alps
2. Urals
3. Andes Mountains
4. Himalayas
5. Carpathians
6. Great Dividing Range
7. Rocky Mountains
8. Appalachians

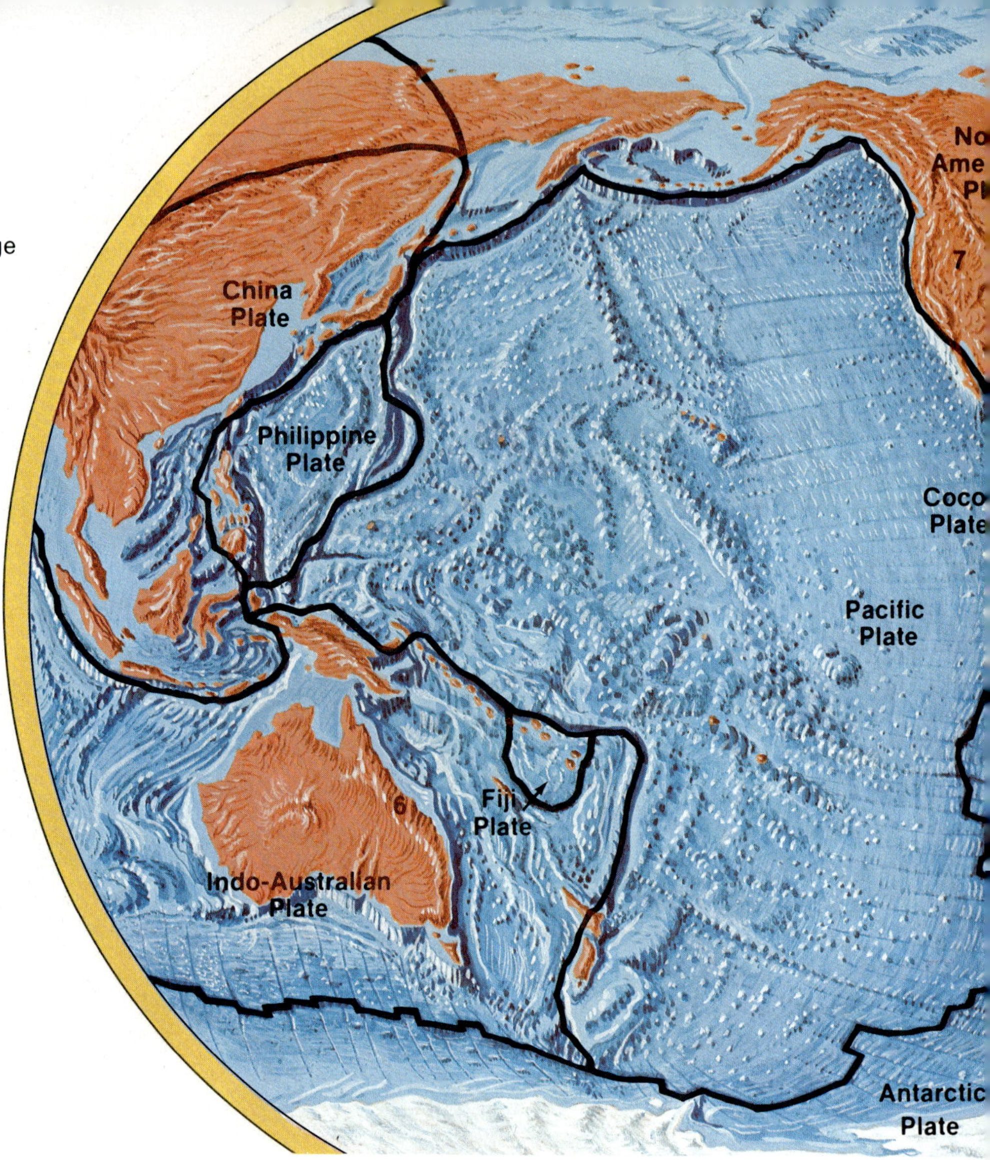

Figure 15.22 Mountain ranges and plate boundaries are closely related.

Look for mountain ranges on the map. Find the Alps, the Urals, the Andes, the Himalayas, the Carpathians, and the Great Dividing Range. What plate boundaries are close to each?

The longest and most recently discovered mountain ranges in the world can't be seen. They lie on the floor of the ocean. The mid ocean ridges on diverging plate boundaries form the longest and highest mountain ranges found on the earth. You can see that these mountains stretch almost unbroken nearly 64,000 km through the ocean basins. Some approach the height of Mount

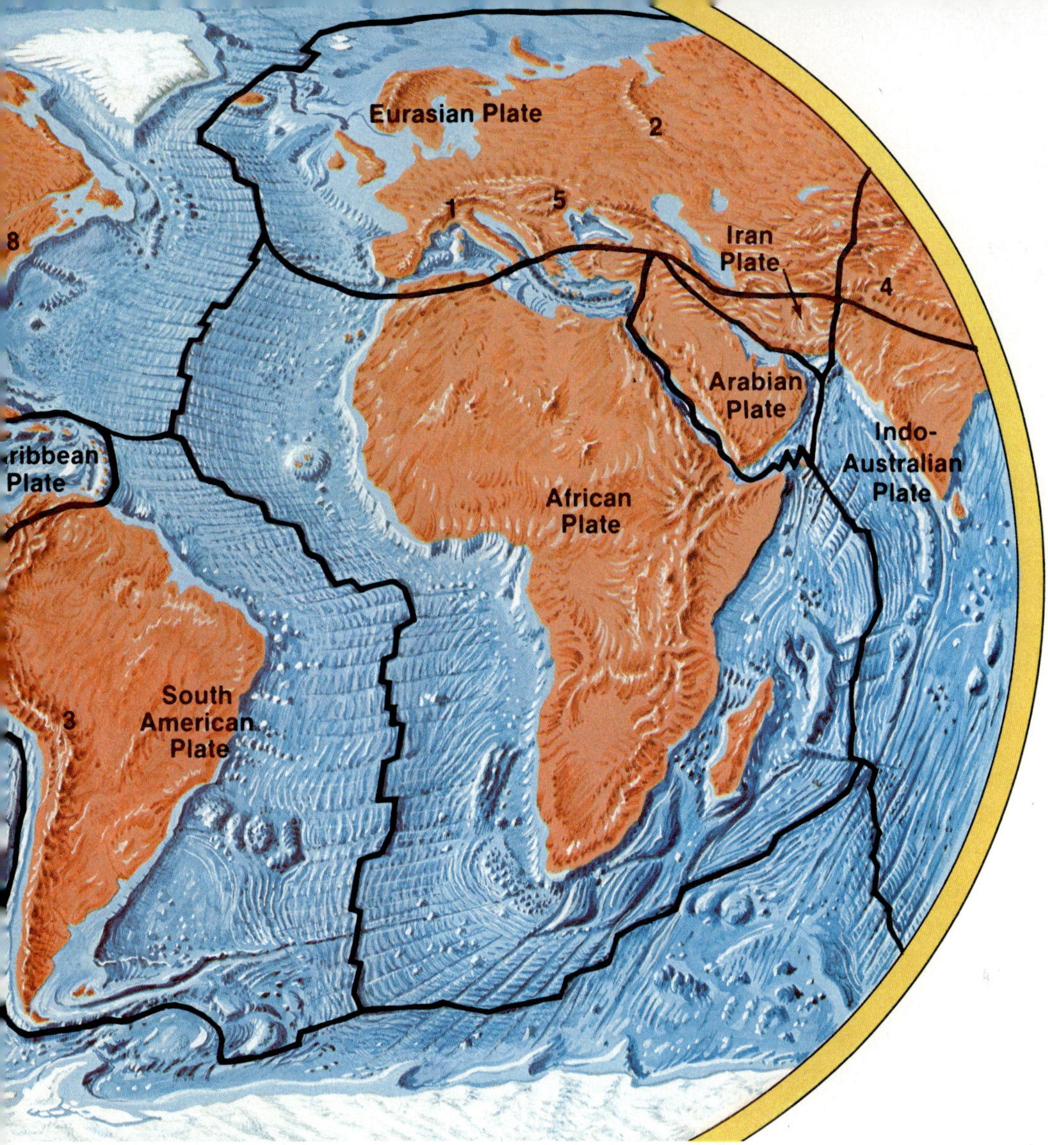

Everest, 8,850 m above their surroundings. Mountain building on the ocean floor or on the land is a worldwide geologic activity that continues through time.

Study Questions for 15.4

1. What mountain ranges are found in North America, South America, Asia, Europe, and Australia?
2. Where are oceanic mountains located?
3. What are the major geologic processes making the geologic cycle?

What Did You Learn?

- A mountain is any part of the earth's surface that stands high above its surroundings.
- Geologists believe that many of the highest rocks on mountains came from below sea level or deep in the earth's crust.
- Rocks deep in the earth's crust can be bent or folded.
- An anticline forms when the layers of rocks are bent upward into an arch.
- A syncline forms when the layers of rocks are bent downward to form a trough.
- A fault is a plane along which rocks slip.
- The three kinds of faults are called reverse, normal, and horizontal faults.
- Faults are recognized (1) when you see tilted blocks, (2) as straight lines on the surface when seen from the air, and (3) as stream valleys that are offset.
- Volcanoes result from the pileup of igneous rock material around an opening in the crust.
- Lava is a liquid rock coming from a volcano.
- Volcanoes occur along subduction zones.
- The three types of volcanoes are called composite or stratovolcanoes, cinder cones, and shield volcanoes.
- The movement of magma within the crust forms batholiths, dikes, sills, and volcanic necks.
- A fault-block mountain is a landmass higher than its surroundings and bounded by one or more faults.
- Geologists believe the fault-block mountains of Nevada formed over plumes which arched the earth. Later the crust broke and settled as uneven blocks.
- Folded or complex mountain ranges form long, narrow zones of folded sedimentary rocks alongside a region of metamorphic or igneous rocks.
- A series of mountains makes a mountain range.
- The longest and most recently discovered mountain range lies on the floor of the ocean.
- The formation of mountains is related one way or another to movements of plates in the earth's crust.
- The geologic cycle is a model of the processes that build mountains and wear them down.

Key Terms

mountain
folding
anticline
syncline
faulting
fault
reverse fault
normal fault
horizontal fault
volcano
magma
lava
composite or stratovolcano
cinder cone
shield volcano
batholith
dike
sill
volcanic neck
fault-block mountain
folded or complex mountain
mountain range
mountain system
mountain chain
geologic cycle

Biography

Dr. Marie Morisawa

Dr. Marie Morisawa, a Japanese–American, is a professor of geology at the State University of New York, Binghamton. How she got to that position is as interesting as what she does as a geologist.

Dr. Morisawa started as a math major with a chemistry minor, graduating with honors from Hunter College. She discovered geology when a friend twisted her arm to take an introductory geology course. Although she liked the course and took several more courses after graduation, she entered Union Theological Seminary. After earning an M.A. degree, she taught two years in Hawaii before accepting a job as the director for geology laboratory work and equipment at Hunter College. This experience led her to pursue graduate studies in geology at the University of Wyoming and Columbia University.

At the time she received her Ph.D., women were not so well accepted in geology as they are today. She encountered a number of interesting experiences as her career developed, such as waiting aboveground while her male students went underground at a mine because women were not allowed to go into mines. She summarizes her experiences by saying, "I discovered that those who like you and respect you as a person are more numerous and more important than those who do not. It is one's inner self that determines success or failure, whether one is a man or woman."

Dr. Morisawa is currently pioneering a new field of aesthetics and environmental geology.

TO THINK ABOUT AND DO

Vocabulary

Copy the terms and definitions and match the term with its definition.

Terms	Definitions
1. cinder cone	shortens the crust
2. composite volcano	folds form an arch
3. reverse fault	made of volcanic fragments
4. sill	extends the crust
5. dike	lies parallel to other rocks
6. normal fault	folds form an upside-down arch
7. syncline	cuts across other rocks
8. volcanic neck	left over after volcano is eroded
9. anticline	made of layers of ash, rock, and lava

What Do You Remember?

1. Explain what a mountain is in your own words. Give an example.
2. Draw a cross-sectional sketch of an anticline, a syncline, a normal fault, and a reverse fault.
3. Describe how folded sedimentary rocks are formed.
4. Explain the cause of volcanoes along plate boundaries.
5. Describe a folded or complex mountain range.
6. What does a curved pattern of rock layers on the surface tell you?
7. What kind of fault causes the crust to be shortened?
8. Describe three clues that indicate faulting has occurred.
9. The Hawaiian Islands are an example of which type of volcano?
10. Which kind of mountain formation may not be on a plate boundary?

Applying What You Have Learned

1. Explain why the faults forming fault-block mountains are normal faults.
2. Draw a sketch showing a dike and a sill. Label your sketch.
3. Describe how a volcanic neck forms.
4. List three folded or complex mountain ranges.

Research and Investigation

1. Visit the city or town planning office to find out if any faults are running through your area.
2. Based on library study, make a written report of mountain-building features in your area.
3. Divide the class into groups to organize and present a travelogue about a mountain range. Each group is to prepare a geologic history and a pictorial review of the mountain range that the group is describing.

A cable car travels between mountains in the Swiss Alps.

Unit Five

The Rock Story

16 Weathering and Erosion

The Grand Canyon is an awesome sight.

Imagine a gorge in the earth 1.6 km deep, more than 16 km wide in some places, and 450 km long. Cutting through thick beds of colorful rock, the canyon is a monument to erosion by the Colorado River. If you were standing on the canyon's rim, wouldn't you wonder about the events that formed this great canyon? Could that little river at the bottom really cut that great big canyon? Do streams carve their own valleys or are valleys formed separately in other ways?

If the only process working were the stream, valleys would be only as wide as the streams that formed them. Other earth processes must have helped the stream. This chapter is about the processes that break up and move rock.

Figure 16.1 The incomparable Grand Canyon.

... and ... Rocks

You will find out
- what weathering is;
- what three processes cause weathering;
- what erosion is;
- how gravitation moves weathered rocks and soil.

Figu... gs escape the effects of wea... g.

Exposed rocks on the earth's surface are continuously being broken down into smaller bits and pieces. This process of being broken down is called **weathering.** But rocks are not the only things that weather. Buildings, fences, bridges, highways, and automobiles also weather. Weathering can be caused by several different processes depending upon the conditions. The three basic weathering processes are **chemical weathering, mechanical weathering,** and **biological weathering.**

In the process called chemical weathering, substances in the atmosphere combine with materials on the earth and slowly change. You see the results of this process every time you see a rusty nail. Oxygen in the air combines with iron in the nail to form rust. The combination of a substance with oxygen is called **oxidation** [AHKS-uh-DAY-shuhn]. Carbon dioxide in the air combining with water to form a weak acid can also cause weathering. The acid dissolves some of the minerals in the rocks. Chemical weathering gradually changes the materials in rocks to something else. Often the new material is more easily dissolved by water and is slowly carried away.

Chemical weathering sometimes occurs underground where you cannot see it. Rainwater, which is a weak acid, dissolves some of the limestone rock as the water seeps downward. Caves are formed where the limestone has been dissolved and carried away. In some places the roofs of these caves and underground passages collapse. **Sinkholes** are funnel-shaped depressions caused by the collapse of underground caves. In central Florida many of the lakes are really large sinkholes which have filled with water.

Figure 16.3 This sinkhole formed when underground limestone dissolved and collapsed.

In mechanical weathering, rocks are broken apart. They are not changed into something new. Mechanical weathering occurs when water freezes in cracks of rocks. The freezing water expands and opens the crack even further. If you live where it is cold, you've seen this process at work in the street pavement. Large holes develop because repeated freezing and thawing of water breaks up the pavement. In Figure 16.4 you can see that this same thing happens to rocks.

Plant material and organisms can also break up rocks. This process is called biological weathering, and it is really a little of both chemical and mechanical weathering. The roots of plants weather rocks by reacting chemically with the rock. Roots also mechanically weather rocks by enlarging the cracks in which the roots are growing. Every time you see something growing in the crack of a rock or on its surface, you know biological weathering is occurring.

Figure 16.4 *above:* Water expands when it freezes, causing rocks to crack and crumble over time.

Figure 16.5 *left:* Over long periods of time, plants like these will weather the rock away.

Do You Know?
At times weathering extends deep into the earth. Mining operations have exposed thoroughly weathered bedrock to depths of 100 m.

ACTIVITY What Happens When Something Weathers?

Materials
steel wool
vinegar
jar with water
moist sand
container

Procedure

1. Wash the protective coating from the steel wool with the vinegar.

2. Divide the steel wool into 3 pieces.

3. For 24 hours place one piece in water with little pieces sticking out, another in the air outside, and the third piece in moist sand.

Questions

1. What evidence do you see of weathering?

2. Which piece of steel wool weathered the most?

3. Do you see any stains from weathering?

4. If you found a rock that contained iron, what weathering features might you see?

Figure 16.6 *above:* Sometimes weathering causes the outer surface of a rock to flake off in layers.

The rate at which rocks weather depends upon three factors: the kind of rock, the climate, and the surface area exposed to the atmosphere. Hard rocks weather more slowly than soft rocks. A climate that has much freezing and thawing increases the rate of weathering. The more area of a rock that is exposed, the faster weathering can take place. In Figure 16.7 you can see why rocks tend to weather faster at their corners and edges than on their flat faces. A good place to see this effect is in very old graveyards. You can see the rounded-corner effect caused by weathering as well as which kinds of rocks weather most rapidly.

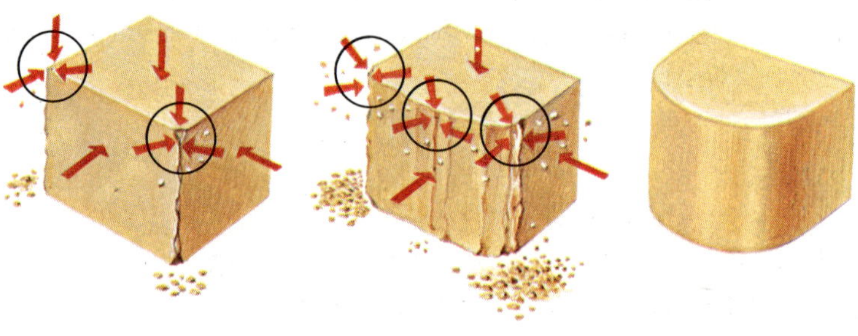

Figure 16.7 *right:* The corners of rocks become rounded because they have more surface exposed to weathering forces.

Figure 16.8 *left:* Landslides and mudflows cause quick and severe changes in local surface features.

Figure 16.9 *below:* Soil creep occurs gradually but can do damage over time.

In addition to the breaking down and decomposition that occurs in weathering, materials on the surface are also loosened and carried away by the action of water and wind. This process of loosening and of carrying away the material is called **erosion** [ee-ROH-zhuhn]. But before you look closely at what causes erosion, you should understand what happens to weathered and loosened rock materials.

In general, weathered rock materials move downhill because of gravity. This downward movement of earth materials is called **mass movement.** You've heard about mass movement when the news reported a **landslide** or **mudflow.** Landslides occur when there is a mass movement of sliding rock materials. A mudflow exists when the mass movement is composed of much finer material and usually occurs after heavy rainfall. In Figure 16.9 you can see evidence of very slow mass movement. A slow mass movement such as this is called **soil creep.**

Gravity also can cause the rock materials to fall vertically. Whenever a stream undercuts a bank, the materials fall into the stream. Sometimes rocks fall from cliffs and collect around the base of the cliff. Whenever there are loose rock materials, gravity will act to move them downhill.

Study Questions for 16.1

1. Describe what is meant by weathering.
2. List three basic kinds of weathering.
3. Tell what causes mass movement.
4. What is meant by erosion?

16.2 Erosion by Running Water

You will find out
- ways that running water erodes the land;
- what some of the landscape features produced by running water are.

Figure 16.10 The features of a streambed can reveal many things about the action of the stream.

To see how water can cause erosion, you need to look at what happens to the water that falls on the earth. Think about what happens in your backyard. The first rain that falls is taken up by the soil and plants. The next thing you notice is water collecting in low spots, and some water may even be running off in mini streams across the yard. You also see that the runoff water is dirty. It is carrying material that it is eroding away.

If the rain is coming down hard, larger pieces of material begin to move as the mini stream grows and runs faster. The streambed of your mini stream may even curve back and forth as it crosses the yard. If you look closely, you can see that the water in the mini stream doesn't always move at the same speed. Drop a small stick in your mini stream, and you'll see it swirl, twist, speed up, slow down, and sometimes get hung up on an obstruction. Notice that the water flows fastest where the mini stream is narrowest. The more you study this mini stream, the more you learn. Your backyard mini stream is almost a perfect scale-down model of a large stream. This mini stream does a lot of the same things a large stream does as it moves along.

Large streams cut streambeds and pick up material loosened by the rolling, sliding, and bumping of pieces of rock along the streambed. Also the materials carried by the stream rub against the channel and against each other. You can think of these materials carried by the stream as cutting or grinding tools. As the water moves along, it uses these tools to carve and shape the earth. Some material such as soil and clay are transported as fine solid particles in the water. Other materials such as salts are dissolved and then carried away.

Do You Know?
Rivers carry off enough sediment to lower the whole surface of the United States about one meter every 27,000 years.

The amount of material carried by the stream changes as the stream slows down or picks up speed. As a stream slows down, it deposits heavier material. A faster-moving stream picks up material and erodes its bed more rapidly. During floods, part of the streambed is carried away. As the stream slows again, sandbars and gravel beds are deposited.

Drainage patterns are one of the landscape features produced by streams. Drainage patterns are distinctive forms of streams and tributaries as viewed on a map. As you can see in Figure 16.11, different drainage patterns can provide you with clues to the kind of land the stream flows over. A branched pattern like a tree generally occurs in flat areas, and the branches are very random in their location. If the drainage pattern appears to radiate in all directions, then there is usually a central high point such as a mountain peak. Sometimes the branches in the pattern appear to be long and run parallel to the main stream. In these cases the land usually consists of parallel ridges or folded mountains. Stream patterns can tell you much about how the land looks.

Figure 16.11 These are drainage patterns produced by streams. *left:* Branched, *above:* Radiating, and *below:* Parallel.

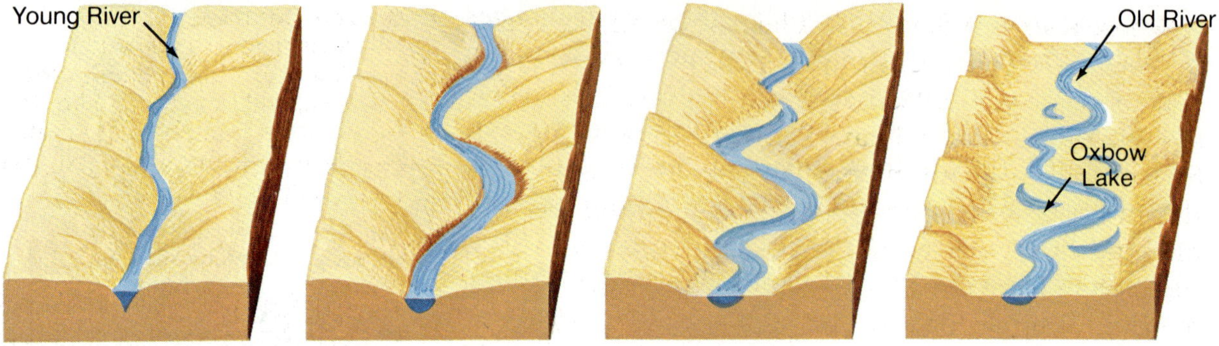

Figure 16.12 *above:* Compare the sizes of the valley floor cut by a meandering stream as time passes.

Other drainage pattern features also develop as streams wear down the land. In level areas streams often wind back and forth in many S curves or bends. These bends in a stream are called **meanders** [mee-AN-duhrs]. Streams widen the floor of the valley through which they run by cutting away on the outside of their bends or meanders. The cut-away material is then deposited on the inside of another bend downstream. In this way the stream works its way slowly across the land. Sometimes one bend in the stream meets another bend and cuts off a meander leaving a long curved lake called **oxbow lakes.**

Streams, along with mass movement, develop valleys. If the land is slowly uplifted, the stream cuts into the earth leaving terraces like you see in Figure 16.13. Within the channel itself streams sometimes develop waterfalls and rapids. In extreme cases the stream meanders cut a deep gorge like the Grand Canyon.

Figure 16.13 *right:* Terraced streambeds usually occur over large, flat surfaces which are gradually uplifted.

Where a stream flows into a lake or ocean, it deposits material. These deposits at the opening of a stream to a lake or ocean are called **deltas** [DEL-tuhs]. Large rivers such as the Mississippi have large and extensive deltas. Streams also deposit material on land. For example, in the dry regions streams may come rushing out of a mountain range as a flash flood after a heavy rain. As the stream spreads out and slows down, it deposits the material it has been carrying in a large fan. This large, fan-shaped deposit is called an **alluvial** [uh-LOO-vee-uhl] **fan.** Alluvial fans are most noticeable in dry regions even though there is no water flowing at the time, but they are also found in other regions.

Do You Know?
The Mississippi Delta began forming near Cairo, Illinois, and has advanced nearly 1,600 km to the south. A section of the delta near New Orleans, Louisiana, has been built in the last 500 years.

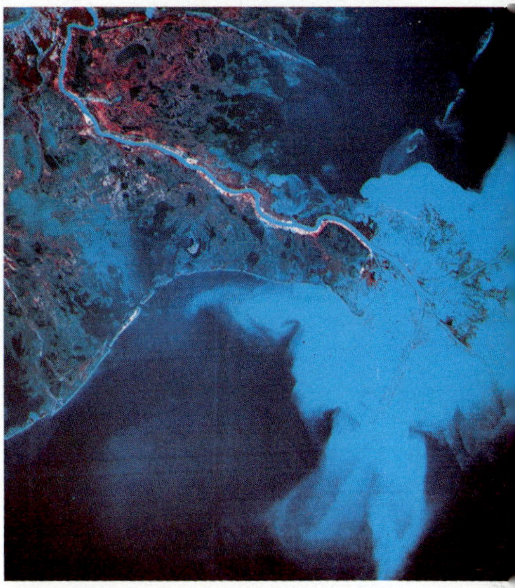

Figure 16.14 *above:* The Mississippi River delta extends into the Gulf of Mexico.

Figure 16.15 *left:* What do alluvial fans such as this one tell you about water flow in an area?

Study Questions for 16.2

1. Explain how a stream erodes the land.
2. Name six landscape features developed by running water.

16.3 Erosion by Wind

You will find out
- how the wind erodes the land;
- what landscape features are produced by the wind;
- why loess is an important wind deposit.

Figure 16.16 This sand dune will change shape and position when wind patterns change.

As the wind moves across the surface, small hard pieces of sand may be moved with it. The sand particles are really small pieces of rock and minerals. These sand pieces moving with the wind at speeds of as much as 100 km per hour or more tear into buildings, cars, rock formations, and people. Over long periods of time the small pieces literally blast and tear away paint, dig holes in rocks, and shape the land. With the exception of dust storms the wind moves sand-sized or smaller particles in an area close to the ground. As with water, when the speed of the wind is slowed, materials being carried along stop moving and are left in a deposit.

The best-known wind deposits are **sand dunes** found in desert and semiarid areas or near beaches. The sand accumulates in piles that shift and flow with each change in wind direction.

Two major processes that make up wind erosion are called **deflation** [dee-FLAY-shuhn] and **abrasion** [uh-BRAY-zhuhn]. Deflation occurs when the wind picks up material and moves it. Sometimes deflation makes shallow bowl-like features on the earth's surface. See Figure 16.17. Abrasion occurs when the particles carried along by the wind polish or break up the things hit by the particles. Deflation and abrasion are the work of the wind. One process picks things up, and the other process polishes and breaks up the things hit.

Do You Know?
Winds blowing over the Mississippi basin have over 1,000 times the transporting power of the river itself.

Figure 16.17 The shallow bowls left from wind deflation are called blowouts.

Figure 16.18 *left:* Flat faces are sculpted onto rock by wind erosion.

Figure 16.19 *right:* Unusual rock sculptures result as wind erodes materials in the rock unevenly.

Figure 16.20 *below:* Dust storms can deposit large amounts of loess but can also blow it away.

Some of the landscape features produced by wind are unusual forms or land sculptures. Some rocks in the desert, as in Figure 16.18, show flat faces almost like you see on a gemstone. Another way wind shapes rock formations is shown in Figure 16.19. When the wind-blown particles strike rock formations, the particles remove the softest material first.

The most important wind deposit is called **loess** [LOH-uhs]. Loess is fine dust deposited by winds over a large area. These deposits can be several meters thick and cover enormous areas. The fertile soil that develops on loess is rich in nutrients needed by grain crops. Rich farmland in parts of Nebraska, Kansas, Iowa, Illinois, and Missouri are loess deposits. Figure 16.20 shows a dust storm in New Mexico that deposited fine material over thousands of square kilometers in the Southwest. Storms such as this can travel for several hundred kilometers.

Study Questions for 16.3

1. Name the two processes of wind erosion.
2. Name three landscape features developed by wind action.
3. Where are loess deposits found and why are these deposits important?

16.4 Erosion by Ice

You will find out

- what a glacier is;
- how glaciers erode the land;
- what landscape features are produced by glaciers.

Figure 16.21 *above:* A continental glacier spans flat areas.

Figure 16.22 *right:* Valley glaciers are common in high mountain areas.

In some high mountains and in the polar areas there are places where the snow does not completely melt in summer. The edge of these places where the snow accumulates is called the **snow line.** The snow line moves back and forth with the seasons. The ice and snow that accumulate above the snow line make up a **glacier** [GLAY-shuhr].

The ice in the glacier is not like the ice in your freezer. Glacier ice forms gradually through a change from snow to granular snow and ice. As the granular ice gets buried deeper and deeper, the weight of the snow and ice above causes the bottom granules to recrystallize. This glacial ice is tough, is hard, and melts slowly.

Glaciers that cover a large area of a continent are called **continental glaciers.** If the ice is confined to valleys, it is a **valley glacier.** Gravitation constantly acts on the glacier to move the ice downhill.

It's hard to visualize how glaciers move. They don't slide along as one mass. The lower part of the glacier flows and continuously changes its shape by recrystallization. The upper part of the glacier acts like a brittle solid and cracks and slides over the lower layer.

Glaciers do a lot of work as they move along. Freezing ice breaks loose chunks of rocks. These rock chunks become part of the glacier and move along with it. Over a long period of time the chunks of rock are scattered throughout the glacier, but more rocks are near the bottom. As rocks are pushed along by the glacier, they erode the land underneath. These rocks polish or make huge grooves or scratches in the land under the glacier. At the same time the rocks inside the glacier are also grinding against each other. This grinding puts scratches on the rocks and makes rock powder.

Figure 16.23 *above:* This U-shaped valley was cut by the powerful forces of a glacier.

Figure 16.24 *left:* Rocks moved by glaciers easily make grooves in surfaces they pass.

Gravitation is continuously moving the ice and the rock material to the lower end of the glacier. There the rock material is deposited as the ice melts. This deposit of rock at the farthest point the glacier reached is called a **terminal moraine** [TUR-mih-nuhl muh-RAYN]. Long Island in New York is a terminal moraine that was built up many meters from the seafloor by debris dropped from melting ice.

Do You Know?
If the Antarctic ice sheets were melted at a suitable rate, the melted ice could keep the Mississippi River flowing for 50,000 years or all the rivers in the United States for about 17,000 years.

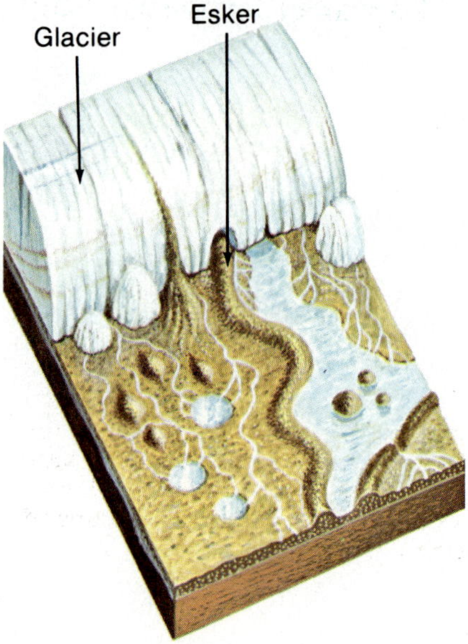

Figure 16.25 *above:* An esker marks the place where meltwater tunneled under a glacier.

The rocks carried by the glacier and deposited in an unsorted way when the glacier melts are called **till.** Glaciers also make deposits that are layered. Large depressions on a glacier may accumulate meltwater. In these pools the rock material settles in layered deposits. At times the meltwater develops huge tunnels that wind their way along the bottom of a melting glacier. Layered deposits then form a snakelike ridge of land in the tunnels. The snakelike ridge shown in Figure 16.25 is called an **esker** [ES-kuhr].

Glaciation is continuously at work changing the landscape in many places today. But the glaciation that formed the landscape features in Yosemite National Park took place thousands of years ago during earlier periods of glaciation. Geologists have found evidence of four major ice ages. The first of these advances of ice took place perhaps more than one million years ago. Figure 16.26 shows the maximum extent of the four ice advances. Following each advance of the ice came a warm period. Will another ice age cause a glacier to cover the northern United States? Nobody can tell for sure.

Figure 16.26 *right:* The combined southern limits reached by all glacial movement over North America.

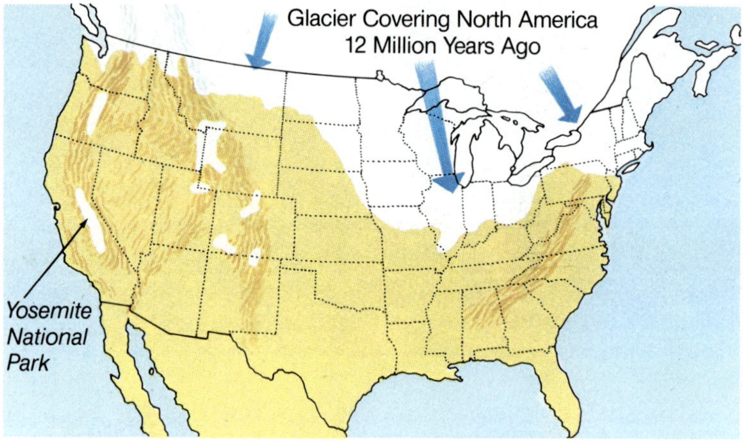

Study Questions for 16.4

1. Explain what a glacier is and why glaciers are found where they are.
2. Tell how a glacier works.
3. Name three landscape features developed by glaciers.

You will find out

- how a topographic map shows the shape and physical features on the land;
- what a contour line is.

Geologists show the shape and features of the land by using **topographic** [TAHP-uh-GRAF-ihk] **maps.** A topographic map shows the contour and elevation of the land. It also shows special features of the land such as streams. On a topographic map points of the same elevation are connected with a **contour** [KAHN-toor] **line.** Where the lines are close together, the land is steeper. Where the lines are farther apart, the land is sloped more gently.

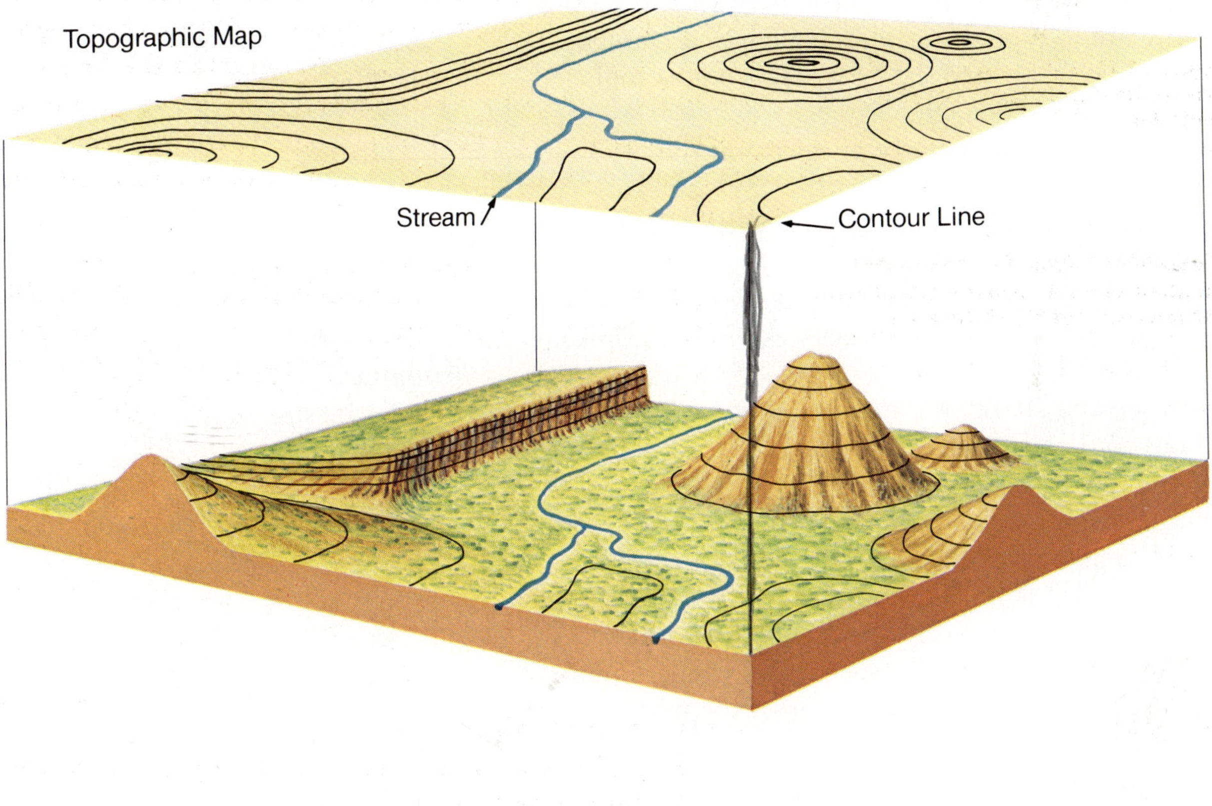

Figure 16.27 Compare this illustration of a land area (*bottom*), with its topographic map (*top*). Notice how the contour lines are spaced to correspond with the slope of the land.

Study Questions for 16.5

1. What does a topographic map show?
2. Describe a contour line.

What Did You Learn?

- The breakdown of rocks is called weathering.
- There are three major weathering processes: chemical weathering, mechanical weathering, and biological weathering.
- The rate at which rocks weather depends upon the kind of rock, the climate, and the surface area exposed to the atmosphere.
- The process of loosening and carrying of material is called erosion.
- Gravity acts to move weathered material downhill.
- The movement of earth material by gravity is called mass movement.
- Mass movements may be fast like landslides and mudflows or slow like soil creep and earthflows.
- Whenever a stream slows down or picks up speed, the amount of material being carried changes.
- The materials carried by streams act as tools to carve and shape the earth.
- Streams form drainage patterns on the land.
- Landscape features formed by streams include meanders, oxbow lakes, wider valley floors, terraces, and waterfalls and rapids.
- Stream deposits include deltas and alluvial fans.
- Two wind erosion processes are deflation and abrasion.
- When the speed of the wind is slowed, materials fall out of the atmosphere to make a wind deposit.
- The most important wind deposit is loess.
- The ice and snow which accumulate above the snow line make up a glacier.
- Glacial ice is not the same as ice in the freezer.
- Gravity causes glaciers to move downhill.
- Recrystallization of ice occurs as glaciers move.
- When they melt, glaciers deposit the material carried as moraines, tills, or eskers.
- There have been four major advances of the glaciers across the northern United States.
- A topographic map shows the shape and physical features of the land.

Key Terms

weathering
chemical weathering
mechanical weathering
biological weathering
erosion oxidation
sinkholes
mass movement
landslide
mudflow
soil creep
drainage pattern
meander
oxbow lake
delta
alluvial fan
deflation
abrasion
sand dune
loess
snow line
glacier
continental glacier
valley glacier
terminal moraine
till
esker
topographic map
contour line

Career

Geomorphologist

A geomorphologist is a person who makes a study of surface landforms and the processes that cause them. Geomorphologists are also interested in the changes in landforms over time. Geomorphologists recognize that every landform is in the process of becoming something else. Hills that are worn down and V-shaped stream valleys that become flatter and wider are two examples of changes in the land surface that geomorphologists study.

Dr. A. Wesley Ward is a geomorphologist working for the United States Geological Survey in Flagstaff, Arizona. Currently he is working on a large-scale map showing the geology in the San Juan Basin. This region is located in northwest New Mexico. Dr. Ward is studying the surface features, soils, and climate relationships involved in the development of landforms in this area. When completed, the map will be used in the search for oil, gas, and coal that are thought to exist in this region.

To make such a map requires a great deal of training. Dr. Ward took many courses in geology, chemistry, physics, engineering, atmospheric science, and mathematics.

Geomorphologists usually have a great interest in and love of the outdoors. Dr. Ward's love of the outdoors began as a child when he collected frogs, turtles, and other living creatures with his grandfather in Michigan. When asked why he liked his work, Dr. Ward replied, "Because it is like solving a great big puzzle. I like the outdoors. I observe the surface and then try to figure out how it got to be that way. I like this kind of detective work."

TO THINK ABOUT AND DO

Vocabulary

On a separate piece of paper unscramble the letters in the words and define each.

1. gnateewhir ________ definition ____________
2. osil eperc ________ definition ____________
3. adtel ________ definition ____________
4. ibnsaaro ________ definition ____________
5. reeks ________ definition ____________

What Do You Remember?

On a separate piece of paper write the answers to these questions.

1. How is chemical weathering different from mechanical weathering?
2. What happens to a rock that is being weathered chemically?
3. What are three factors that affect the rate of weathering?
4. Two rapid mass movements of earth material are ________ and ________.
5. Why would anyone want to study water flowing through or off his or her backyard?
6. Suppose you are watching a pebble being pushed along by a stream. What would you see?
7. What are the tools a stream uses to sculpture the earth?
8. What are deltas?
9. Where is wind action most effective in moving material?
10. Explain this sentence: Deflation and abrasion are the work of the wind.
11. Why is loess considered the most important wind deposit?
12. Name five states that have extensive wind deposits.
13. How is glacial ice different from ice in your freezer?
14. Explain how glaciers move.
15. Why does a glacier deposit material?

Applying What You Have Learned

1. How can mechanical weathering help chemical weathering?
2. What causes a stream to change its speed?
3. Can a tributary cut deeper into the earth than the main stream it is entering? Explain your answer.
4. Explain how a glacier could grow and shrink.

Research and Investigation

1. Make a picture collection of landscape features near where you live. Label and describe weathering processes shown in each picture.
2. Make a library study on a national park and make an oral report on the natural land features of that park that make it unique.

Erosion can produce spectacular natural structures such as those in Arches National Park in Bryce Canyon, Utah.

17 The Rock Record

For thousands of years people have tried to determine what geologic events happened on the earth in the past. Imagine the look on the face of the Greek scholar Herodotus when he discovered a fossil fish on the top of a mountain. There wasn't an ocean in sight, and yet he held a rock with a fossil fish in it. What did he think? How could he explain what he had found?

There are several ways that that fish could have gotten there. Maybe the seas had once been much deeper, and they had covered the mountaintop. Maybe the mountain containing this fossil fish had been pushed up to this high level. If you had found the fossil fish, how would you say the fossil fish had gotten there? What would you look for to determine what really had happened? In every rock and every fossil there are clues that help you read what has happened.

How do you read a rock? That's quite a task, but it's easier if you know what clues to look for. This chapter is all about clues that are found in rocks.

Figure 17.1 A fossil can tell a story if you know how to read it.

17.1 The Rock Record

You will find out

- what the law of superposition is;
- why the law of crosscutting relationships is useful in determining what event occurred first;
- why the principle of uniformity is important.

Figure 17.2 *above:* Sedimentary rock layers follow the law of superposition.

Figure 17.3 *right:* The law of crosscutting relationships can be applied to this sedimentary rock. The rock was faulted after the layers were formed.

The last time you were watching an old black-and-white movie on TV, did you try to guess when it was made? If so, what did you look for? Was it how the cars looked? the clothes the people were wearing? the hair styles? or some other clues? A clue is something that aids in the solution of a problem. Clues can help you learn about the ages of the rocks you see.

The position of the rock layers is one clue to look for. In Chapter 12 you learned that layers of sediment are deposited (laid down) in horizontal layers. These layers form sedimentary rock in time. If the rock layers have not been turned over, the oldest layer will be at the bottom. The youngest layer will be on top. This relationship of youngest layer on top and oldest layer on the bottom is called the **law of superposition.**

What's another clue to look for in the rock record? Look for faults in rocks or magma features that cut across other rocks. Figure 17.3 shows a layer of sedimentary rock cut by a fault. Which came first, the layer of rock or the fault? The layer of rock had to come first because the rocks had to be there before the fault could occur. The fact that younger features cut across older features is called the **law of crosscutting relationships.** Faults, for example, are younger than the rocks they cut.

Geologists believe that the laws of nature that are recognized today have always existed. Water has always flowed downhill and carried sand and silt to the sea. The rock cycle worked the same millions of years ago as it does today. Rocks have weathered and eroded in the same manner throughout geologic time. Earth features formed long ago were made by the same processes operating in the same way as they do today. This is called the **principle of uniformity.**

Figure 17.4 Ripples in ancient sandstone (*left*) were formed by the same process that made the new ripples in the sand (*right*).

Is the principle of uniformity ever questioned? Some geologists believe the rate at which the earth processes work may differ over time. Volcanism may have been more frequent at certain times in the past. There may even have been earth processes operating in the past that are not apparent today. The earth's atmosphere may have been different a long time ago. This would cause different chemical reactions in the weathering process. But one thing seems certain. The physical and chemical laws of nature are unchanging. The earth processes that these laws of nature caused in the past continue to happen today. By studying the present, you can infer the behavior of these processes in the past.

Study Questions for 17.1

1. Name two geologic laws used in reading the rock record.
2. Make a sketch illustrating the two geologic laws.
3. Describe the principle of uniformity.

17.2 Gaps in the Rock Record

You will find out

- what is meant by an unconformity in the rock record;
- how to use correlations to determine if rock layers are related in age.

Figure 17.5 These are examples of unconformity. *top:* Sedimentary layering interrupted by erosion, *middle:* A break between igneous rock and sedimentary rock of different ages, and *bottom:* A change in tilt between sedimentary layers.

By now you may think that reading the rock record is an easy process. You just apply the geologic laws. This would be true if the rock layers in each area were a complete story. Why isn't the story complete? The principle of uniformity gives you some clues about why the story is incomplete. Remember that according to this principle the earth processes that are going on now have probably always been occurring. If rocks are being eroded away today, they were also eroded away millions of years ago. The erosion wore away the rock layers and deposited the rock material somewhere else. As the rock layers eroded away, that part of the rock record also disappeared.

After a period of time new layers of sediments may again have been deposited on top of the eroded area. These sediments in time became sedimentary rock. But an examination of the layers shows a time lapse between the top of the eroded layers and the new layers. These time lapses are breaks in the rock record. You no longer can read a complete story. Something is missing.

When geologists find such a break in the rock record, they say an **unconformity** [uhn-kuhn-FOHR-mih-tee] exists. Unconformities then represent a lapse of time. During that time, the rock record was not recorded or was being destroyed by erosion, creating an unconformity. An unconformity may occur within rocks of the same kind or between different kinds of rocks.

The easiest unconformity to spot is an **angular unconformity.** An angular unconformity is one in which the older layers slope at a different angle from that of the younger layers. In Chapter 16, you learned that there are pressures in the earth's crust that can fold the layers of sedimentary rock. In time the top of a fold or an anticline is worn away, leaving only the sloping layers.

With more time this area may again be covered with a sea, and new sedimentary deposits may be formed. The new rock layers are laid down horizontally on the sloping lower layers. The result is an angular unconformity. In Figure 17.6 you can see how this happened.

Figure 17.6 The four stages in the formation of an angular unconformity.

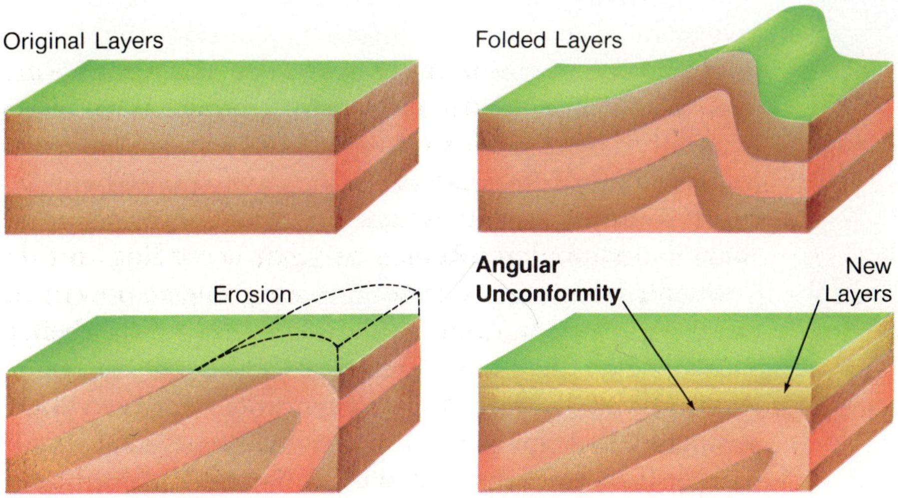

By now you can see that sometimes there are problems in reading the rock record. If you are trying to compare the rock records in two different areas, these problems can make the task very difficult. But geologists have developed some ways of comparing rock layers in different places. One way is to walk along a rock layer and trace it from place to place. You can walk along a rock layer and follow it from location A to location B without a break in the layer. What would you think? The rock layer is obviously the same rock formation in both location A and location B. The rock must be the same age in each location.

Another way is to note the sequence of distinctive rock layers. Examine the rock layers in Figure 17.7. Suppose that 100 km away in another area you also find the same distinctive layers. You find the red shale, gray limestone, and yellow sandstone are laid down in the same order. You can infer that both areas contain rock layers that were laid down at the same time. They were probably joined together in the past.

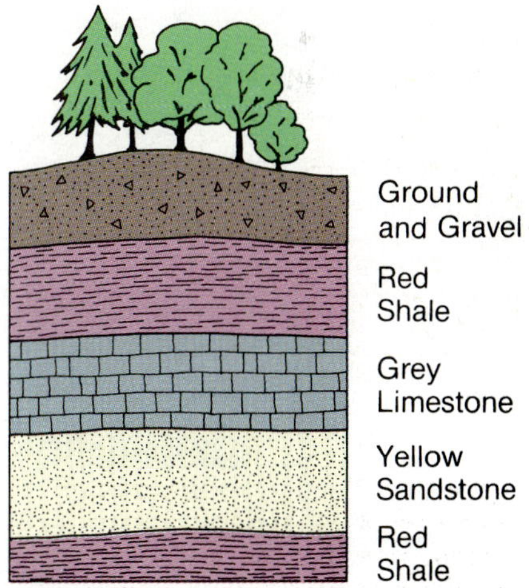

Figure 17.7 If you found these same rock layers in another area, what would they tell you about the age of each area?

A third way uses a **marker layer.** Marker layers provide an even more distinctive characteristic. A volcanic-ash layer can be a marker layer. It would cover a wide area and be deposited at one time. Wherever this layer is spotted, you'll find rocks of a similar age.

A fourth way is the location of **index fossils.** If rock layers in two different locations contain the same groups of fossils, they are thought to be about the same age. Sometimes there is a very distinctive fossil that lives only a brief geologic time. Yet this fossil is found in rocks in many different places. Such a distinctive, single fossil is called an index fossil. Rocks containing the same index fossil are related by age.

When geologists compare the ages of rocks in two different locations, they use a process called **correlation.** In a correlation they compare the rocks in the different locations by looking for comparable features. The geologists look for such things as distinctive layers, a marker layer, and index fossils. If they find any of these in both areas, they have a correlation. The geologists can then say the rocks are related in age. Correlations are much easier to do with rocks separated by only a few kilometers. Yet correlations have been done with rocks separated by oceans.

Figure 17.8 Fossil remains of this organism make it a valuable index fossil. The organism is related to the chambered *Nautilus,* which still exists.

Figure 17.9 Examine the correlation clues shown in the diagram. Can you pick out the related rock layers?

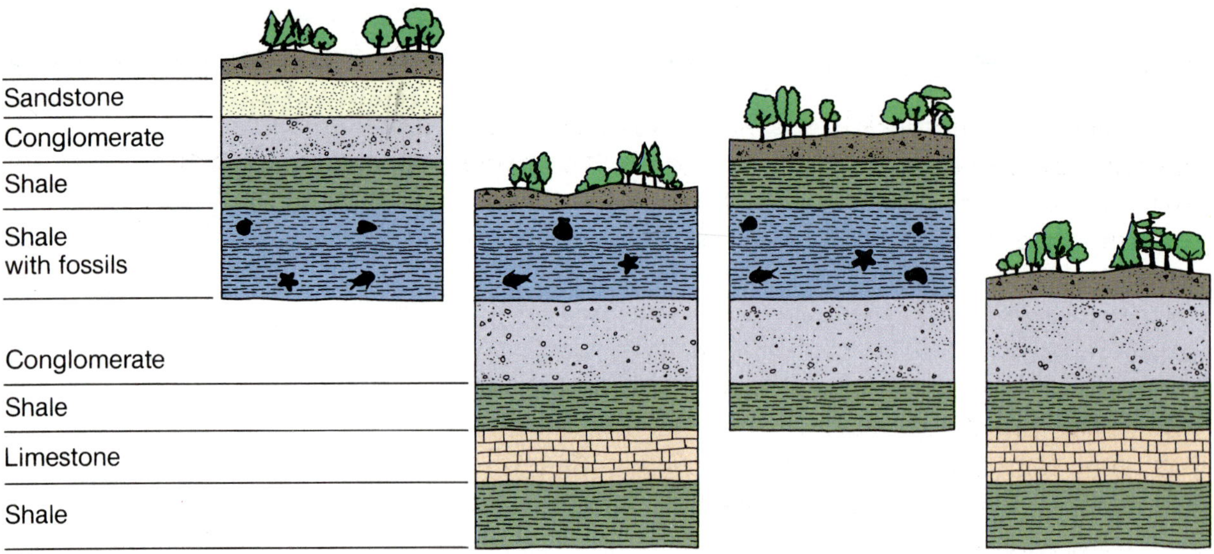

ACTIVITY

Which of These Rock Layers Are Related?

Materials

pencil
paper
colored pencils
scissors

Procedure

The rock layers in four different places were identified as shown below. (Fossil symbols represent groups of fossils.)

1. Copy the rock layers and mark each layer as shown.

2. Cut out the columns of rock layers.

3. Arrange the columns of rock layers side by side by matching layers of the same age.

Questions

1. Which rock layer(s) correlates with layer 2? With layer 3? With layer 4? With layer 9?

2. How many different rock layers are shown?

Just as reading the rock record has its problems, correlations are sometimes difficult to make. The same fossil may be found in different kinds of rocks. Geologists must use as many clues as possible when they correlate rock layers.

Study Questions for 17.2

1. What is the name of the geologic feature that indicates part of the record is lost?
2. Describe what it means to correlate rock layers. How are index fossils and marker layers used in correlations?

17.3 Clues in Sedimentary Rocks

You will find out

- what sedimentary structures are;
- how rock layers indicate a change in sea level.

Figure 17.10 *right:* The imprints of lightly falling rain are preserved in sandstone. What do they tell you?

Figure 17.11 *below:* Four sedimentary structures that give geologists clues about past conditions: (a) ripple marks, (b) crossbedding, (c) graded bedding, (d) mud cracks.

In learning how to read the earth's history, you must look everywhere for clues. Remember, a clue is something that helps you find a solution to a problem. What clue do you see in Figure 17.10? What could make these circular imprints in this piece of sandstone? What you see in the sandstone are raindrop imprints. These imprints tell you that the sand was above water and a light rain was falling when the imprints were made. From the raindrop imprints the geologist also knows which side of the rock layer is the top side. The imprints were made more than 100 million years ago. They are a reminder of the principle of uniformity. The same earth processes have operated throughout time. Rain fell upon the earth 100 million years ago as it does today.

You can find many clues to the conditions that existed when sedimentary rocks were deposited. Figure 17.11 shows four **sedimentary structures** that are good clues. Sedimentary structures are patterns formed in the sediment before it becomes solid rock. Ripple marks are formed in the sand, which later becomes rock. They indicate that water or wind flowed back and forth or in one direction. Crossbeds show shifting water or wind currents flowing in one direction. Graded bedding indicates rapid settling of particles of different sizes in still water. What conditions exist when mud cracks form? Mud cracks when it is exposed to air and dries out. Each clue in a sedimentary structure helps us to learn the conditions existing at the time the sediments accumulated.

By using clues, geologists can usually tell land deposits from ocean deposits. Most land deposits occur in the beds of streams, on the flat areas of valleys, in lakes, in swamps, and in deserts. Each kind of deposit will contain clues that help in reading the earth's history. For example, the roundness of pebbles is usually related to how far they have traveled.

Ocean deposits occur below sea level. Since the level of the sea changes because of tides, there is a transition zone between land and ocean deposits. When you go to the beach, you play in this zone. The next time you go to the beach, look carefully and you'll see both land and ocean deposits.

Figure 17.12 Notice the order of deposits of sediment shown here. Compare the pattern with those in Figure 17.13.

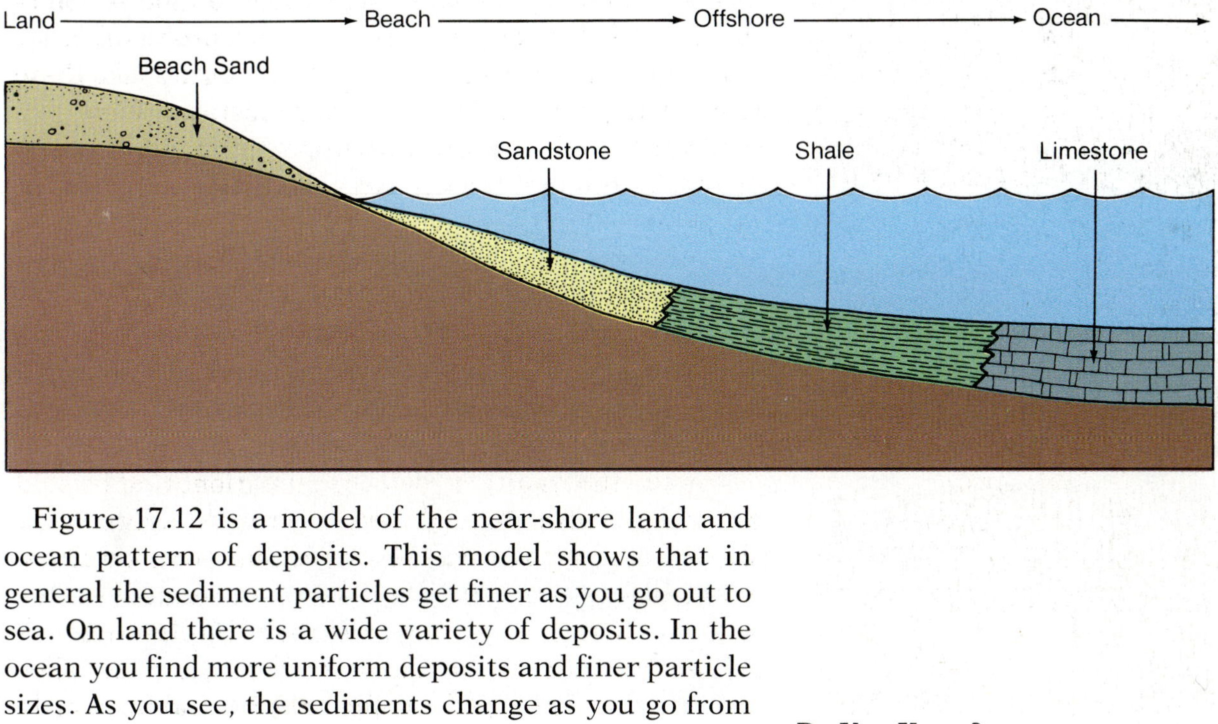

Figure 17.12 is a model of the near-shore land and ocean pattern of deposits. This model shows that in general the sediment particles get finer as you go out to sea. On land there is a wide variety of deposits. In the ocean you find more uniform deposits and finer particle sizes. As you see, the sediments change as you go from shore out to sea. Deposits near shore are mostly sand and form sandstone. Farther out to sea the fine deposits of silt form shale. In the deeper seawater the deposits form limestone. If you live in an area that is largely limestone, you know it was once under a sea. If your area is largely shale, you can guess that it was an area nearer shore than the limestone area was.

Do You Know?
The total thickness of all sedimentary rock has been used to estimate the age of the earth. If the average rate of deposition was 3,000 years per meter, 128 km would require at least 384,000,000 years to accumulate. Geologists believe this figure is far too low.

How do geologists use this information? By examining the layers of rock, they can tell that an area was submerging and the sea was moving onto the land. They can also tell if the land was rising and the shoreline was moving outward. In Figure 17.13A you can see how layers were formed in a submerging area. As the land submerged, the shoreline with its deposit of beach sand moved in over the former land area. The fine silt that would form shale was deposited over sand deposits. Farther at sea, deposits that would form limestone were laid over the area that was shale. If the water rose even higher, you would find these patterns repeated again farther toward land. Suppose you are looking at three layers of rock with limestone on top and then shale with sandstone on the bottom. This pattern of deposit tells you that the land submerged or the sea level rose. This caused the shoreline to move inland.

Figure 17.13 *top:* A submerging land area; *bottom:* An emerging land area

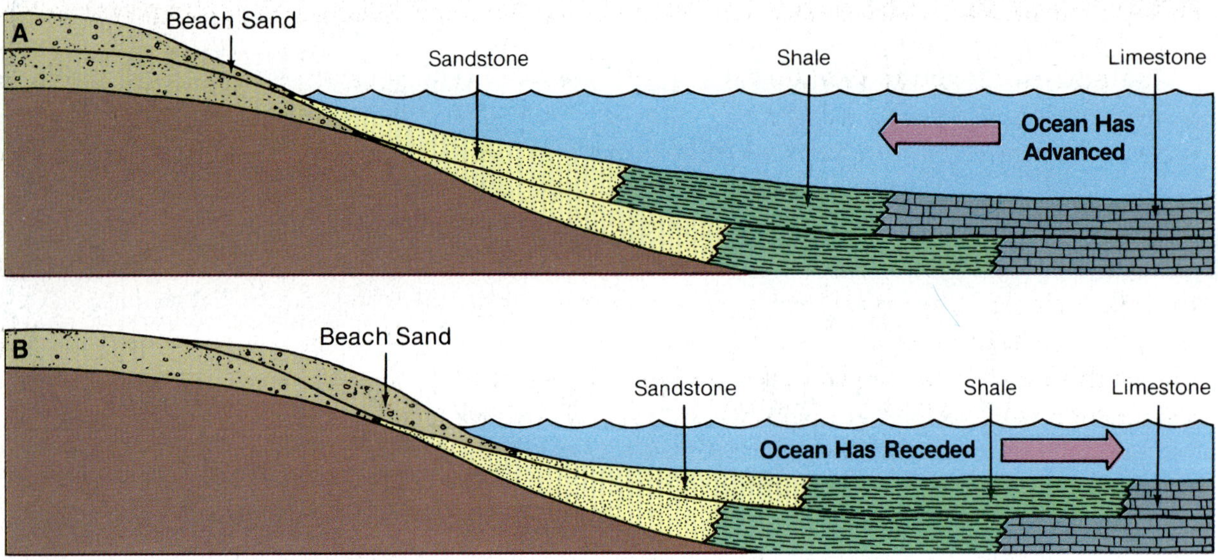

Suppose the three layers were in this order—sandstone (top), shale, and limestone (bottom). Does that change the earth's history at that spot? Geologists would say that sea level had been falling when these layers were deposited. You can see how this occurred in Figure 17.13B.

Figure 17.14 Exposed rock layers like these provide clues about sea level changes.

The next time you see some rock layers, look for the order of the sandstone, shale, and limestone layers. The order of deposits in an area indicates changes in sea level.

By now you should have learned the rock record is not always clear and easy to read. Sometimes it is necessary to look for more clues and the interpretation becomes more difficult. In fact, you may not be able to interpret the earth's history as well as you would like. This is true for geologists as well.

Study Questions for 17.3

1. What is a sedimentary structure?
2. What conditions produced the following clues: ripple marks, mud cracks, and graded bedding?
3. What sequence of sedimentary layers would indicate a sea moving over the land?

17.4 Interpreting What You See

You will find out

- how geologists use cross sections to read the earth's history;
- what information can be found on a geologic map.

Interpreting what has happened in the earth's past is somewhat like deciding what happened in building your school. You see the finished building, but what was the first event in building it? What were all the events in between the first and the last event? In what order did these events occur? These are the kinds of questions geologists ask about the earth's history.

Figure 17.15 Can you read the history written in this road cut? The illustrations below show the order of events that happened before the road cut.

In Figure 17.15 you see a road cut. You can think of it as a **geologic cross section.** There are several sedimentary layers, and the layers have been cut by a fault. Now think about what happened and in what order. What evidence do you have? There is evidence of (1) folded sedimentary rock layers, (2) a fault, (3) a period when this place was underwater, (4) a period when it was above water, (5) erosion of the surface, and (6) a recent road cut. In what order did these things occur?

The first thing to occur was that the region was underwater. While the region was underwater, the sedimentary rock layers were deposited and folded. The next event was the faulting. Remember the law of cross-cutting relationships? Geologists then assume that the uplift of the land, or the fourth event, which raised the land out of water, occurred during faulting. The last natural event was the surface erosion, followed by the event of the road cut being made by people. In interpreting the earth's history, then, geologists look for geologic features and the sequence in which these features occurred.

ACTIVITY

Material

pencil

What Is the History of the Rocks in These Geologic Cross Sections?

Procedure

1. Make a list of the things that occurred in cross section A.

2. Write down the order in which these things occurred.

3. From the evidence you should have found these things:

two sets of sedimentary layers
area underwater two times
angular unconformity
erosion two times
uplift two times
tilting

You should have listed the order in which things happened as (1) area underwater; (2) sedimentary layers C, D, E, F; (3) tilting; (4) uplift; (5) erosion; (6) underwater; (7) sedimentary layers K, J, I; (8) uplift, (9) erosion.

4. Now repeat steps 1 and 2 for cross section B.

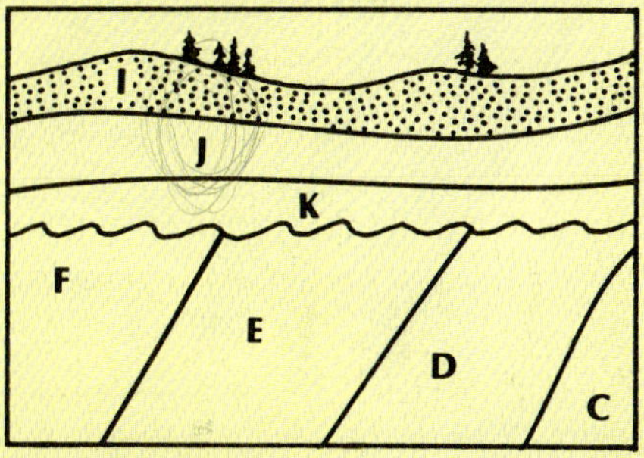

Cross-section A

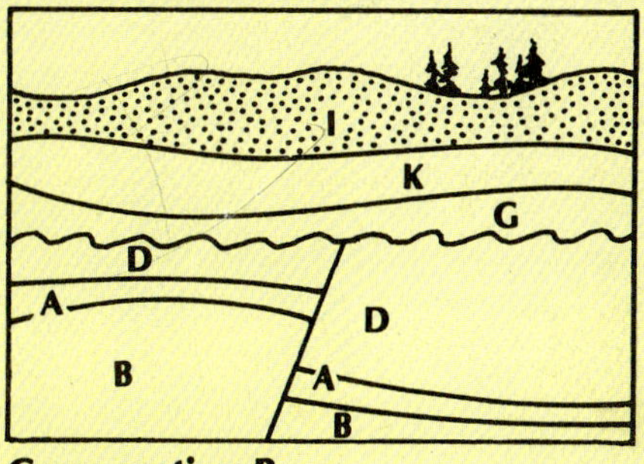

Cross-section B

Questions

1. List the evidence you find.

2. In what order did these things occur (first to last)?

Road cuts or exposed cross sections are not always available. Geologists must rely on other sources for the information they need. For many areas special maps have been prepared. These are maps that show rock types on the surface, ages of rocks, topography, and how the rocks lie. A map showing geologic features of the land is called a **geologic map.** Geologic maps show a vertical look at the land as if you were looking straight down from an airplane.

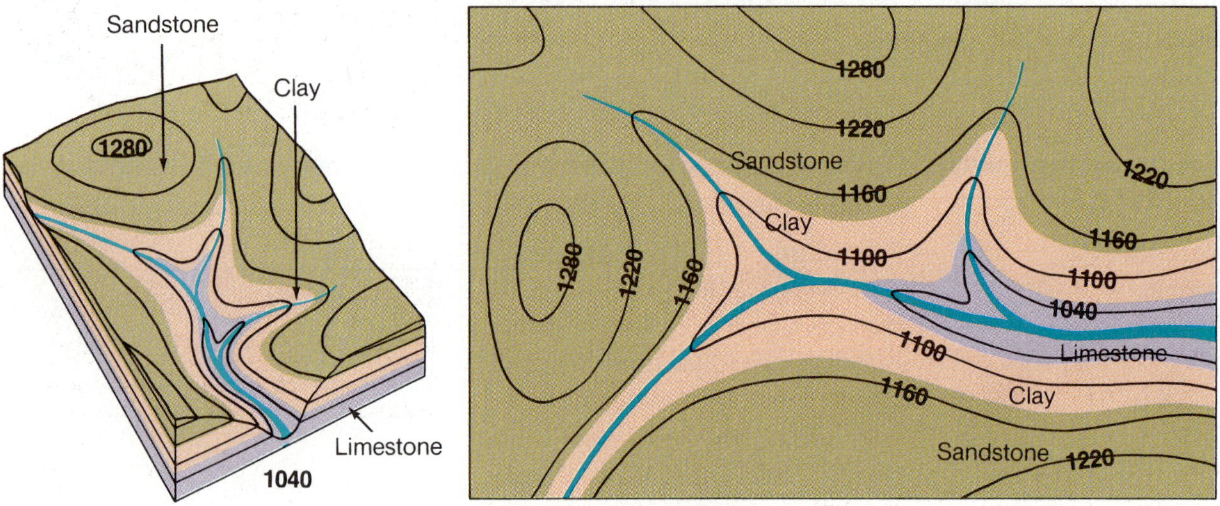

Figure 17.16 The geologic features of an area (*left*) are shown on the geologic map with contour lines added (*right*).

In Figure 17.16 you see a part of a geologic map. It shows three different kinds of rock. You can also see that there is a creek. The direction of the stream flow can also be determined. Where is the highest land? Where is the lowest land? The contour lines tell you how high the land is in each area. Now look at the rock layers. Which layer is the oldest? Remember what the law of superposition tells you. Which layer is the youngest? The rock layers are also horizontal. What on the map tells you that? Did you notice that the contour lines and rock layer lines run side by side?

You can read the history of the earth in this area from the map. Remember what you have learned about what the sequence of deposits tells you. Since there are sedimentary rocks, the area was once underwater. The sequence of sandstone, clay bed, and limestone tells you that the area was uplifted. As the area rose, the shoreline moved out away from this area. Finally, from the contours and the stream flow, you know the area has been eroded. You can read this history because you now understand the earth-shaping processes.

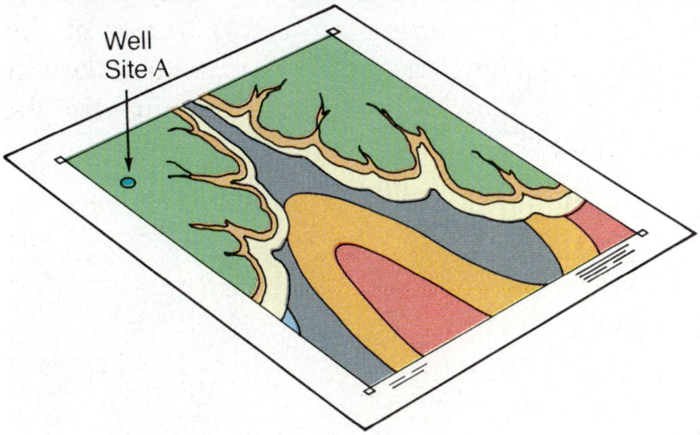

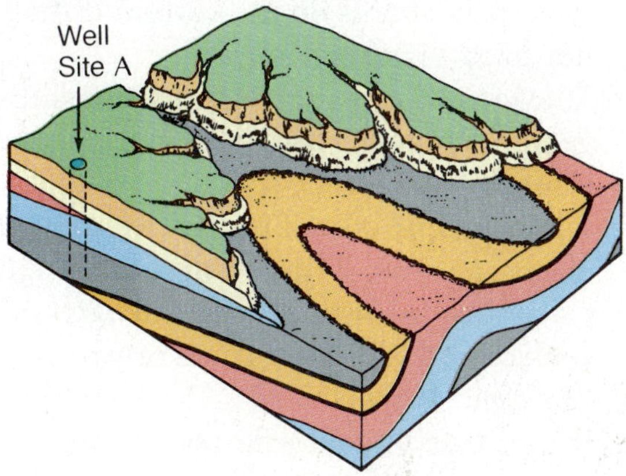

Figure 17.17 *left:* A geologic map showing surface features of an area.

Figure 17.18 *right:* A cross section of the same area based on the surface patterns.

Geologists also use geologic maps to tell them what rock layers are found underground. Look at the geologic map shown in Figure 17.17. You can see the pattern of the rocks on the earth's surface. Suppose you are going to drill a well at location A. What rock formations will you find when you drill the well? Now look at Figure 17.18. This figure shows the surface and an underground view. This view is based on the geologist's interpretation of the geologic map. A geologic map shows the patterns of the rocks on the earth's surface. From these patterns a geologist usually can interpret what rock layers are underground and how the rock layers lie. By now you should be able to see that there are many clues in the rock record that are used to interpret the earth's history. You must be able to use them all.

Study Questions for 17.4

1. What two things should you look for in making an interpretation of a geologic cross section?
2. What are some features shown on a geologic map?
3. Draw a geologic map showing three horizontal rock layers. What is the earth's history of the place you have drawn?

What Did You Learn?

- Geologists believe that sedimentary rock layers are deposited (laid down) horizontally.
- The law of superposition says that the oldest sedimentary rock layer is on the bottom if the rock layers have not been disturbed.
- The law of crosscutting relationships says that the geologic feature that cuts across other geologic features is younger.
- The principle of uniformity says that the physical and chemical laws of nature are unchanging.
- Unconformities represent a length of time when the rock record was not formed or was being destroyed by erosion.
- In angular unconformities the older rock layers are sloped at a different angle than the younger layers.
- The process of comparing features found in rocks in different locations to find the age relationship between different layers is called correlation.
- Correlation can be accomplished by (1) walking along a rock layer, (2) looking for a distinctive sequence of rock layers, (3) using marker layers, and (4) using fossils.
- Four sedimentary structures providing good clues to conditions at time of deposition are (1) ripple marks, (2) crossbeds, (3) graded bedding, and (4) mud cracks.
- Ocean deposits become finer as they go toward deeper water.
- The order in which layers of sandstone, shale, and limestone appear is a clue to whether the sea was moving toward or away from land at the time of deposition.
- In making an interpretation of the rock record the geologist tries to decide what happened and in what order.
- The earth's history can be interpreted from geologic cross sections and geologic maps.
- By using a geologic map the geologist can determine how the rocks lie below the surface.

Key Terms

law of superposition
law of crosscutting relationships
principle of uniformity
unconformity
angular unconformity
marker layer
index fossil
correlation
sedimentary structures
geologic cross section
geologic map

Biography

Dr. David Lopez (1949–)

Dr. David Lopez, a field geologist, was born in El Rito, New Mexico, in 1949. While he was still a small child his family moved to Grand Junction, Colorado. He grew up loving the outdoors and has enjoyed fishing and hunting all of his life. His mother says he was a great rock collector as a child.

He started his career in geology when a college advisor suggested that he change from a forestry major to a geology major. He did very well in his studies in the new area. However, a summer job in the field really convinced him that work in geology was to be his future. Dr. Lopez was an outstanding student and received a number of scholarships during his college years. He completed his Ph.D. in 1981 at the Colorado School of Mines, Golden, Colorado.

Dr. Lopez's work experience includes several years as a field geologist doing research for the United States Geological Survey in Idaho and Montana. This work was primarily a mapping study that included an appraisal of the mineral and petroleum potential in this area. Currently he is working for Montana Power Company, Billings, Montana, exploring for petroleum in southwest Montana, Idaho, and northern Utah.

His love of field geology is based on a desire to know how things fit together. He enjoys figuring out the three-dimensional relationships between and within rock formations. He says that he has to develop a model of how things fit together before he feels right.

TO THINK ABOUT AND DO

Vocabulary

In your notebook unscramble the letters to form the word indicated by the explanation.

1. deposit covering a wide area
   R K A M E R   R L Y E A
2. comparison of the ages of rocks in different locations
   O L A E C R T I O R N
3. distinctive fossil
   D N I X E   S L I F S O
4. lapse of time
   C F Y U O N R I T O N M

What Do You Remember?

1. State the law of superposition in your own words.
2. Find an example of something in your classroom that is arranged according to the law of superposition.
3. Explain this sentence: "The youngest event cuts across older events."
4. Describe an example that illustrates the principle of uniformity.
5. Why has so much of the rock record been lost?
6. Describe how rock layers can be correlated.
7. Explain what is meant by the fact that the eruption of Mount St. Helens laid down a marker layer of volcanic ash.
8. Describe an index fossil.
9. Whenever you see sedimentary rock layers on land, you know three occurrences in the earth's history at the location. What are they?
10. List four types of geologic clues found in the rock record.

Do You Know?
On May 18, 1980, Mount St. Helens erupted. A marker layer of volcanic ash 60 cm deep covered more than 1,500 km^2. This ash added a time line to the rock record.

Applying What You Have Learned

1. Describe how an unconformity could occur between two adjacent layers of sedimentary rock.
2. Interpret the cross section shown below left.

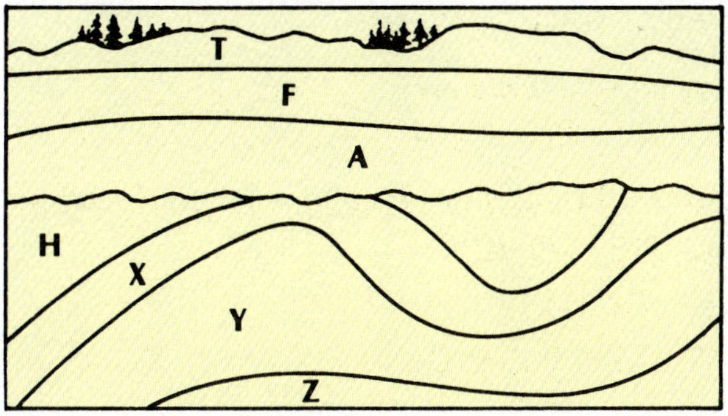

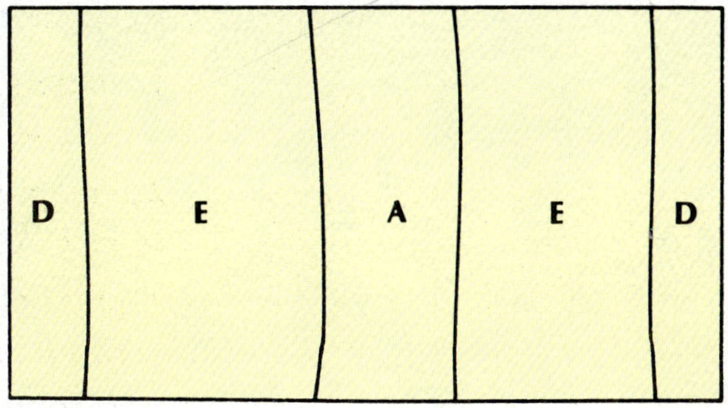

3. Interpret the geologic map shown above right.
4. Draw a geologic map showing an unconformity. Label the age of two rock bodies.
5. Suppose you found the following sequence of sedimentary rock layers—shale, limestone, shale, limestone, shale. What was happening to cause this pattern of rock layers?

Research and Investigation

1. Use references in the library to find a description of the earth's history where you live.
2. Obtain a geologic map of your community. Make a list of the rock layers that lie under your school and your home.
3. If you live in an area of sedimentary rocks, try to find a sequence of sedimentary layers in your community that indicates the movement of the sea toward or away from land.

18 Dating and Geologic Time

Time has no natural divisions. There is nothing in nature that rings or buzzes to tell when an hour, a year, or a century has gone by. The ancients created elaborate observatories to mark the passage of time. They wanted to know when to plant crops or when to prepare for winter. They were concerned with the seasons and two basic units of time, the day and the year.

Does nature keep a record of what has happened on the earth in the past? If so, what kind of record does nature keep? What is found in nature that will help determine the age of a rock, the number of years it takes to build a mountain, or even the age of the earth itself?

In this chapter you'll learn about geologic time and how geologists mark the passing of time.

Figure 18.1 Stonehenge is the mysterious remains of an ancient calendar, built in England over 3,000 years ago.

18.1 The Concept of Geologic Time

You will find out
- what the requirements for a geologic clock are;
- how the age of the earth has been estimated in the past;
- how the age of the earth is estimated today.

Figure 18.2 There are 21 sedimentary layers exposed in the Grand Canyon. Stacked one above the other, these strata give a view of the earth that stretches back a billion years.

A key difference between geologists and other scientists is their attitude toward time. Physicists and chemists make measurements in their laboratories over fractions of a second, a few minutes, or a few weeks. Geologists make measurements of the earth processes that take up to millions and even billions of years. For example, assume that the Colorado River deepens its channel one tenth of a centimeter per year. In 1,000 years it would cut down one meter. To cut the Grand Canyon (1.7 km deep), it would take only about two million years. Even that is only a brief instant in geologic time. In reality it is believed the Colorado River began its task about 6 million years ago. How do you measure time over millions and billions of years? What kind of clock would you use?

This was the kind of problem facing early geologists. They wanted to know the age of the earth and the length of time for a given geologic event. For example, how long did it really take for the Colorado River to make the Grand Canyon in Arizona? How could these early geologists determine geologic time? They wanted a clock for geologic events.

In a large clock you can see certain things happening. The pendulum swings back and forth with great regularity. The hands of the clock count the number of swings to tell how much time has passed. Geologists have looked for things that have happened with regularity on the earth. This regularity would compare with the pendulum in the clock, and each occurrence could then be counted to measure geologic time. The search for such a clock focused on finding a series of actions or happenings that met certain requirements. First, the actions had been operating since the earth began. Second, the actions had never varied in their rate of operation. Third, the rate of the actions had to be measurable. Early geologists thought of three earth processes that could meet each of these requirements. The processes were (1) the rate at which salt was added to the oceans, (2) the rate of deposition of sediments in the sea, and (3) the rate the earth loses heat.

In 1899 John Joly estimated the age of the earth by using the rate at which salt was being added to the ocean. He estimated the amount of salt in the ocean and the rate salt was being added to the ocean each year. He divided the amount of salt in the ocean by the rate it was being added. From this he obtained an age for the earth of about 90 million years. But really all he was measuring was the age of the ocean. Most scientists today do not think that the ocean salt is a good clock for measuring the true age of the earth.

The rate at which sediments were being deposited was another process used for determining the age of the earth. In the late 1800's five scientists made estimates of the age of the earth using this method. Each of these scientists estimated the total thickness of sedimentary rock found on the earth. Each used his experience and knowledge to estimate the rate of deposition. By dividing the total thickness by the rate of deposition, each estimated the age of the earth. There were wide differences in their estimates, ranging from 17 million years to 1,584 million years. The differences occurred because there was no accurate way to measure maximum thickness or rate of deposition.

A third idea for a clock to estimate the earth's age was based on the rate the earth was losing heat. Lord Kelvin, a physicist, assumed the earth had been losing heat from its beginning. He measured the rate at which the earth was losing heat. He also estimated how hot the earth had to be for rocks to be melted. Using these two estimates, Lord Kelvin calculated how long it would take the earth to cool. His estimates for the age of the earth ranged from 20 to 100 million years.

Figure 18.3 Finding the right clock takes time.

Figure 18.4 *below:* One half of the amount of radioactive material present will decay when the half-life passes.

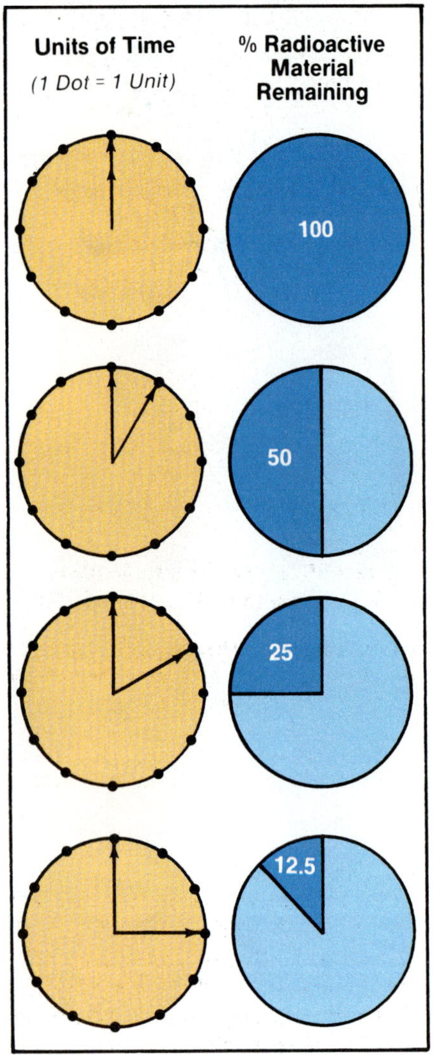

Geologists now use **radioactive decay** [ray-dee-oh-AK-tihv dee-KAY] processes to estimate the age of the earth. Radioactive decay is a natural, regularly occurring process in which certain radioactive elements change into different elements. This happens as their atomic nuclei break down and simpler atoms are formed. The rate of change for different elements varies. Scientists base their measurements on the length of time it takes half the atoms in a sample to change. The amount of time it takes for half the atoms in a sample to change into simpler unchanging atoms is very regular. This period of time is called the **half-life** for that element. The half-life of a radioactive element, then, because it is always the same, is like a clock. It is an atomic clock that can be used to measure the age of rocks and also the age of the earth. Atomic clocks are not affected by changes in temperature, by weathering, or by erosion.

How an atomic clock works is shown in Table 18-1. This table shows that the number of atoms of A and B changes through time. Compare the amount of A atoms to the amount of B atoms. Study the changes taking place in the ratios. Which type of atom is decreasing in number? By how much is it decreasing? What is happening to the A atoms? They change to B atoms. This process could continue for a long time.

Table 18-1 Radioactive Decay

Ratio of A Atoms to B Atoms	1:0	1:1	1:3	1:7	1:15	1:31	1:63
Time in Half-life Units	0	1	2	3	4	5	6

■ = A Atoms
□ = B Atoms

To see how the age of a rock is estimated, imagine a rock which contains both uranium 238 and lead 206. The numbers 238 and 206 refer to forms of these elements. It takes 4.5 billion years for half of the uranium 238 atoms to turn to lead 206. If a rock has equal amounts of uranium 238 and lead 206, how old is it? It is 4.5 billion years old. If a rock contains three quarters of uranium 238 and one quarter of lead 206, how old is it? The answer is marked on Figure 18.6. Since only one quarter of the uranium 238 has changed to lead 206, the age is one half of the half-life, or about 2.25 billion years. The regular rate of decay of radioactive elements in a sample has provided geologists with the accurate geologic clock they were looking for.

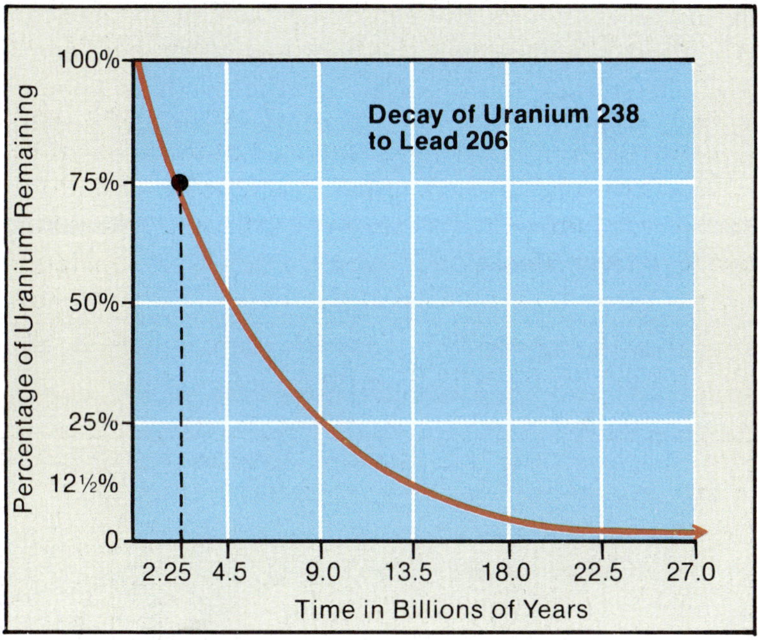

Figure 18.5 *above:* The radioactive decay of uranium 238 into lead 206

Figure 18.6 *left:* During radioactive decay, uranium 238 changes to lead 206 in the ratios shown here.

Do You Know?
Antoine Henri Becquerel discovered radioactivity in 1895. It wasn't until 10 years later in 1905 that Ernest Rutherford suggested that radioactive minerals could be used to date rocks.

Study Questions for 18.1

1. What three requirements must be met by an earth process before it can be used as a geologic clock?
2. List three ways the age of the earth has been calculated in the past.
3. What method is used today to estimate the age of the earth?

18.2 Geologic Dating

Figure 18.7 Atomic measurements indicate that this rock is over 3.9 billion years old. It is one of the oldest earth rocks ever found.

You will find out

- what methods are used today to date rocks;
- what the difference is between relative and radiometric, or absolute, dating.

How can you measure the age of the earth? One way geologists use is to treat the whole earth as a great big rock. Estimates of the present ratio of uranium 238 to lead 206 is about 1 to 1. In other words, about half of all the uranium 238 atoms have changed to lead 206 atoms. From Figure 18.6 you can see that the time it would take for this to occur is 4.5 billion years. This is one half-life of uranium 238 and makes the age of the earth about 4.5 billion years old. How does this age of the earth compare with the age found by measuring the salt in the seas? The earth's sediments? The heat loss? Since the age of the oldest rock found so far on the earth is about 4 billion years old, the atomic clock seems to provide more accurate information.

ACTIVITY

How Old Is This Model Rock?

Materials

boxes labeled *model rock,* each containing 80 pieces of paper (some of the pieces of paper are marked *atom A*, some *atom B*)
pencil

Procedure

1. Select one of the model-rock boxes.

2. Count the number of *A* atoms.

3. Count the number of *B* atoms.

4. Divide the number of *B* atoms by the number of *A* atoms. This is the ratio of *B* atoms to *A* atoms.

5. Use Table 18-1 to find the estimated age of your model rock.

Questions

1. How many atoms *A* and atoms *B* do you have?

2. What is your ratio of *B* atoms to *A* atoms in step 4?

3. How old is your rock?

When you determine the age of a rock using an atomic clock, you are using **radiometric** [ray-dee-oh-MET-rihk] **time.** For example, if you find that the amount of radioactive decay of atoms in a rock would require 100,000 years, then the rock has existed for 100,000 years of radiometric time. When you place a date or length of time on an event by using an atomic clock, it is called radiometric, or absolute, dating.

Do You Know?
Radiometric dating of rocks from the moon range from 3.1 to 4.5 billion years old. There are none younger.

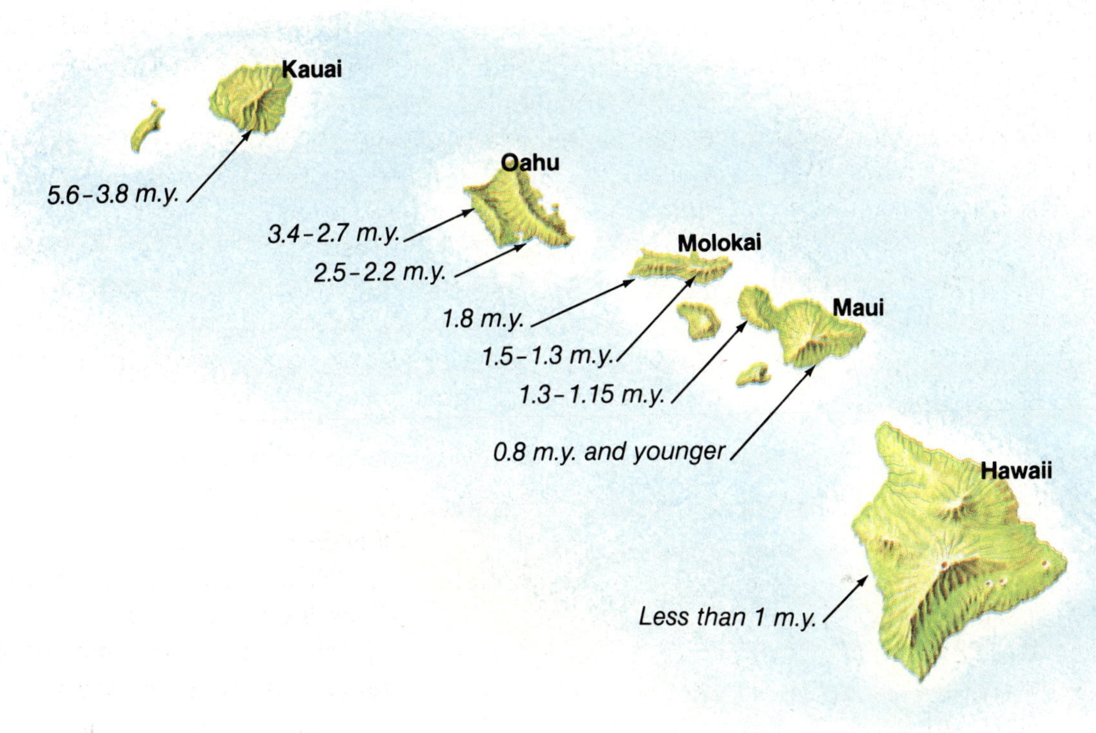

Figure 18.8 The Hawaiian Islands formed individually as their crustal plate gradually drifted northwest over a hot plume. The islands in the northwest erupted first.

The most widely used atomic clock is the potassium 40-argon 40 clock. Small amounts of potassium 40 are found in many rocks. It takes 1.3 billion years for half of the potassium 40 atoms to change to argon 40 atoms. In other words, the half-life of potassium 40 is 1.3 billion years. The amount of potassium 40 atoms that have changed can be used to date rocks as young as a few thousand years as well as the oldest rocks known. The potassium 40-argon 40 atomic clock was used to determine the age of the rocks making up the Hawaiian Islands in Figure 18.8.

Figure 18.9 *above:* Radiometric dating can identify how long ago once-living organisms existed.

An atomic clock for determining the age of material that was once part of a living organism is called **radiocarbon** [ray-dee-oh-KAHR-buhn] **dating.** Radioactive carbon 14 is created in the atmosphere. Carbon 14 atoms enter into all living things and slowly decay to nitrogen 14 atoms. As long as life exists, there is a ratio maintained in the organism between carbon 14 atoms and ordinary carbon 12 atoms. New carbon 14 atoms are added as fast as some of the old ones decay to nitrogen 14 atoms. When death occurs, carbon 14 atoms are no longer added and the clock is started. Then the ratio between unchanging carbon 12 atoms and carbon 14 atoms begins changing. In 5,730 years, the half-life of carbon 14 atoms, one half of the carbon 14 atoms will be changed to nitrogen 14 atoms. During this time carbon 12 atoms do not change. The amount of carbon 14 remaining in a dead object can be compared to the amount of carbon 12. If you find one half the ratio of carbon 14 to carbon 12, then half of the carbon 14 has changed to nitrogen 14. You then know that 5,730 years, the half-life of carbon 14, have passed since the organism died.

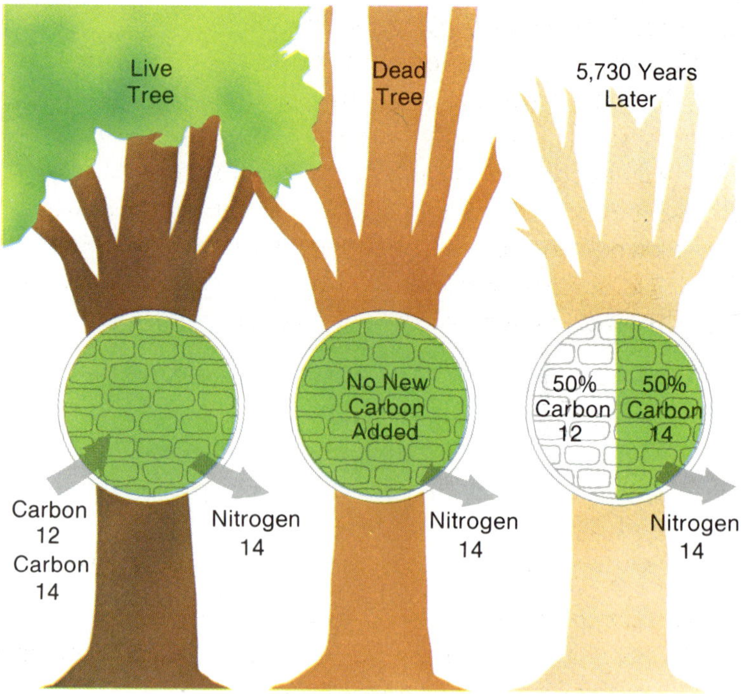

Figure 18.10 *left:* In a live organism, carbon 14 decay is balanced by newly added carbon. *middle:* When the organism dies, carbon 14 continues to decay, but no new carbon enters the organism. *right:* The percentages of carbon 12 and carbon 14 remaining reveal the age of the organism.

Another way of looking at geologic time is to compare two events by asking which event came first. When you do this, you are thinking of **relative time** and relative dating. When you apply the law of superposition, you are using relative dating of events. You don't know when the events happened or how long they lasted, but you do know in what order they occurred.

Relative dating is used to find the "before" and the "after" time of a rock. Radiometric dating can be used to tell the age of a rock in years. In Figure 18.11 you can see how relative and radiometric time can be used to figure out the earth's history. You know the relative age of the rocks: first, rock A was deposited, and then rock B cut across it, followed by the deposition of rock C, and finally the deposition of rock D. Suppose your atomic clock has shown that rock B is 90 million years old and rock D is 15 million years old. How old are rock A and rock C? You won't know the age of rock A, but it must be older than 90 million years. Now you also know that rock C is younger than 90 million years and older than 15 million years. By using both radiometric time and relative time, you can add dates and time spans to the geologic history of that cross section.

Why couldn't the ages of rock layers C and A be measured by the atomic clock? Unfortunately, not all rocks contain radioactive substances. Sometimes too much of the radioactive substance has been lost since the rock first formed to permit atomic measurement. Atomic clocks work only if radioactive substances are present in the rock. You must use every clue you can find to make your interpretations of the rock record.

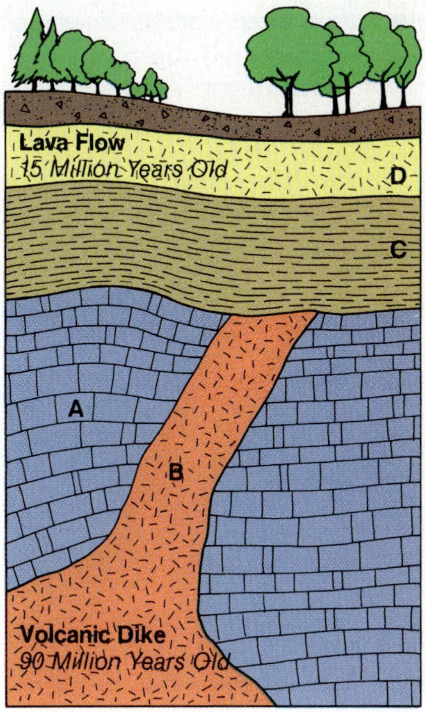

Figure 18.11 How do relative dating and radiometric dating help identify the history of this geologic sample?

Study Questions for 18.2

1. List the methods used to date rocks today.
2. Describe the difference between relative dating and radiometric, or absolute, dating.

18.3 Dimensions of Geologic Time

Figure 18.12 *above:* William "Strata" Smith

Figure 18.13 *below:* An example of a geologic column

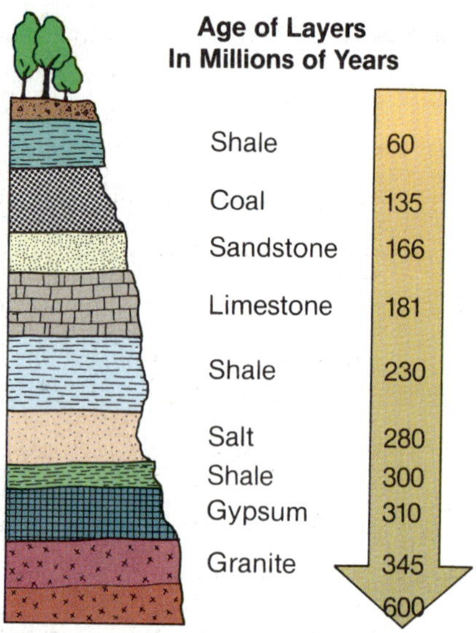

You will find out

- what a geologic column is;
- how the geologic time scale was developed;
- what the names of the eras in the geologic time scale are.

How long is geologic time? It has to be a long, long time, since it began when the earth formed. There is some evidence that it was at least 4.5 billion years ago. Geologic time encompasses such a vast interval that it's too much to comprehend fully. Humans can keep track of events in their lives for about 50 years. After that, they usually rely on some method of recording events to keep track of what happens around them. Geologists can use the information that is recorded in rocks. That is a record of events that have happened on the earth since the beginning of time.

The first relatively complete part of the geologic record was compiled by William "Strata" Smith. Smith made the first geologic map of England. He was able to place the rock layers into a sequence in order from oldest to youngest. Thus he made a type of relative time scale. Later, other geologists working in France, Germany, and Russia also began to order the rocks. At each place the relative time scale of the rock record was worked out. With the discovery that rocks containing similar kinds of fossils were the same age, the relative sequence of the rock layers in all of Europe was put in order. Geologists call such a sequence of layered rocks arranged in order a **geologic column.**

Many attempts were made to divide the geologic column into blocks of time. Finally, by international agreement, the rock record was divided into blocks of time, creating a relative geologic time scale. Later, by using radiometric dating, dates in years were added to the scale. Thousands of rocks were dated. This allowed geologists then to combine the relative geologic time scale with a radiometric, or absolute, scale. All this information has been used to make up the geologic time scale.

The geologic time scale, as shown in Figure 18.14, is divided into **eras** [IHR-ruhs], **periods,** and **epochs** [EP-uhks]. Eras are the largest time blocks of the scale. There are four eras. They are in the left-hand column of the time scale. Periods are blocks of time within an era. The periods are in the middle column. Epochs are blocks of time within a period. The epochs are in the third column from the left. The eras were named to indicate the fossils they contained. **Cenozoic** [see-nuh-ZOH-ihk] refers to the time of recent life. **Mesozoic** [mez-uh-ZOH-ihk] refers to middle life. **Paleozoic** [pay-lee-oh-ZOH-ihk] is the time of ancient life. **Precambrian** [pree-KAM-bree-uhn] refers to the time before ancient life.

Do You Know?
If the geologic time scale was plotted alongside the Empire State Building (381 m high), a dime (1/8 cm thick) would more than equal the span of recorded history.

Figure 18.14 The divisions of the geologic time scale

Geologic Time Scale			Time* (Millions of years)	Duration (Millions of years)
Era	**Period**	**Epoch**		
Cenozoic	Quarternary	Holocene	last 10,000 years (approx.)	
		Pleistocene	2-3	1
	Tertiary	Pliocene	12	10
		Miocene	26	14
		Oligocene	37-38	11
		Eocene	53-54	26
		Paleocene	65	12
Mesozoic	Cretaceous		136	71
	Jurassic		190-195	54
	Triassic		225	35
Paleozoic	Permian		280	55
	Carboniferous	Pennsylvanian	320	40
		Mississippian	345	25
	Devonian		395	50
	Silurian		430-440	35
	Ordovician		500	70
	Cambrian		600	100
Precambrian				3,900

*Estimated ages of time boundaries

Another way of visualizing geologic time is to think of all the time since the earth formed at least 4.5 billion years ago compressed into a year. How would the rock record appear on that scale? If the earth began January

ACTIVITY

What Does a Time Line of the Geologic Scale Look Like?

Materials

5 m of adding-machine tape
pencil
metric ruler

1. 3.9 billion years
2. 50 million years
3. 160 million years

Procedure

1. Place a mark labeled *today* near one end of the tape.
2. Place another mark labeled *beginning* at 4.5 m from *today*.
3. Using a scale of 1 m = 1 billion years, compute the scale distance for the beginning of each era and period. See Figure 18.14 for dates.

$$\text{Scale distance in meters} = \frac{\text{number of years ago}}{1{,}000{,}000{,}000}$$

4. Using the scale values, plot each era and period on the tape.
5. Label each era and period.

Questions

1. How long did the Precambrian Era last?
2. How long was the Devonian period?
3. How long was the Mesozoic Era?

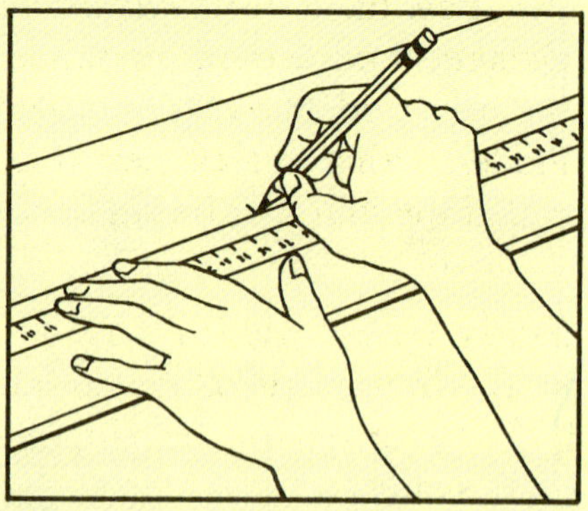

1, living things appeared in the sea in May. Land plants and animals showed up in the rock record in November. The swamps that formed most of the coal existed for about four days in early December. Dinosaurs appeared in mid-December and disappeared on December 26. This is about the time the Rocky Mountains were uplifted. On this scale, the first humans appeared during the evening of December 31. The last of the big glaciers receded about one minute and fifteen seconds before midnight on December 31. Rome ruled the world for five seconds. Columbus discovered America about three seconds before midnight.

This way of looking at geologic time is based on the best inference geologists can make with the evidence they now have. With each new piece of evidence geologists reexamine their models and inferences. Sometimes the new evidence confirms and supports their previous model. Other times new inferences and models are called for that can be supported by the new evidence.

Figure 18.15 One geologic theory states that the age of the earth is about 4.5 billion years. This age can be condensed to show the rock record on a relative scale of one year.

Do You Know?
How long is 5 billion years? If you were to begin counting at the rate of one number a second and continued 24 hours a day for 7 days a week and never stopped, it would take two lifetimes (150 years) to reach 5 billion.

Study Questions for 18.3

1. What is a geologic column?
2. Describe the development of the geologic time scale.
3. List four eras of the geologic time scale.

What Did You Learn?

- To serve as a geologic clock, an earth process must (1) have been operating since the earth began, (2) never have varied in its rate of operation, and (3) have a rate that is measurable.
- In the past the age of the earth was measured (1) by estimating how long it had taken for the oceans to become as salty as they were, (2) by how long it had taken to deposit the total thickness of sedimentary rock, and (3) by estimating how long the earth had been cooling.
- Today geologists use radioactive decay processes to estimate the age of the earth and to date rocks.
- The time it takes for half of the atoms in a radioactive substance to change to simpler unchanging atoms is called the half-life.
- Geologists have discovered that atomic clocks are not affected by changes in temperature, by weathering, or by erosion.
- There is evidence that the age of the earth is about 4.5 billion years. This is based upon the radiometric dating of rocks containing uranium 238, which changes to lead 206.
- When you determine the age of a rock by using an atomic clock, you are using radiometric dating.
- The radiocarbon clock is used to date things that were once living.
- Relative time is determined by determining which event happened before or after another.
- In some situations both relative and radiometric times are used in interpreting geologic history.
- Geologists make measurements of earth processes that take up to millions and even billions of years.
- Geologists call a sequence of layered rocks arranged in order a geologic column.
- Radiometric dating has made it possible to place dates in years on the geologic time scale.
- The geologic time scale has four eras: Precambrian (before ancient life), Paleozoic (ancient life), Mesozoic (middle life), and Cenozoic (recent life).

Key Terms

radioactive decay
half-life
radiometric time
radiocarbon dating
relative time
geologic column
era
period
epoch
Cenozoic
Mesozoic
Paleozoic
Precambrian

Career

Research Engineer Scientist in Radiocarbon Dating

Persons working in radiocarbon dating use radiocarbon dating techniques to date archeological objects, antiques, charcoal samples, bones, shells, and wood samples. Radiocarbon dating techniques are usually applied to objects that are thought to be less than 40,000 years old. A research engineer scientist uses special chemical treatment techniques in preparing the samples for dating. This work requires at least a bachelor's degree in mathematics and chemistry.

Alejandra Gloria Varela is a research engineer scientist working at the Radiocarbon Dating Laboratory at the University of Texas in Austin. She attributes her interest in science and mathematics to her mother, who held daily classes with the children of the family in their early years. In this class her mother would teach the children to read, write, and do mathematics. Alejandra knew the multiplication tables before she entered first grade. In high school she had an outstanding biology teacher who helped develop her math and science talents by setting up special projects and extra assignments for Alejandra.

In college she took as much mathematics and chemistry as possible. This prepared her for dating archeological materials, antiques, charcoal samples, bones, shells, and wood. Each substance must be treated chemically before its age can be determined. Calculating the age of a substance requires the application of her mathematics.

Alejandra has had more than ten articles published in scientific journals. She finds a special reward when she learns that other scientists refer to her work.

TO THINK ABOUT AND DO

Vocabulary

On a separate piece of paper complete the statements below.

1. The change of one substance into another by the breakdown of atomic nuclei is called ________.
2. One event happening before or after another is an example of ________.
3. Determining the age of a rock through an atomic clock is an example of ________.
4. An ordered sequence of layered rocks makes up a ________.

What Do You Remember?

1. Describe how the age of the earth was estimated by using the salt content of the ocean.
2. What method of estimating the age of the earth was proposed by Lord Kelvin?
3. Define *half-life*.
4. What is the most widely used atomic dating procedure clock?
5. Which atom decays—carbon 12 or carbon 14?
6. Define *geologic column*.
7. Which geologic era lasted the longest?
8. Who was "Strata" Smith? What did he do?

Applying What You Have Learned

1. How old is a rock that contains 1/16 potassium 40 and 15/16 argon 40?
2. Explain how the application of the law of crosscutting relationships provides the geologist with relative time.
3. In this drawing, about what age is layer 1? Layer 2? Layer 4? Layer 5? Layer 7?

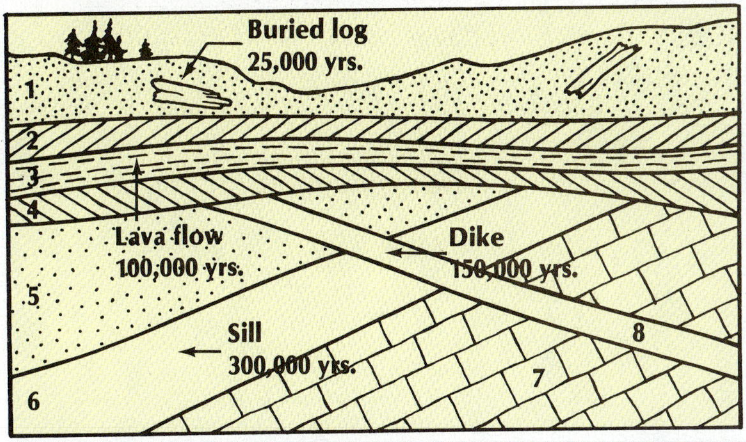

Research and Investigation

1. Have the students use the library to find the rocks making up the geologic column in their area.
2. Plot the geologic time scale in a stairwell of your school by determining the number of steps for each era and period.
3. Assign an oral report on the problems of radioactive dating.

The oldest moon rock ever found is dated as being over 4 billion years old.

19 The Fossil Record

Can you read the story of these footprints? They were made by a human more than 6,000 years ago. The rock in which the footprints were found is volcanic ash. Look at the footprints carefully. Look at the toe prints. Look at the heel prints. What was this person doing? Why was this person making these footprints? Was the person a woman or a man or a child? How big was the person? How much did the person weigh? Paleontologists say that these footprints show a human being running through muddy volcanic ash. The person may have been trying to escape an ash fall from a nearby volcano. Do you think this is a good explanation?

These fossil footprints as well as every other fossil have a story to tell. The fossil record is really the story of the history of life on planet Earth. In this chapter you'll be able to read some of the fossil record of the animals and plants that have lived on the earth.

Figure 19.1 Was this person running from a volcanic threat?

19.1 Types of Fossil Preservation

You will find out

- what the five types of fossil preservation are;
- what the most likely conditions are for making a fossil?

The record of life on the earth is preserved in **fossils** [FAHS-uhls]. A fossil is any evidence of past life. Fossils come in many different sizes. A fossil may be a piece of bone, a shell, a tooth, or a footprint. It can even be a piece of animal flesh frozen in ice. How can material that was once alive be preserved for millions and millions of years? How are fossils made?

Figure 19.2 Fossils provide pictures of living things as they looked long ago.

Fossils can be made in several ways. The five types of preservation are (1) preservation without change, (2) replacement by a mineral, (3) the filling of a hollow space in a shell or bone with a mineral, (4) formation of a thin carbon film, and (5) formation of an imprint or the filling of an imprint. How did each of these ways of preservation occur?

When an organism dies, parts of it may be preserved as a fossil without change. Bones, shells, and teeth are examples of such fossils. Sometimes the whole organism becomes a fossil. An example is an insect trapped in tree sap. As the tree sap ages and hardens, it forms a transparent yellow-orange fossil, which is called **amber** [AM-buhr]. The insect trapped in the tree sap is now clearly seen embedded in the transparent amber. These fossil insects often have nearly perfect forms. Animals that have been frozen in ice can also be preserved with hair and flesh intact. In some fossils even the food the animal was eating is preserved.

Sometimes the once-living organism is replaced by a mineral. **Petrified** [PET-ruh-fyd] **wood** is made in this way. The original wood parts of a tree are removed cell by cell, and the mineral silica is substituted. This process can preserve remarkable detail. Notice you can count the tree rings in Figure 19.4.

Figure 19.3 *below:* An insect is perfectly preserved in amber.

Do You Know?
Thomas Jefferson, in the spring of 1797, carried some bones from a cave floor in western Virginia to Philadelphia. There he wrote a memoir describing the specimens. The memoir, published in 1799, was the first study of American fossils by an American.

Figure 19.4 Wood can become petrified when conditions preserve the cell materials from rapid decay.

At times small open spaces in a shell or bone are filled with a mineral. If you pick up a fossil bone, you'll notice it is heavier than you might expect because of the added mineral material. The mineral matter hardens and preserves the bone as a fossil.

Figure 19.5 *right:* A bone fossil preserved with added mineral matter

Some fossils appear as thin films of carbon in a rock. Fish, leaves, and insects are found preserved in this way. After the organism is buried, it is pressed by forces within the earth. Under this pressure the organism leaves a thin film of carbon in the rock. When the rock is cracked open, the thin carbon fossil is discovered. See Figure 19.6 and Figure 19.7.

Figure 19.6 *below:* A carbon fossil within this rock is a lucky find for a geologist.

Figure 19.7 *below right:* Another example of a carbon fossil

Traces of past life may be preserved as an imprint in rock or as a filling of an imprint. Footprints, shell imprints, and worm burrows are fossils of this type. **Molds** and **casts** are examples of these kinds of fossils. When water dissolves a shell or bone around which sand or mud has already hardened, a cavity is left in the rock. This cavity is called a mold. If the mold is later filled with mud or mineral matter, the filling is called a cast. A cast can be identified because it has no inner structure of the original organism.

Figure 19.8 *above left:* The imprint of a shell is a pretty reminder of a once-living sea animal.

Figure 19.9 *above:* The imprint on the left is the cast. The one on the right is the mold.

Generally, fossils form best under certain conditions. First, hard parts of the plant or animal are more easily preserved. Second, the plant or animal should be buried or frozen soon after death. As a result, only a few of all the plants and animals living in the past have become fossils. After death the soft parts of plants and animals are removed by scavengers, bacteria, and chemical decay. The hard parts are not so easily destroyed and are more likely to become fossils. Shells, bones, teeth, and skulls are the parts of animals most likely to be preserved. An animal such as a clam has hard parts and lives in the mud or sand on the bottom of a lake or an ocean. When the clam dies, it is already buried. The soft parts may be destroyed, but the shell remains for a long time and may become a fossil.

Study Questions for 19.1

1. What are the five types of fossil preservation?
2. What are the most likely conditions for making a fossil?

19.2 Recognizing and Identifying Fossils

You will find out

- how to recognize fossils;
- how to identify fossils.

Figure 19.10 *top:* The modern horseshoe crab is sometimes called a living fossil. *bottom:* Compare its resemblance to an ancient horseshoe crab.

Figure 19.11 Geologists identify unknown fossils by matching them to organisms which are familiar.

In order to tell whether something you've found is a fossil, you must be able to tell whether or not it was alive. Look for pieces of bone or shell or for other evidence that tells you what you found was once alive.

Geologists sometimes say, "I know this is a fossil because it looks like . . ."—a whole or part of an organism that was once living. Geologists identify a fossil by the features it has. A fossil horseshoe crab looks much like the horseshoe crab living today. A fossil snail and a modern snail also have a similar appearance. The major key to whether or not something is a fossil is its appearance. Generally it looks like something you know that is living today. The more you know about living things, the easier it is to recognize a fossil.

Once you know something is a fossil, how do you find out what kind of fossil it is? How is it identified? In Figure 19.11 you see three fossils. Opposite shows three sketches of what the organisms these fossils came from are believed to have looked like. Can you match the fossils with the sketches? One way you can identify a fossil is by matching a fossil with a sketch of a once-living organism.

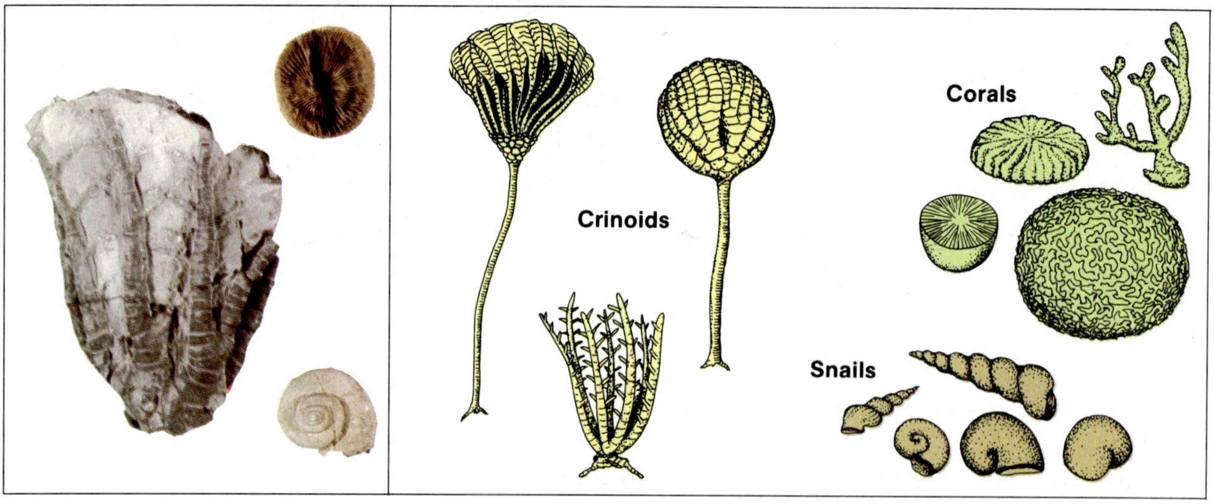

ACTIVITY

Which Molds Look like Once-Living Things?

Materials

plaster of Paris
plastic dish
water
two objects: something once alive and something that was never alive
petroleum jelly
paper towels
stirring rod
pencil
containers

Procedure

1. Coat each object with petroleum jelly.

2. Make a thick paste of the plaster of Paris.

3. Place the objects in the plaster of Paris and set aside for 24 hours.

4. Remove the objects and clean the molds that you have made.

5. Exchange your molds with another student's. Don't tell what your molds are.

Questions

1. Which mold looks like something that was once living?

2. What is it and how do you know?

Study Questions for 19.2

1. What do you look for to tell whether you've found a fossil?
2. Describe a procedure for identifying a fossil.

19.3 Dating Fossils

You will find out

- how groups of fossils can be used in dating another fossil;
- how index fossils are used in dating other fossils.

Figure 19.12 *top:* Can you determine when this cephalopod lived?

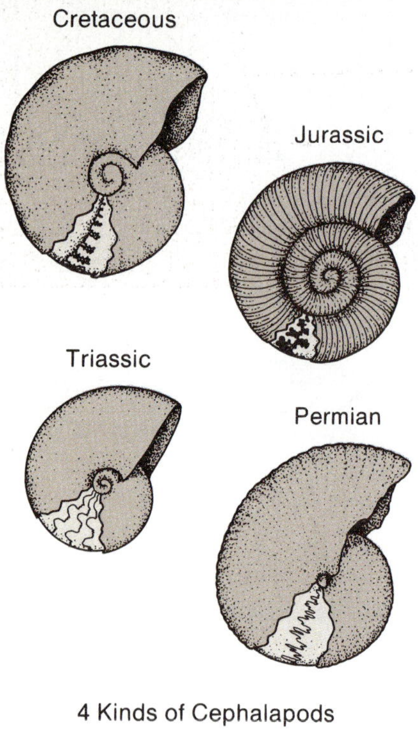

Figure 19.13 *bottom:* Match the unknown fossil in Figure 19.12 to one of these four cephalopod types.

Fossils do not occur with any signs telling their ages. Yet paleontologists are able to determine the age of a fossil. How do they do it? They must rely on many pieces of related evidence. Even then, the age is an estimate and may be off by many millions of years.

Imagine looking at an old photograph of your family. You may recognize relatives, but they look somewhat different from how you know them. You may want to know when the photos were taken. Suppose your grandfather examines the photo carefully and then says it was probably taken at the 1952 family reunion. What evidence does he find in the picture for his answer? The ages of the people when the photo was taken may help identify the time. The styles of clothes, hair, and car may also help. Using the evidence in the picture, your grandfather can date it. Any other photos of the same group of people who look the same way must also have been taken at about the same time.

Paleontologists do almost the same thing. Whenever they find a group of fossils, they compare these fossils with other groups of fossils. If the fossils in one of the other groups are largely the same, they both probably lived about the same time. Each block of geologic time has its own set of plants and animals that lived at that time.

How can you use this information to date the fossil shown in Figure 19.12? The paleontologist would start by finding in which group of fossils it lived. In Figure 19.13 are four coiled, shelled animals called **cephalopods** [SEF-uh-luh-PODS]. Each of the cephalopods lived at a different period of time. Paleontologists guess this by the fossil groups with which each of these cephalopods has been found. After examining the fossil carefully, you can see that it is like the cephalopod that lived in the **Jurassic** [juh-RAS-ihk] age. Your fossil then is a

Jurassic cephalopod. Now you can determine its approximate age.

Geologists also look for other pieces of evidence called **index fossils.** Index fossils represent organisms that lived only a short time, but whose remains are found in many areas. Because index fossils are unusual, they are special markers to geologists. Figure 19.14 shows how index fossils can be used to date other fossils. If widely separated rock layers contain the same index fossil, the rock layers are probably the same age. Therefore, other fossils in the rock layer would also be the same age.

Figure 19.14 Notice how the rock layers shown below can be matched using index fossils.

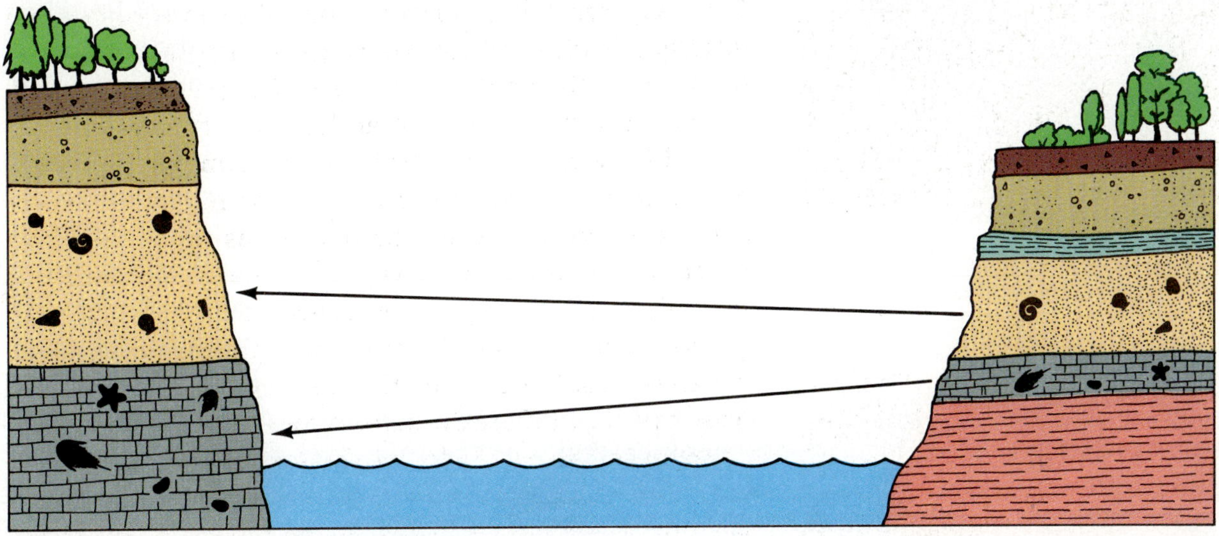

Study Questions for 19.3

1. Describe how the age of one fossil can be determined by using groups of other fossils.
2. Describe how index fossils are used in dating other fossils.

19.4 The Fossil Record at Different Times

You will find out

- why the age of fossils and of the rocks they are found in is the same;
- what the dominant forms of life in the past were;
- why eras came to a close.

Figure 19.15 Soft-bodied fossils decay too quickly to leave fossil evidence.

In Chapter 18 you learned about the geologic time scale and how rocks are dated. Figure 19.16 shows the record of life at different times on this scale. What do geologists believe about life that existed 600 million years ago? What is the source of their information?

Fossils can be dated by the rocks they are found in because the organisms were deposited at the same time the rock was laid down. Thus, the forms of life that existed in Cambrian time are revealed by the fossils found in rocks of Cambrian age. Not all forms of past life are found in the fossil record. The earliest life forms were largely soft-bodied. In addition, much of the fossil evidence may have been destroyed by heat and pressure in older rocks. However, the best evidence of what life existed at any time in the past is found in the fossil record. The fossils provide the best picture of life in the past, and finding new fossils adds to that picture.

Look carefully at the time span for each of the eras

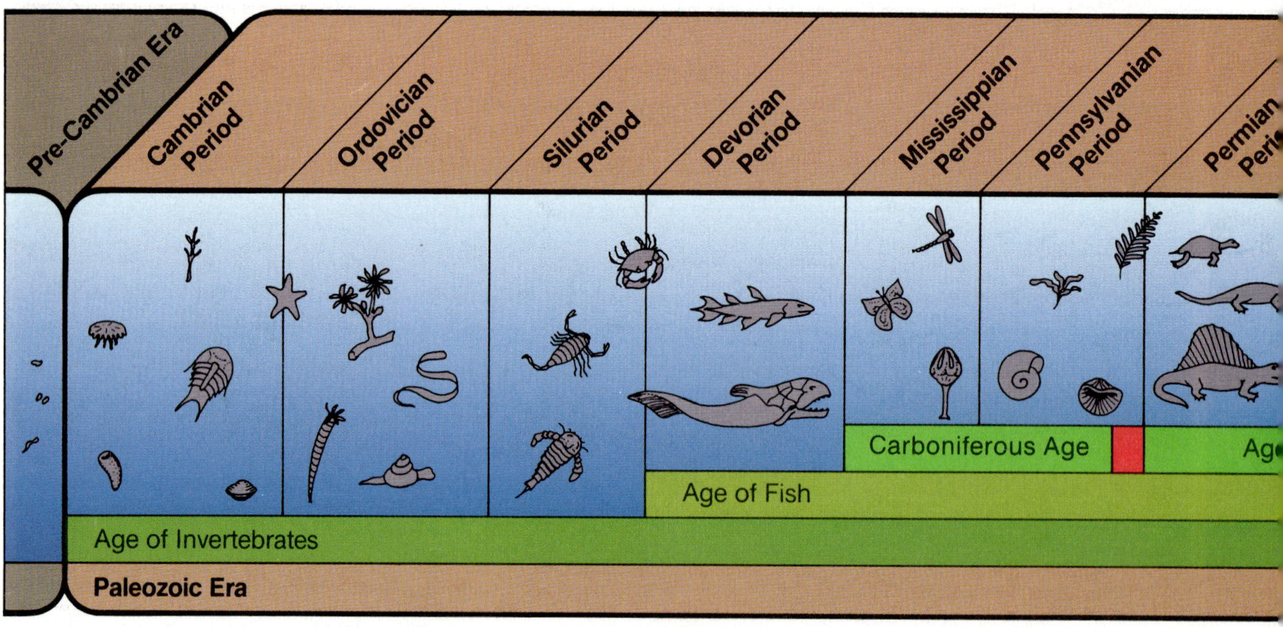

Figure 19.16 The geologic time scale indicates when various life forms appeared on the earth.

and periods of the geologic time scale. There seems to be no order or regularity to this system of dating the eras. What evidence did the geologists have for choosing such random dates? Why did they close out one era and begin another?

The evidence the geologists used is found in the fossil record. Although the record is incomplete, it is the best evidence available. Geologists have been able to make several inferences from this evidence. The first inference is that there were blocks of time when certain groups of animals and plants were the major groups then living. The second is that there were distinct times when many of the plants and animals disappeared. The boundaries of the geologic eras were set to correspond with those times when many of the plants and animals became extinct.

On the geologic time scale you see that the earliest fossils are from the Cambrian Period at the beginning of the Paleozoic Era. During the early periods of the Paleozoic Era nearly all animal types were without backbones. Animals without backbones are called **invertebrates** [in-VUR-tuh-bruhts]. Geologists call this time the Age of Invertebrates. Fossil **trilobites** [TRY-luh-byts] and **brachiopods** [BRAY-key-oh-pods] provide extensive evidence of these early invertebrates.

Figure 19.17 *top:* An ancient trilobite; *bottom:* A brachiopod

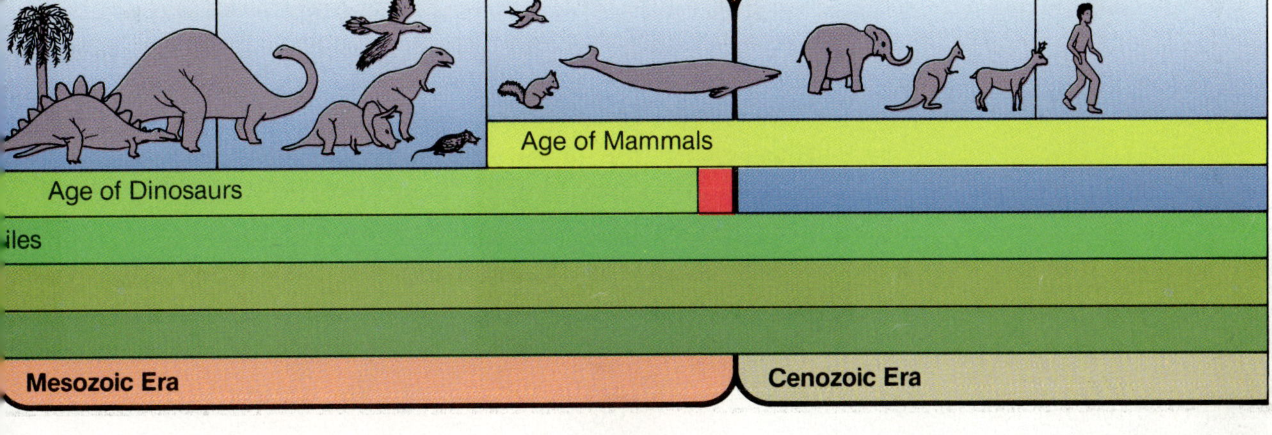

Figure 19.18 The structure of a once-living fish is clearly seen in this fossil.

Figure 19.19 *below:* Dinosaurs ruled the earth for 90 million years before being mysteriously wiped out.

The fossil record in the rocks of the Devonian Period shows that the major forms of life were fish. This period, lasting about 50 million years, is called the Age of Fishes.

The Mississippian Period and Pennsylvanian Period are called the Carboniferous Age. This is the time when plants lived that later became coal deposits. The Paleozoic Era concludes with what geologists call the Permian extinction. For some reason over 90 percent of the shallow-water animals died out.

For the next 160 million years, the Mesozoic Era, the dominant forms of life were the **reptiles** [REP-tyls]. Reptiles are cold-blooded, air-breathing animals that lay eggs. This era is known as the Age of Reptiles. The reptiles came in many sizes and many shapes. Most were plant eaters, but a few were meat eaters. It was during this era that the reptiles known as dinosaurs roamed the world. But about 65 million years ago something occurred that wiped out the dinosaurs and many other marine creatures. This major extinction closed the Mesozoic Era and opened the Cenozoic Era.

For the last 65 million years, the Cenozoic Era, the dominant form of life has been the **mammals.** Mammals are warm-blooded and have hair, and the females nurse their young.

Will the Cenozoic Era also be closed in time with a mass extinction of many organisms? The fossil record suggests this may happen. Geologists don't really know why the earlier eras ended. Why did those times of major extinctions occur? From geologists' knowledge of the earth's history, they can make some inferences. Perhaps the drifting crustal plates and continental changes

ACTIVITY

During Which Geologic Periods Did These Fossils Live?

Materials

pencil
paper
metric ruler

Procedure

1. Look at the shaded column *Late Cambrian time* on the chart. Note that fossil groups A, B, and C lived at that time.

2. Now find the time when fossil groups A, B, C, and D lived together.

3. Find the time when fossil groups A and B, fossil groups C and D, and fossil groups A, C, and D lived together.

Questions

1. During which geologic time(s) did fossil groups A and B live together?

2. When did fossil groups C and D live together?

3. When did fossil groups A, C, and D live together?

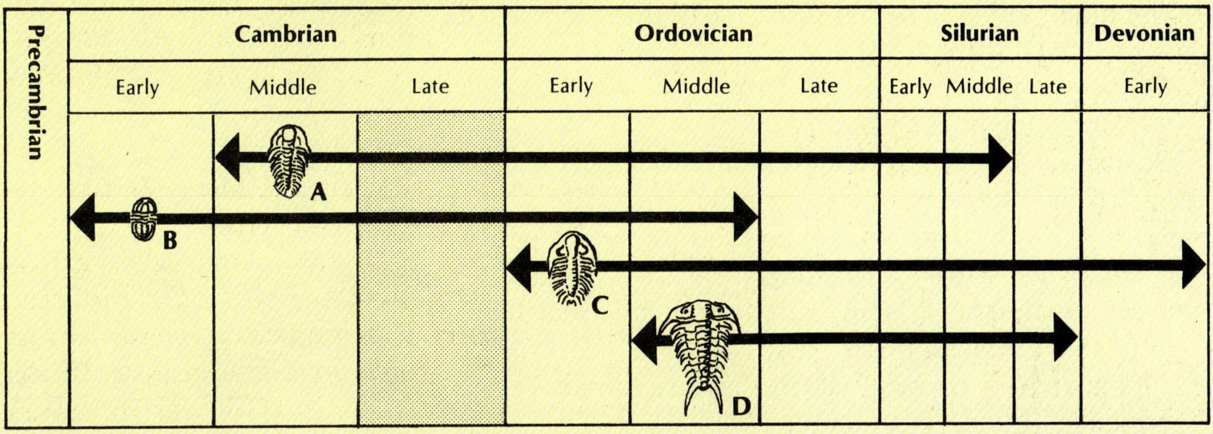

caused major climatic changes. Perhaps a comet or an asteroid collided with earth and caused the great change at the end of the Cretaceous Period. The fossil record leaves many questions yet to be answered. Perhaps you may be the one who solves the mystery of why the dinosaurs disappeared.

Study Questions for 19.4

1. Why is the age of a fossil and of the rock it is found in the same?
2. What were the dominant forms of life in each era?
3. What event caused the close of the Paleozoic and the Mesozoic eras?

What Did You Learn?

- The record of life on the earth is preserved in the fossils found in rocks.
- When an organism dies, it may be preserved without change and become a fossil.
- Sometimes the once-living organism is replaced by a mineral to become a fossil.
- At times small open spaces in a shell or bone are filled with a mineral to become a fossil.
- Some fossils appear as thin films of carbon in a rock.
- Traces of past life may be preserved as a mold or as a cast.
- The most likely conditions for making a fossil are hard parts and immediate burial.
- The major key to whether or not something is a fossil is its appearance.
- One way to identify a fossil is to match it with a sketch of a once-living organism.
- Geologists have found that each block of geologic time has its own set of plants and animals.
- Paleontologists date a fossil by finding out in which group of other fossils it belongs.
- Index fossils are the remains of organisms that lived for a short period of time in a wide area.
- Fossils and the rocks in which they are found are the same age.
- A major extinction of plants and animals occurred to close the Paleozoic and Mesozoic eras.
- Invertebrates are animals without a backbone, and these animals were the dominant life forms in the Paleozoic Era.
- The fossil record in the rocks of the Devonian Period shows that the dominant life forms were fish.
- In Mississippian and Pennsylvanian time, about 80 million years, the dominant life forms were plants.
- The dominant life forms in the Mesozoic Era were the reptiles.
- Mammals have been the dominant life forms during the Cenozoic Era.

Key Terms

fossil
amber
petrified wood
mold
cast
cephalopod
Jurassic
index fossil
invertebrate
trilobite
brachiopod
reptile
mammal

Career

Vertebrate Paleontologist

Vertebrate paleontologists study ancient life forms that had backbones. Fish, amphibians, reptiles, birds, and mammals are animals with backbones. Examples of ancient vertebrate life forms include such animals as dinosaurs, mammoths, mastodons, and pterodactyls. A paleontologist requires a strong background in geology, paleobiology, zoology, and chemistry.

Dr. Douglas A. Lawson, a vertebrate paleontologist, found the fossil bones of the world's largest flying creature, a huge pterodactyl. This flying reptile had a wingspan of more than 16 m and lived in southwest Texas about 65 million years ago. Dr. Lawson was walking up a dry streambed when he noticed some unusual looking bits of rock. These turned out to be fossil bone. Near these fossil bone pieces he found a larger fossil bone, nearly 1 m long. He took some of the fossil bones back to the lab where he and others discovered the bones had been hollow. This information, plus more of the fossil bone and many comparisons with other animals having hollow bones, left no doubt that the bones were that of a huge pterodactyl.

Dr. Lawson made his discovery of the pterodactyl while studying geology at the University of Texas in Austin. Currently Dr. Lawson is working at a petroleum research center in Bartlesville, Oklahoma.

TO THINK ABOUT AND DO

Vocabulary

Unscramble these words.

1. *ssfoli* A (An) ________ is evidence of past life.
2. *eepritl* A (An) ________ is cold-blooded, is air breathing, and lays eggs.
3. *ieeerrattnvb* A (An) ________ is an animal that has no backbone.
4. *mmmlaa* A (An) ________ is warm-blooded and has hair, and the females nurse their young.

What Do You Remember?

1. Can a fossil be a piece of bone? Explain.
2. Give an example of an organism preserved without change.
3. How is petrified wood made?
4. What makes a fossilized bone heavy?
5. Describe how a leaf might become a fossil.
6. List three fossils that are traces of past life.
7. What are the most likely parts of an animal to be preserved?
8. How can you tell whether or not an object you picked up is a fossil?
9. You can identify a fossil by ________________.
10. Paleontologists can date a fossil by finding out in which ______________________________.
11. If widely separated rock layers contain the same ________ fossil, they are the same ________.
12. Explain this statement: Fossils and the rocks in which they are found are the same age.
13. What kinds of fossil evidence are found in the very old rocks?
14. When did the fossil record first show animals with hard parts?
15. What happened at the end of the Age of Reptiles?
16. In which block of geologic time is the fossil record most complete?

Applying What You Have Learned

1. Describe why only a few of all plants and animals living in the past have become fossils.
2. Explain this statement: Whether or not a plant or animal becomes a fossil depends upon what happens after it dies.
3. The major key to whether or not something is a fossil is ________. Explain.
4. How do paleontologists know what forms of life existed in the past?

Research and Investigation

1. Make a collection of the fossils found in your area. Label and mount the collection for display.
2. Add the fossil record to the time line you plotted in Chapter 18. Use sketches to show the forms of life at different times.
3. Make a library report on the reasons scientists give for the disappearance of the dinosaurs.
4. Write a description of Precambrian fossils based on information found in an encyclopedia.

Can this be a footprint of an ancient dinosaur?

Do You Know?
One of the largest Devonian fish was about 9 m long with an enormous bite. The fish probably was a poor swimmer, and so it waited for other fish to swim within reach. Then its enormous mouth opened and snapped shut on the victim.

20 Observing Land Formations

The name for this landform is Camel Rock. You knew it had to have that name, didn't you? Camel Rock is located near Santa Fe, New Mexico.

It took a long time for Camel Rock to get to look like this. Years ago it was part of a series of sedimentary layers. Some layers were harder than others. Erosion of the soft layers and erosion along large cracks in the sedimentary layers produced the distinctive camel shape of the rock. What will it look like 50 years from now? 100 years from now? 500 years from now?

Camel Rock is like every other landform. It's a place where two earth processes are interacting. Erosional processes are wearing it away, but forces inside the earth are raising the rock layers. In time Camel Rock will change. It will probably lose its distinctive shape. Observing the changes in the land that take place through time is the focus of this chapter.

Figure 20.1 Camel Rock is one of several unusual rock structures in the southwestern United States.

20.1 Geologic Features Found on Maps

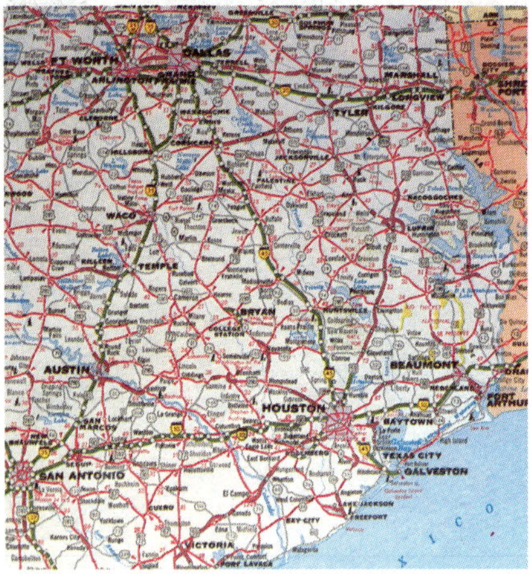

Figure 20.2 *above:* Have you ever read a road map?

Figure 20.3 *below:* River drainage patterns of the continental United States

You will find out

- what geologic features can be found on a road map;
- how to recognize geologic features on a topographic map;
- why geologists find geologic maps useful.

Whenever you take a long trip, a road map comes in handy. By reading the road map, you can usually get to where you want to go without getting lost. What information other than roads can you find on a road map? Depending on where you are looking, you can sometimes find the height of mountain passes and mountains. Rivers and lakes are usually shown on road maps too. If you know what to look for, you can also interpret other geologic information from road maps.

Look at the red-shaded section of Figure 20.3, covering the states of Georgia, South Carolina, and North Carolina. Notice that the rivers all run side by side in the same direction. This parallel river drainage pattern develops on long, gently sloping land surfaces. Can you find any other examples of parallel drainage?

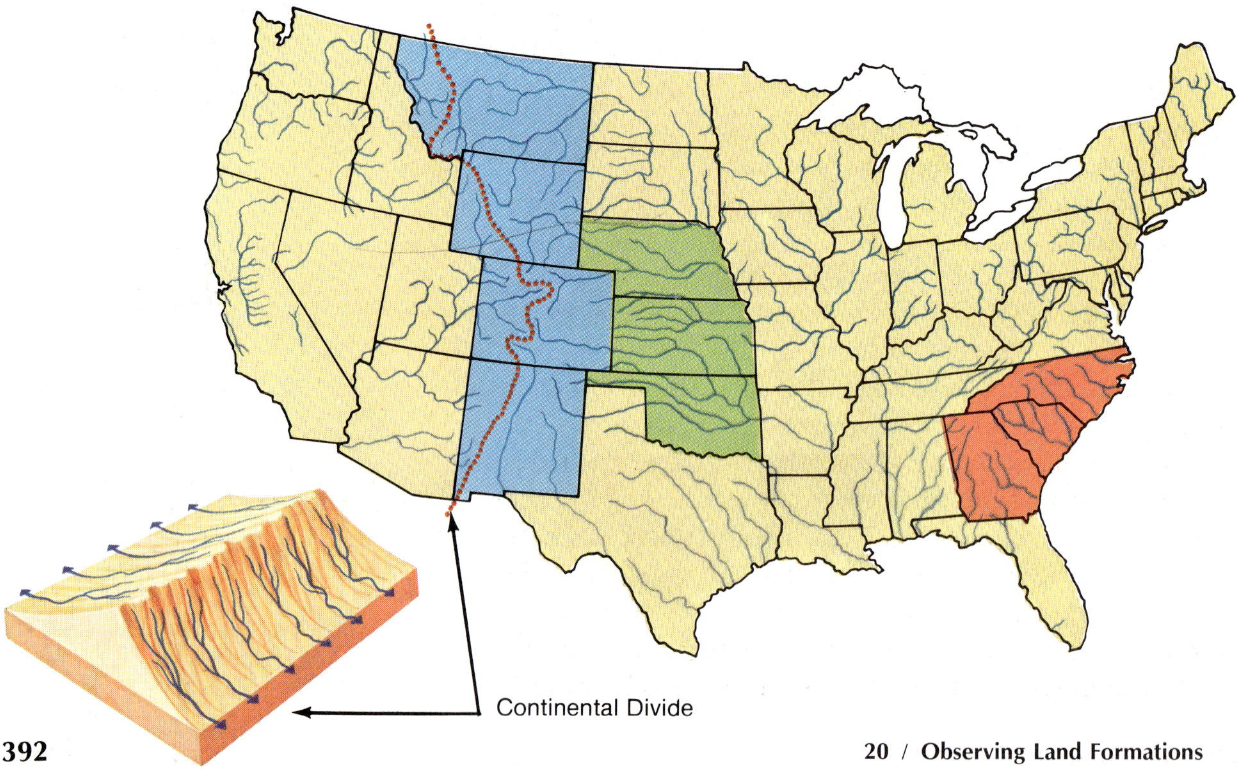

Look at the green-shaded area of Nebraska, Kansas, and Oklahoma. What do you think the land looks like in these areas? This area is also gently sloping land. Now look for other areas that are similar. Once you are familiar with the idea, you can see many other places where gently sloping land surfaces are found.

Another major geologic feature that can be identified on a road map is a **continental divide.** A continental divide separates regions that drain into one ocean from regions that drain into another. Notice the direction that the streams flow in the blue-shaded section of the map in Figure 20.3. Some streams flow to the Gulf of Mexico, some flow to the Gulf of California, and others flow to the Pacific Ocean. If you draw an imaginary line on the land separating these rivers that flow in opposite directions, you have the continental divide. You'll find your line will curve up through New Mexico, Colorado, Wyoming, a corner of Idaho, and Montana. Can you find another continental divide in the United States?

Do You Know?
The United States has two continental divides, both running almost north-south. These divides separate huge drainage basins from each other.

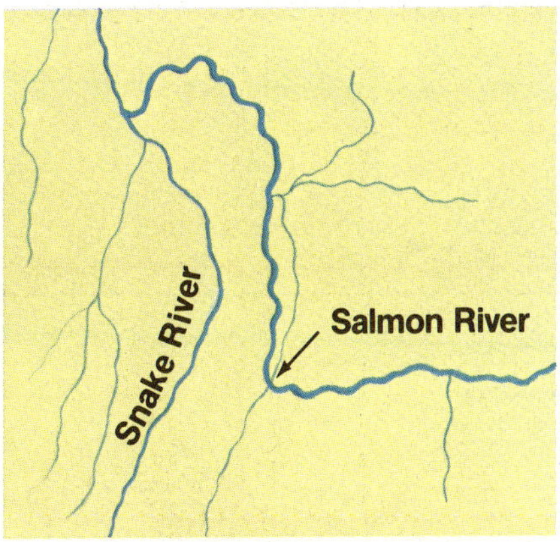

Now look at the stream patterns in Figure 20.4. In some parts of the country, the streams seem to make many right-angle bends. These right-angle turns are evidence of geologic features. One likely reason that these turns occur is that the streams flow toward high, resistant, rock layers and are diverted by them. This kind of flow is particularly true in mountainous areas. Another likely cause of right-angle turns is that the stream changes course when it is interrupted by a fault.

Figure 20.4 *left:* Notice the right-angle turns made by the river shown on this map.

Figure 20.5 *right:* Mountains can cause rivers to bend at right-angles.

Although you can find geologic features on a road map, there are other maps more useful for locating such features. In Lesson 16.5, page 329, you learned that topographic maps show the physical features of the land. Figure 20.6 is a topographic map. Examine the map carefully and see what features you can find. You'll notice that the contour lines fit around a feature. What land feature do you find where contour lines are close together? What do you find where the contour lines form *V*'s? Can you find a gentle slope?

Figure 20.6 *left:* A topographic map.

Figure 20.7 *right:* A geologic map and its corresponding cross section *below*

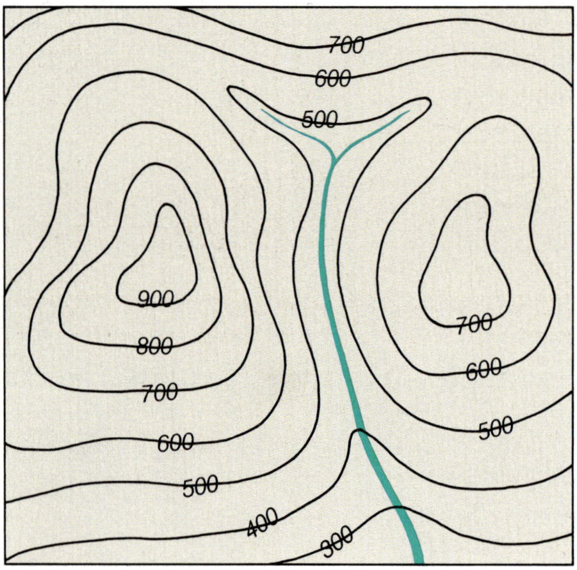

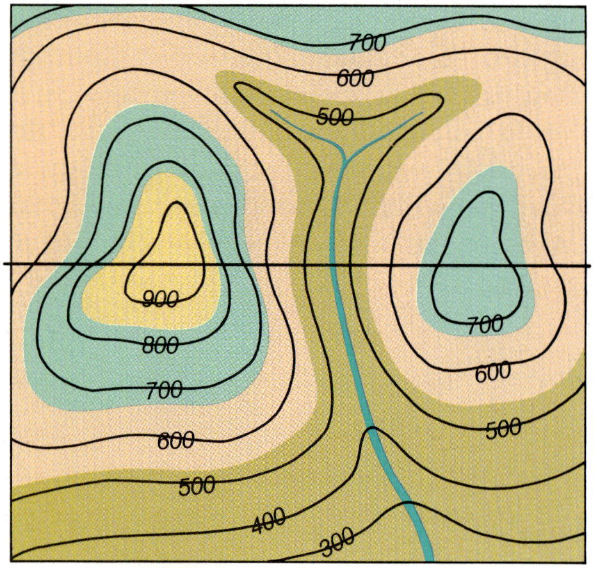

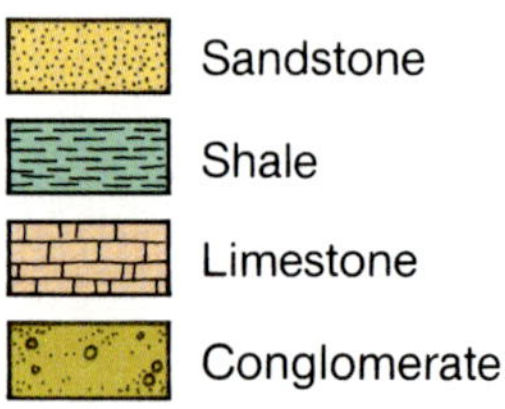

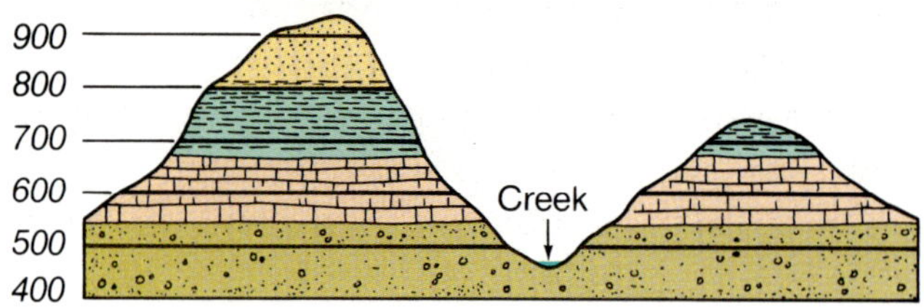

In Chapter 17, page 348, you learned that geologists use maps that also show geologic features of the land. These geologic maps combine both topographic and geologic information. In Figure 20.7, the geology has been added to a topographic map. Look at how the pattern of the rock types and contour lines relate to each other. What does the map tell you about how the rocks lie? Are they tilted or level? If you said "The layers are

Do You Know?
Much or most of the landscape features you see today have developed during the last 30 million years.

ACTIVITY

What Geologic Features Are Shown on These Topographic Maps?

Material

Pencil

Procedure

1. Study each of the topographic map sections.

2. Decide what land features are shown in each.

Questions

1. Which of these maps shows a cliff? Isolated hills? Gentle slope? An alluvial fan?

2. How do contour lines that show a cliff differ from those that show a gentle slope?

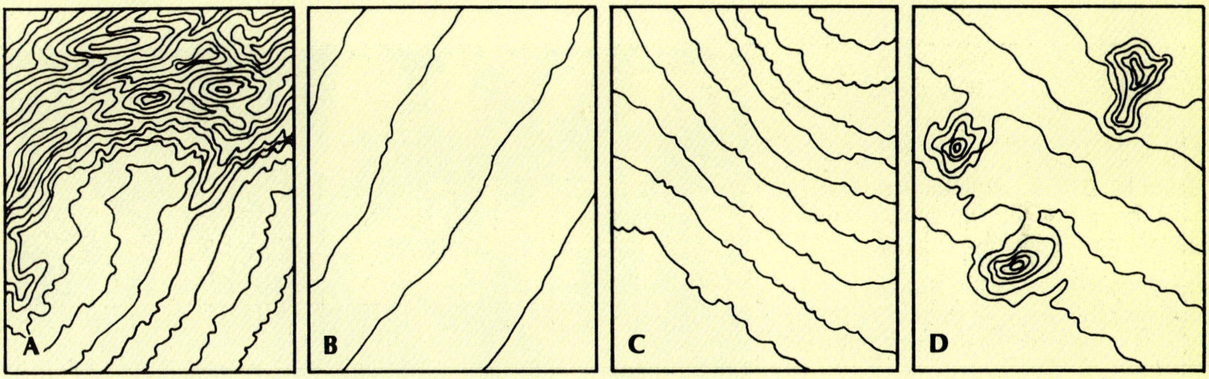

flat or level lying," you'd be correct. A cross section is also shown in Figure 20.7. Notice that the flat-lying beds can be seen on both sides of the hill. A geologic map is very useful because it not only shows elevations and surface features but also shows the kinds of rocks and ages of the rocks underneath the surface.

Maps of this type are very useful. Imagine the information a contractor needs in order to build a road across unlevel country. A combination of topographic maps and geologic maps would be a good choice. Then the contractor could dodge the cliffs and perhaps locate the easiest rock formations through which to cut.

Study Questions for 20.1

1. Describe three things to look for that help you to recognize geologic features on a road map.
2. Sketch how contour lines look to show a hill, a cliff, and a gentle slope.
3. What information is found on a geologic map?

20.2 Geology from an Airplane

You will find out
- how to spot geologic features from the air;
- how to interpret these features.

From high in the air, the land looks almost flat. As you can see in Figure 20.8, patterns of roads, fields, and streams show on the surface. Do you see any mountaintops? They are difficult to spot because their tops don't look high when seen from above.

Figure 20.8 *right:* Mountains do not have the same lofty appearance when viewed from the air.

Figure 20.9 *below left:* From an airplane, the drainage pattern of this river is clearly seen as a branching network.

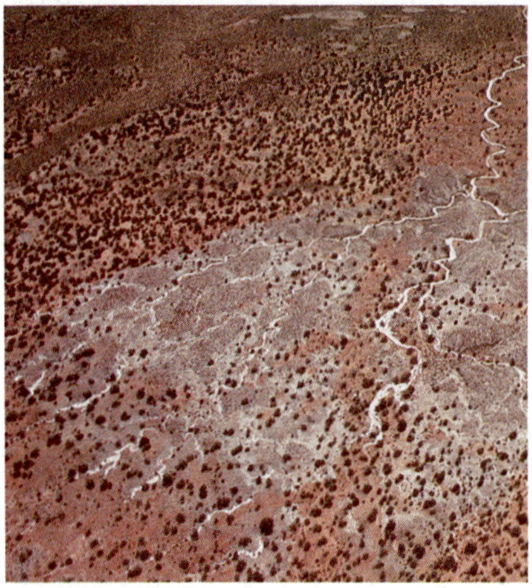

Figure 20.10 *above center:* A pattern of oxbow lakes helps to locate the old, abandoned channels of a stream.

Figure 20.11 *above right:* An anticline is easily spotted from the air because it forms a curved pattern of rocks on the surface.

Other geologic features that are frequently visible from the air include stream drainage patterns, abandoned stream channels, some anticlines, and some faults. Each geologic feature has its own characteristic pattern when viewed from above. Once you know these patterns, you can easily identify the geology of the area you are flying over.

ACTIVITY

Are the Sedimentary Rocks in the Aerial Photo Flat Lying or Tilted?

Material
pencil

Procedure

Look at the photo. The light and dark tones indicate different rock layers. Contour lines are shown at A and B.

2. Examine the rock pattern.

Questions

1. If there was a complete contour map of this area, would the rock patterns match the contours?

2. Are the rocks tilted or flat lying? Explain your answer.

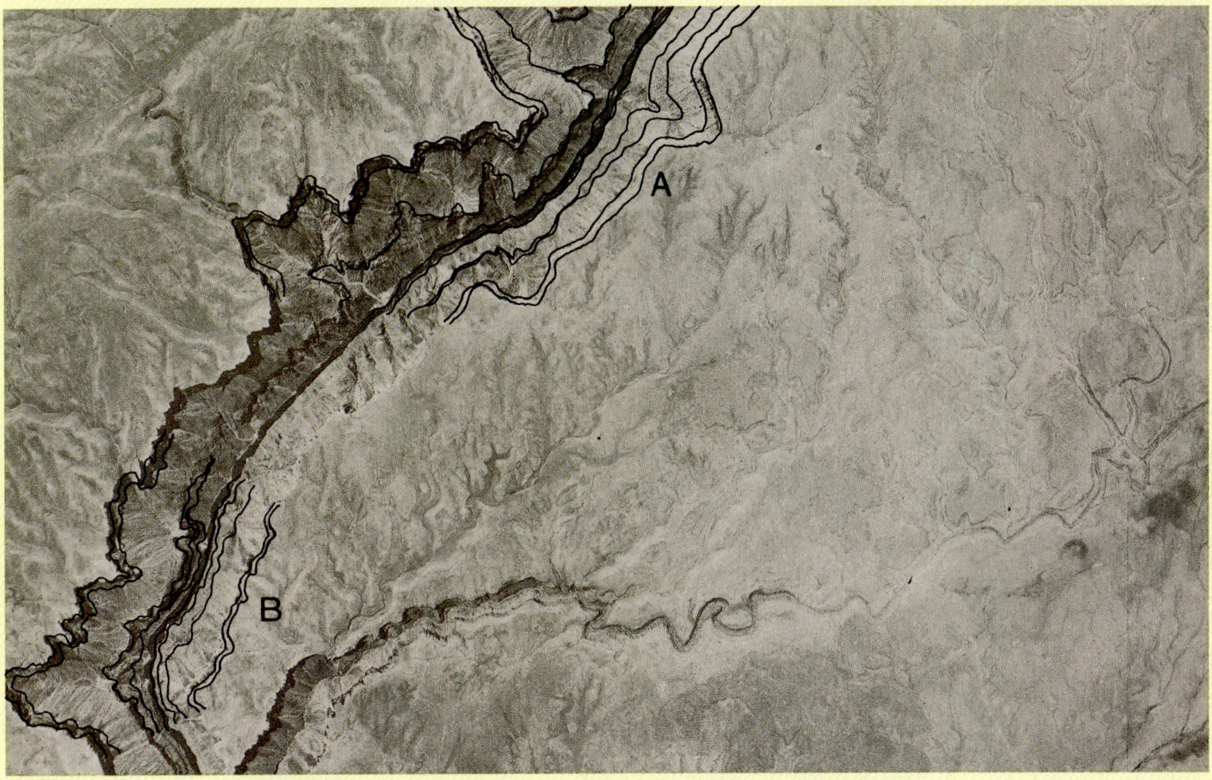

Study Questions for 20.2

1. What do you look for to spot a stream drainage pattern, an abandoned stream channel, an anticline, and a fault from the air?
2. Choose one of the features mentioned in question 1 and explain how this feature was made.

20.3 The Major Landform Regions of the United States

You will find out

- what makes a major landform region;
- what the major landform regions of the United States are.

The United States can be divided into eight **major landform regions.** A major landform region is a large area which contains similar or related landforms, soils, rock types, and rock structures. The eight regions of the continental United States are the Coastal Plain, Appalachian Highlands, Central Plains, Rocky Mountains, Colorado Plateau, Basin and Range, Columbia Plateau, and Pacific Mountains.

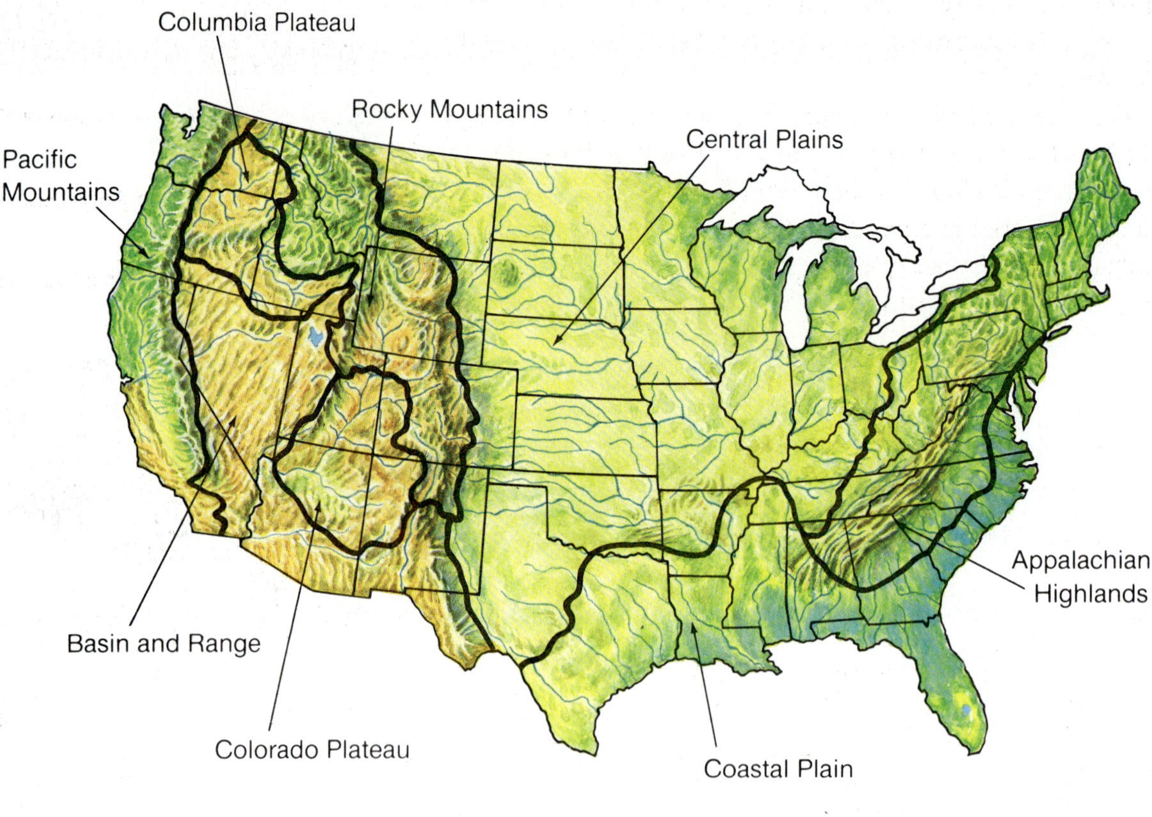

Figure 20.12 The major landforms of the continental United States

The major landform region in the United States in which the most people live is called the Coastal Plain region. A section of the Coastal Plain region is photographed in Figure 20.13. The Coastal Plain is low in elevation and, in general, consists of a gently sloping surface of rocks dipping toward the ocean. The rock types are sedimentary and 135 million years old or younger. The land is favorable for farming and supporting large populations. Many large rivers run parallel courses across the Coastal Plain to the ocean, where they form large deltas.

West and north of the Coastal Plain lies the Appalachian Highlands. See Figure 20.14. Included in this region are the Appalachian Mountains and the land next to those mountains. In general, the land has been heavily eroded by streams. Rivers are numerous and large, and many lakes exist. Many different rock types of many different ages are associated with the Appalachian Mountains.

The Central Plains is another major landform region. This is the largest region, and part of it is photographed in Figure 20.15. It has low elevation. Most of the rocks in this area are sedimentary rocks. It has great rivers and large lakes. Soils are thick, and the farmland is rich. It is the major agricultural area of the United States.

Figure 20.13 A typical section of the Coastal Plains

Figure 20.14 *left:* The Appalachian Highlands

Figure 20.15 *right:* The Central Plains region

Figure 20.16 The San Juan Mountains in the Rocky Mountain Region

Further west is the Rocky Mountain landform region, such as that shown in Figure 20.16. Here are some of the highest elevations of the United States. Igneous, metamorphic, and sedimentary rocks are found. The region is the principal headwaters for many rivers. Soils are thin, and there are large areas of barren rock.

South and west of the Rocky Mountain region lies the Colorado Plateau, as you see in Figure 20.17. Here are found extensive areas of nearly horizontal sedimentary rocks, with some igneous rocks occurring as volcanic rocks. The whole region is high in elevation and has a shortage of water.

The region photographed in Figure 20.18 is the Basin and Range region. Drainage is confined to local areas in the Basin and Range. There are no through streams. Many alluvial fans border fault-block mountains. The land is mostly barren. Igneous, metamorphic, and sedimentary rocks are found.

Figure 20.17 *left:* Part of the Colorado Plateau region

Figure 20.18 *right:* The east side of the Sierra Nevadas in the Basin and Range region

North of the Basin and Range region is the Columbia Plateau. Part of this area, shown in Figure 20.19, is covered by great thicknesses of lava flow. Except for a few areas, soils are poorly developed. The better soil areas are from windblown deposits. The Columbia River and the Snake River canyons are found in the Columbia Plateau.

Figure 20.19 *left:* Lava flows in the Columbia Plateau region

Figure 20.20 *right:* The northern Cascades in Washington, part of the Pacific Mountain region

To the west and south of the Columbia Plateau is the Pacific Mountain region, shown in Figure 20.20. This area has a great variety of landforms, rock types, and rock structures. In the north are volcanic mountains; the mountains somewhat in the center are the Sierra Nevada granites; near the ocean lies the Coast Mountain range. The great valley of the Pacific Mountain region, the San Joaquin, is a vast farmland.

In which region do you live? Would you like to live in one of the other regions?

Study Questions for 20.3

1. Define a major landform region.
2. Name the eight major landform regions of the United States.
3. In which landform region do you live? In general, what is this region like?

What Did You Learn

- You can find evidence of geologic features on a road map.
- Parallel river drainage develops on long, gently sloping land surfaces.
- A continental divide separates water that flows into one ocean from water that flows into another.
- Right-angle bends in streams may be caused by a resistant rock layer diverting the stream, or a fault.
- Contour lines show land features.
- A geologic map not only shows elevations and surface features but also shows the kinds of rocks and ages of rocks underneath the surface.
- From high in the air, the land looks almost flat.
- A major landform region of the United States contains similar or related landforms, soils, rock types, and rock structures.
- The major landform region in which most people of the United States live is the Coastal Plain region.
- The Appalachian Highland region includes the Appalachian Mountains and the land next to those mountains.
- The Central Plain region is the largest major landform region of the United States.
- The Rocky Mountain region contains some of the highest elevations in the United States.
- The Colorado Plateau is high in elevation and has a shortage of water.
- The Basin and Range region does not have through streams, as all drainage is local.
- The Columbia Plateau is covered by a series of thick lava flows.
- The Pacific Mountain region contains a great variety of landforms, rock types, and rock structures.

Key Terms

continental divide
major landform region

Career

Exploration Petroleum Geologist

An exploration petroleum geologist studies surface and subsurface rocks in search for petroleum. The work involves the evaluation of potential areas as a source of petroleum and the analysis of the rock material that is removed when a well is being drilled. The exploration petroleum geologist is the first to know whether or not drilling is successful.

Marsha Findlay is an exploration petroleum geologist. She attended Walt Whitman High School in Huntington, New York. Later she studied geology and earth science at Vassar College. Her master's degree was earned at the University of South Carolina. Her thesis study on the effects of tectonic events on carbonate sediments of the Jurassic Period provides her with the kind of information that helps her locate potential sources of petroleum.

Her interest in geology began in elementary school when she learned about the glacial features near her home on Long Island. In the ninth grade she took an earth science course that she liked very much. This interest has continued and led to her present position.

An exploration petroleum geologist must be able to interpret surface landforms and how they relate to subsurface rock formations that may hold petroleum. The scale of events that must be studied range from the microscopic to global. Marsha Findlay finds the study of geologic changes in surface and subsurface rocks exciting and the challenge of using this knowledge in the search for petroleum important.

Marsha Findlay is a member of the Scholarship Committee for the American Geological Institute's Minority Participation Program. This group encourages minority students to study the geologic sciences.

TO THINK ABOUT AND DO

Vocabulary

Unscramble the words below. Match the words with the descriptions.

a. cooleggi apm
b. jraom ndalrmfo neroig
c. aoclinntten iiddev

________ **1.** separates water flowing into one ocean from water flowing into another ocean

________ **2.** large area of the United States which contains similar or related landforms, soils, rock types, and rock structures

________ **3.** shows the distribution of rock types on the earth's surface

What Do You Remember?

1. What gives evidence of geologic features on road maps?
2. What types of drainage patterns develop on long, gently sloping land surfaces? Make a sketch showing this pattern.
3. In what direction does the continental divide run in the northwest corner of Wyoming? In southwestern New Mexico?
4. What landscape feature would be shown by contour lines that are spaced equally apart but widely separated? Spaced equally apart but close together?
5. How can you tell rocks are flat lying on a combination topographic and geologic map?
6. What does a geologic map show?
7. List four geologic features that are frequently visible from the air.
8. Which states lie partially or completely in the Coastal Plain region?
9. How many states are partially or completely in the Central Plain region?
10. In which major landform regions would you find continental divides?

Applying What You Have Learned

1. What does the land surface look like in northeastern Iowa?
2. The cause of a stream making a right-angle bend may be a fault. Draw and label a sketch showing this geologic feature.
3. Explain this statement: The pattern of flat-lying rocks on the surface is related to the pattern of the contour lines. (You may use a sketch.)

Research and Investigation

1. Find a map of your state. Lay a piece of tracing paper over this map. Try to divide your state into landform regions. Start by locating mountainous areas and low, flat areas. Then subdivide the state into areas of similar landform character.
2. Determine the major landform region you live in. Use the library to find out what landforms, rock types, and rock structures are found in your region. A geography text is usually a good reference.
3. Use the library to find a geologic map of the area in which you live. List the types of rocks found near where you live.
4. Plan a trip from your home to the Grand Canyon. Make a list of the geologic features you would expect to see on this trip.

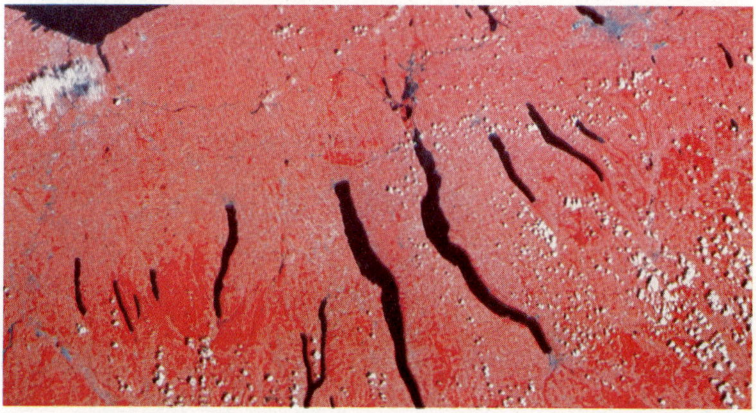

A satellite photograph shows the pattern of the appropriately named Finger Lakes in New York State.

Unit Six

The Earth's Bounty

21 The Earth's Resources

The space shuttle is a marvel of engineering and science. But have you ever thought about all the different materials the space shuttle is made of? Many different kinds of materials were needed to make the computers and electric systems which help control the craft. Other materials were required for the special heat-resisting tiles to keep the craft from burning up upon reentry into the earth's atmosphere. Vast amounts of still other materials are burned in the rockets that lift the shuttle into space. Would you believe that all of the materials needed for the space shuttle come from the earth in some form? In fact, just about everything you use, wear, or eat comes directly or indirectly from the earth's crust. Till now the earth has been able to supply all the materials people have wanted. Will this always be true?

In this chapter you'll learn about some of the materials people take from the earth, where the materials are found, and how much of them there is.

Figure 21.1 *opposite:* Materials to build the space shuttle come from many of Earth's natural resources.

21.1 Metallic Resources of the Land

You will find out

- what a natural resource is;
- what some renewable resources are;
- what some nonrenewable resources are;
- what an ore is.

The earth has a large supply of materials people use. They are called **natural resources.** Natural resources are the supplies taken from the earth and include water, food, building materials, fiber for clothing, minerals, and energy.

Figure 21.2 A new crop will be planted to replace this one after it is harvested, but the machinery is made of resources that can be taken from the earth only once.

Natural resources fall into two types, renewable and nonrenewable. **Renewable resources** add a new supply each growing season or will become available again sometime soon in the future. Wood, food, and wool are examples of such resources. A **nonrenewable resource** is one that is not available again from the earth once it is taken and used. Examples of nonrenewable resources are metallic substances such as copper, iron, and aluminum. Nonmetalic substances such as oil, coal, clays, and chemical fertilizers are also nonrenewable resources. Although paper made from wood and many metals can be recycled, neither wood nor metals are considered to be renewable.

Metallic resources used for making things are found in the earth as the elements that make up minerals. **Ores** are minerals from which metals can be taken at a reasonable cost. Table 21-1 shows some of the widely used metals and their most common ore minerals.

Iron is the most widely used metal. It accounts for over 90 percent of all metals mined. The major ore minerals from which iron is taken are hematite and magnetite. These ores contain from 15 percent to 60 percent iron. In the United States the location of the major iron deposits is the Lake Superior region. Here the ore is found in huge sedimentary deposits. Other metals are added to iron to make special steels. The strength of iron may be increased by adding nickel, carbon, manganese, and other metals. Steel's resistance to rusting is due to the presence of added chromium.

Table 21-1. Ten Common Metals and their Ore Minerals

Metal	Ore Mineral
Iron	Magnetite Hematite
Copper	Native copper Chalcocite Chalcopyrite
Aluminum	Bauxite
Lead	Galena
Zinc	Sphalerite
Gold	Native gold
Silver	Native silver
Chromium	Chromite
Mercury	Cinnabar
Nickel	Pentlandite

Figure 21.3 Steel is one of the most important building materials used in modern industry.

The procedure for obtaining a metal from an ore is done in four steps. The first step is to extract the ore from the earth. Next, the ore must be enriched by removing waste materials found with the ore. The third step is to process the enriched ore to obtain the metal that is wanted. The fourth step is to use the metal to manufacture a final product.

Figure 21.4 *right:* A copper mine

Figure 21.5 *below:* Copper may be fashioned into a variety of useful forms.

Copper is another widely used mineral. You might think a copper ore would be rich in copper. That is not true. Most copper ores contain very little copper. They may have as little as 4 kg of copper for each metric ton of ore. That means for every metric ton mined there is 996 kg of waste material. Because a great deal of copper is used in the electrical industry, some copper mines are enormous.

One of the metals used for its lightness and strength is aluminum. It is used in many household products, in airplane and automobile construction, and in the electrical industry. Most new electrical transmission lines in the United States are made of aluminum. The number of different uses for this metal grows every day.

Gold is a desirable metal for its beauty and the fact that it is almost indestructable. It occurs in small amounts in deposits that are located in present or old stream channels. It is used in making jewelry, in electri-

Do You Know?
The United States is the world's leading producer of copper.

cal circuits, and in gold work by dentists.

Metals play an important role in your life. Imagine how your surroundings, your activities, and even your clothes would be different if there were no metals. Forms of transportation from bicycles to spaceships would be unknown. There would be no telephones, televisions, or stereos. Tools, modern appliances, jewelry, and some kinds of clothing would be impossible without metals. Metals are essential to your way of life.

ACTIVITY

What Happens When an Ore Is Smelted?

Materials

safety goggles
crucible
ring stand
galena (composed of lead and sulfur)
Bunsen burner
pliers
iron ring

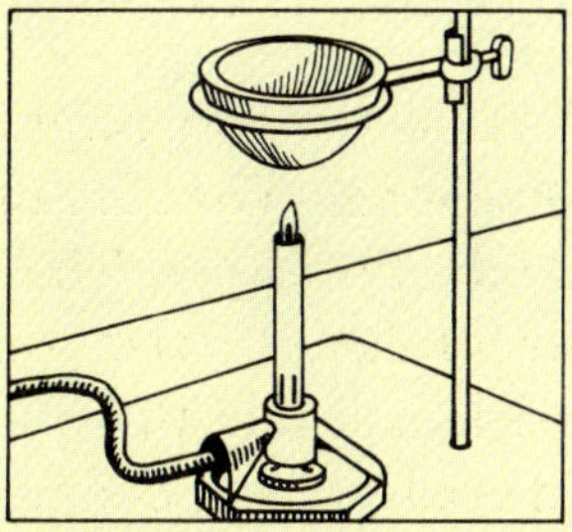

Procedure

Wear your safety goggles at all times and provide some ventilation in the room while heating this material.

1. Place one piece of galena in the crucible. CAUTION: Wear your safety goggles.

2. Heat the galena as hot as possible.

3. Observe the changes.

4. When the galena has changed, remove the heat. Let the mass cool and then squeeze it with pliers. CAUTION: Do not handle the mass until it is completely cool.

Questions

1. Did you smell anything coming from the galena while it was being heated? What did you smell? Remove flame before testing for smell.

2. What happens when you squeeze the material remaining in the crucible with pliers?

3. What metal is in galena?

Study Questions for 21.1

1. Define *natural resources.*
2. Give three examples of a renewable resource. Explain why they are renewable.
3. What are three nonrenewable resources? Explain why they are nonrenewable.
4. What is an ore and what are the basic procedures followed to obtain metal from an ore?

21.2 Nonmetallic Resources of the Land

You will find out

- what some nonmetallic resources are;
- how nonmetallic resources are used.

Would you prefer to own a gold mine or an oil well? A coal mine or an iron mine? A copper mine or a clay pit? Most students say they would prefer to own a metal mine because they think metallic resources are more valuable than nonmetallic resources. More than 10 billion dollars yearly is spent on metals, but the amount spent on nonmetals is more than three times that amount.

What are the major nonmetallic resources that cost so much? The major nonmetallic resources are (1) the fossil fuels—petroleum, natural gas, and coal; (2) minerals for fertilizers; (3) minerals for the chemical industry—sulfur and salt; (4) abrasive minerals; and (5) building materials. Today's industry runs because there is a continuing supply of nonmetallic resources.

Figure 21.6 Which do you think is more valuable: an oil well (*left*) or a sulfur mine (*right*)?

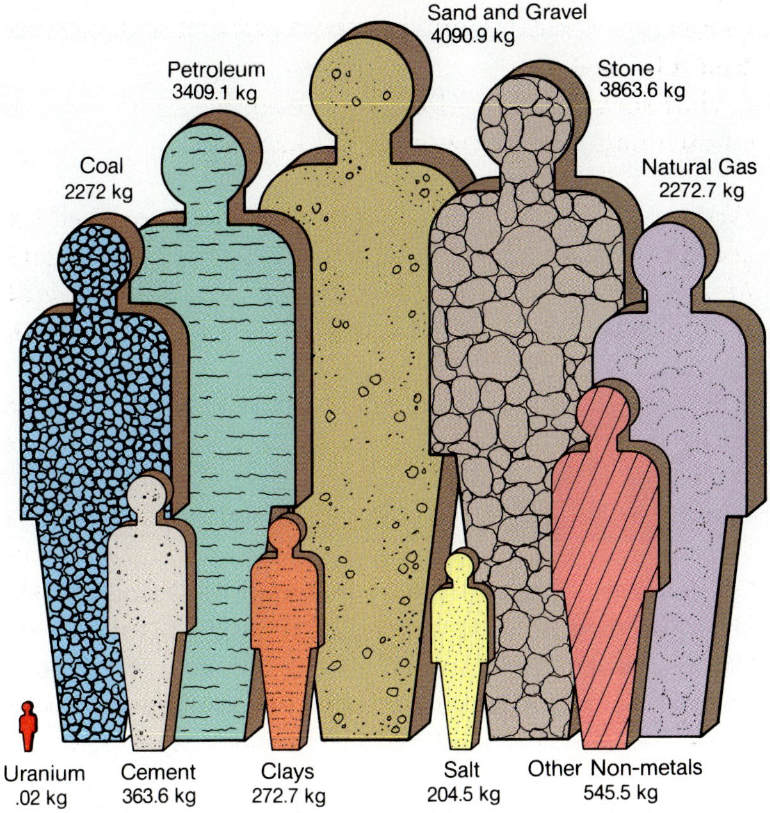

Figure 21.7 *left:* The consumption of nonmetallic resources in the United States per person

Figure 21.8 *below:* Fossil fuels are used to manufacture a wide variety of products. How many fossil fuel products do you use in one day?

Look at Figure 21.7. This figure shows the amount of some nonmetallic resources used for each person in the United States each year. The numbers were obtained by dividing the total amount of each nonmetallic resource used each year by the population of the United States. You probably didn't realize how much of the earth's crust you use each year. You use such a great amount of nonmetallic resources each year because they are used in so many different ways. Nonmetallic resources are used for fuels, fertilizers, chemicals, foods, medicines, abrasives, and building materials to name just a few of the ways.

The most valuable of the nonmetallic resources is the fossil fuels. Their primary use is to produce energy for power and heat, and lubricants. However, many other things can be made from the fossil fuels. Did you know that nylon, Styrofoam, polyesters, aspirin, and more than 3,000 other things can be made of chemicals obtained from fossil fuels?

Figure 21.9 *top:* Crop-dusting enables farmers to add important fertilizer minerals to large areas of crops.

Figure 21.10 *bottom:* What does a diamond have in common with these other materials?

One of the most necessary nonmetallic resources for the United States and the rest of the world is the fertilizer minerals. Fertilizers are essential for growing crops to feed the world's population. The big three in fertilizers are nitrogen, potassium, and phosphorus. The demand for these three elements doubles every ten years. Fortunately, there appear to be large reserves of these elements.

The chemical industry is a large user of nonmetallic resources. Sulfur is one of the most widely used chemical nonmetallic resources. Sulfur is used in making insect and weed killers. It is also used in making sulfuric acid, which is used in many manufacturing processes. Sodium chloride, or salt, is another chemical nonmetallic resource. It is used as a food, in medicines, as a food preservative, and in packing houses. Many other nonmetals are used by the chemical industry but, with the exception of the fossil fuels, only in relatively small amounts.

The abrasive minerals include substances that are hard. Diamond is the hardest natural substance and the most important abrasive. It is used in grinding wheels, as a powder, and on the tips of oil-drilling bits. Other abrasive materials are pumice—used in soaps and scouring powders—and garnets and quartz used to make sandpaper.

ACTIVITY

Which Cleanser Contains Quartz? Which Contains Feldspar?

Materials

2 kinds of scouring powder (1 made of feldspar and another made of quartz)
4 glass microscope slides
gloves
water

Procedure

1. Put a small pinch of scouring powder near one end of a glass microscope slide.

2. Add a couple drops of water.

3. Place another glass microscope slide on top of the first slide and rub the powder between them. Wear your gloves while doing this.

4. Repeat steps 1–3 for the other scouring powder.

5. Carefully examine the surface of each set of slides.

Questions

1. Which powder scratched the glass slides?

2. Why didn't both powders scratch the glass slides?

3. Which powder contains quartz? Which contains feldspar? Explain your choices. *Hint:* Quartz has a hardness of 7; feldspar has a hardness of 6.

Building materials are another valuable nonmetallic resource. Stone is used for building structures and for ornamental display. Cement for buildings and roads is made from limestone and clay. Sand and gravel are widely used for roads. They are also mixed with cement to make concrete for roads and concrete buildings. Clay is used to make bricks used in the building industry. Pure quartz-sand is made into glass.

Look around your school to see how nonmetallic resources were used in building it. How many nonmetallic materials can you spot in your home? You'll be surprised how much use you make of these materials.

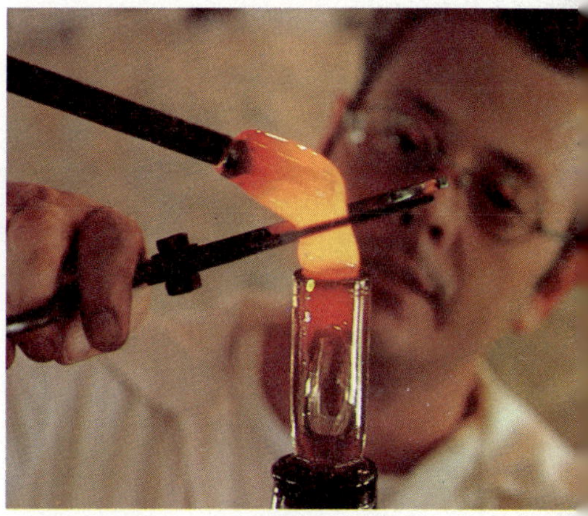

Figure 21.11 Ordinary quartz-sand is used to produce glass.

Study Questions for 21.2

1. List five major nonmetallic resources.
2. Select one of the major nonmetallic resources and describe its use.

21.3 Resources of the Ocean

You will find out

- what mineral resources are obtained from seawater;
- what some of the mineral resources are in or on the ocean floor;
- what some of the biological resources of the ocean are.

Figure 21.12 "Okay, Mule, all we have to do is bring up those elements, and 1.6 billion dollars will be all ours!"

Do You Know?
The abundant elements formed in seawater are also abundant in rocks with the exception of chlorine, which is much more abundant in the ocean. Scientists are not sure why the ocean has so much chlorine.

The thing that impresses everyone who has ever taken an ocean voyage is the ocean's vastness. There's a lot of it, more than 1,370,000,000 km³. This enormous volume contains more than just water. In every cubic kilometer there are 5×10^{12} metric tons of dissolved elements. The ocean is said to contain enough gold to make everyone a millionaire. Each cubic kilometer is worth about $1.6 billion if all the elements could be recovered and sold.

Seven elements found in seawater and their dollar values are shown in Table 21-2. The elements are worth a lot of money but are costly to remove. Imagine an industrial plant that could remove all of the elements from seawater flowing through it. At the rate of flow of one million liters per minute, it would take almost two years to process one cubic kilometer of seawater. It is worthwhile, however, to extract only a very few substances from seawater. The table salt you eat on your food may have come from the sea. Salt is made from a combination of two elements, sodium and chlorine. About 6 million tons of salt are taken from the seawater each year. Magnesium is also removed in useful amounts.

Table 21-2. The seven most abundant elements in seawater

Element	Estimated value in combined form ($ per 4×10^6 liters)
Chlorine	$924
Sodium	$378
Magnesium	$4,130
Sulfur	$101
Calcium	$150
Potassium	$91
Bromine	$190

The most valuable thing taken from seawater is fresh water. Fresh water is the world's most important resource. Whether or not fresh water is available determines where people live, where they can farm, and what kinds of manufacturing can be done. The recovery of fresh water from seawater is important for countries that border on the ocean and have a desert climate. Figure 21.13 shows one type of plant used to obtain fresh water. The cost of such fresh water is as low as \$.50 for 4,000 liters. Costs are expected to drop in the future as more efficient plants are developed. Fresh water from the sea could change many deserts into fertile farms.

Figure 21.13 Desalinization plants can recover large volumes of fresh water from the ocean.

ACTIVITY

How Will This Water Taste?

Materials

2 Styrofoam cups
plastic sandwich bag
3 soda straws
6 toothpicks
1 jar lid, 6.5 cm in diameter, painted black
plastic tape
water and salt

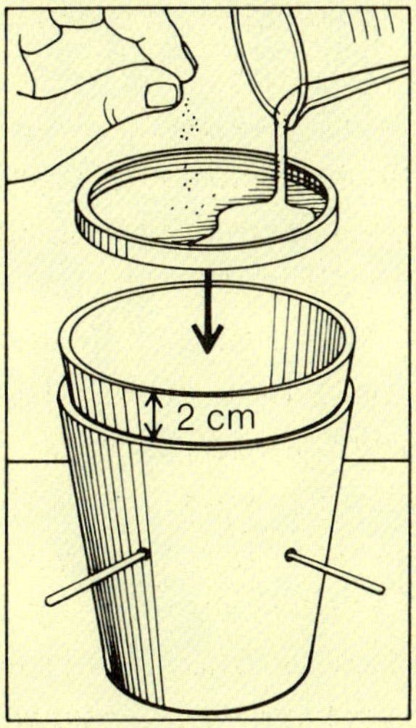

Procedure

1. Set up the apparatus by first placing one cup inside the other. The inside cup should be raised enough to give a 2-cm clearance between the two. Hold the inner cup in this position with four toothpicks.
2. Place the jar lid in the inner cup and fill halfway with water. Add a pinch of salt and stir. Now taste the water.
3. Fasten the soda straws together with tape to form a tripod and set into the jar lid.
4. Use two toothpicks to make a support at the top of the tripod.
5. Place a plastic bag over the tripod with the bottom edge inside the outer cup. Use plastic tape to secure the bag in place and hold it away from the inner cup.
6. Place the apparatus in sunlight for a day.
7. Dismantle the apparatus and taste the water that has collected in the outer cup.

Questions

1. Where did the water in the outer cup come from?
2. How did it taste?
3. Describe how you could make fresh water from seawater.

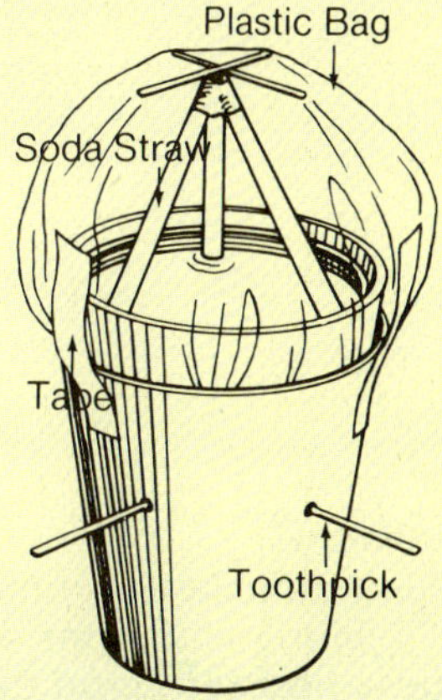

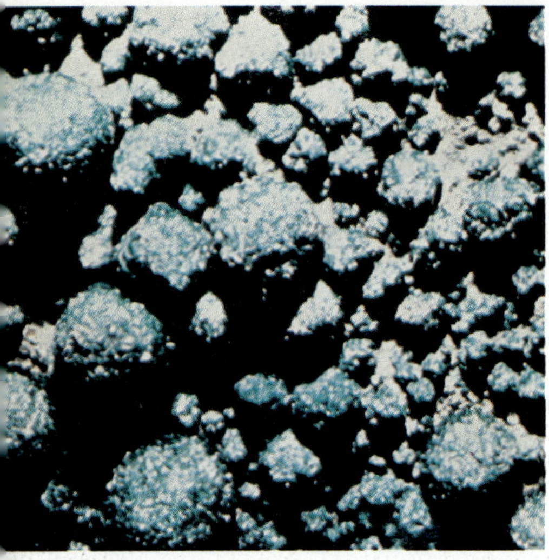

Figure 21.14 *above:* Manganese nodules on the ocean floor

Figure 21.15 *right:* Devices such as this one will be used for many sea-mining operations in the future.

The ocean floor is also a rich source of minerals. Some of these minerals are found as small lumps called **nodules** [NAHJ-ools]. These nodules, found on the ocean floor, are rich in manganese, copper, nickel, and cobalt. These are important minerals that have wide use in manufacturing. Will the ocean floor ever be mined for these minerals? Some people believe it will. In Figure 21.15 you see the design of a device proposed to recover the nodules from the ocean floor.

Beneath the ocean floor is another mineral resource. Huge drilling platforms can be located offshore to drill through the ocean floor for oil and gas. Geologists expect that more and more of the oil you use in the future will be taken from the ocean floor.

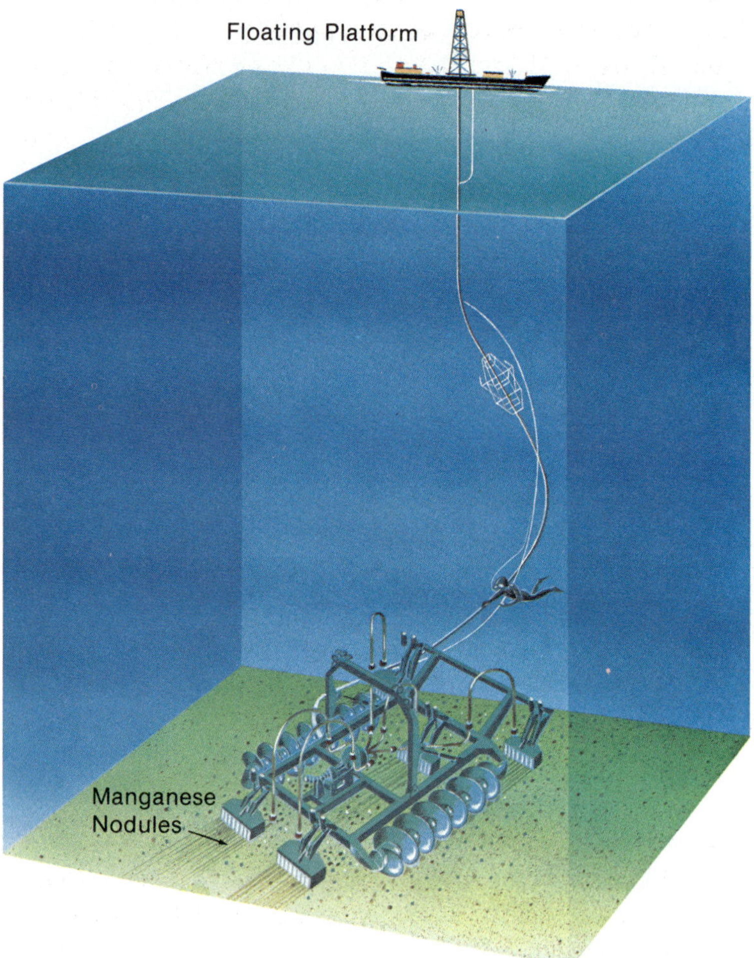

Do You Know?
It has been estimated that more than 1.5 billion metric tons of manganese nodules lie on the Pacific Ocean floor.

An even more promising resource found in the ocean is biological. The ocean is an excellent source of food, especially **protein.** Proteins are substances that the body uses for the growth and repair of tissues. Fish are rich in proteins. About 90 percent of the ocean harvest is fish; the rest consists primarily of whales and shellfish.

Not all fish end up directly on someone's dinner table. About half of the fish landed in the United States go to feed animals in the form of fish meal. The poultry industry uses large amounts of fish in this form. Fish meal is rich in minerals that are essential to animal and plant growth.

Although fishing will continue to provide an important food resource from the ocean, it is believed by many that even more food can be obtained by farming the ocean. The process of farming the ocean is called **aquaculture** [AH-kwa-KUL-chur]. The Japanese grow seaweed, oysters, and shrimp under farming conditions in the sea. This makes the ocean an important resource because people can grow the kinds of things they like to eat. The ocean is one of the earth's treasures. In the future it will play an even more important role in supplying people with minerals and food.

Figure 21.16 These chickens are being fed a nourishing diet of fish meal. The chickens will then produce food for people.

Do You Know?
Farmers used powdered fish to feed their livestock as early as 325 B.C.

Figure 21.17 The ocean is becoming a rich resource for food because of modern techniques of aquaculture.

Study Questions for 21.3

1. Name the mineral resources taken from seawater.
2. What resources are found in or on the ocean floor?
3. How can the biological resources of the ocean be increased?

21.4 Where the Resources Are Found

You will find out

- where mineral resources are found and used worldwide;
- where fossil fuels are found and used;
- how future supplies of mineral resources will be met.

Everyone who studies about the uses of natural resources sooner or later asks the questions, "Are there enough? Will the earth's resources run out? What's going to happen in the future?"

The earth's natural resources are being used up faster than they are being produced in the earth. This might lead you to think that the earth's resources will someday be all used up. That's not exactly true. Instead, the materials that people use will just become increasingly difficult to obtain. They will cost more, and it will require more energy to take the material from the ore.

Figure 21.18 The location of the world's major mineral resources

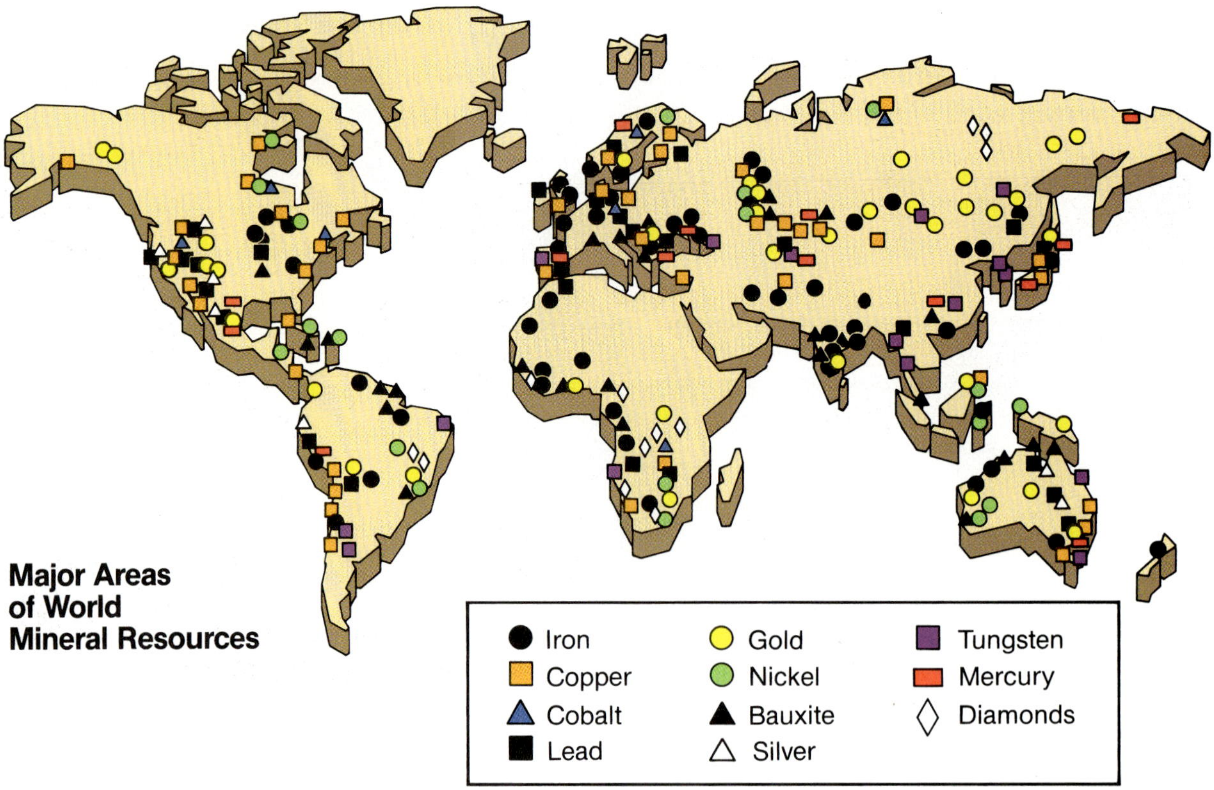

One clue that the earth's natural resources will not run out all at once is shown in Figure 21.18. The main sources of minerals used by people and industry are shown in this figure. Each kind of mineral is found in many different places around the world. All of these places won't run out of ore at the same time, and some new sources of ore may be discovered.

Many of the earth's resources, however, are not found where they are used. It is necessary to move some of the minerals used in manufacturing long distances from where they are found. You will see why this is so if you compare Figure 21.20 with Figure 21.18. The industrial countries are big users of minerals. Some minerals, such as cobalt and nickel, are found in locations that are not highly industrial. That means these materials, which are important in making steel, must be moved from one part of the world to another. The buying and selling of mineral resources is part of international trade.

Figure 21.19 *above:* Mineral resources are necessary for manufacturing and industry.

Figure 21.20 The major manufacturing regions shown on this map draw their raw materials from the areas shown in Figure 21.18.

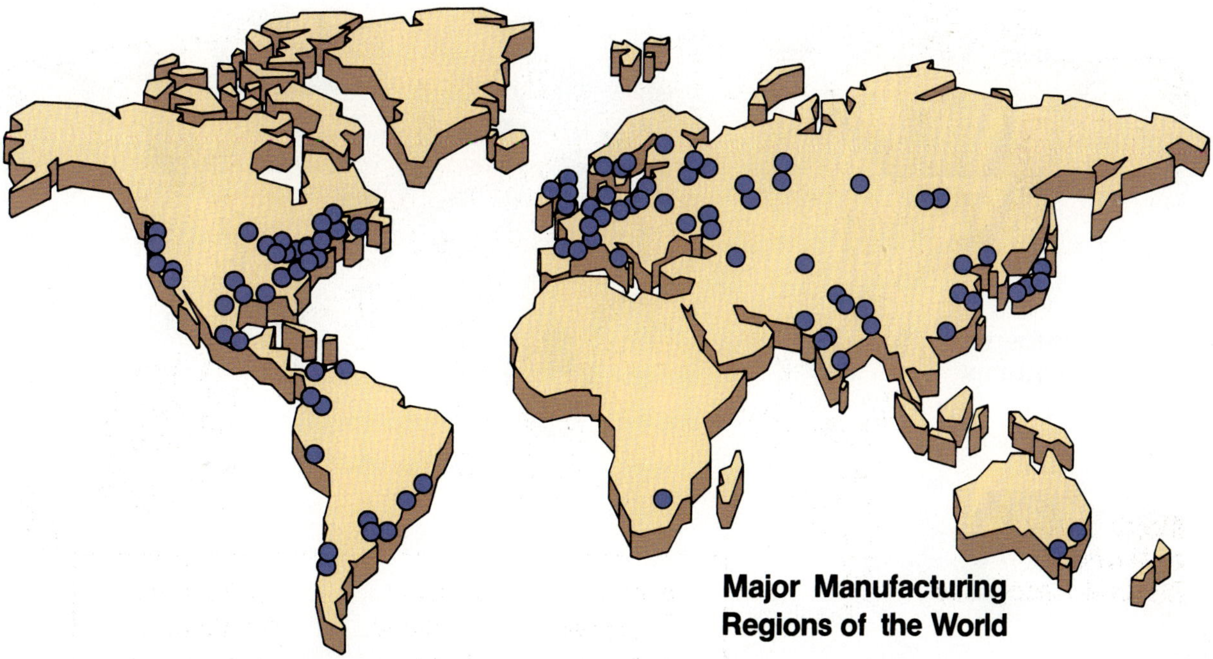

Figure 21.21 *above:* Oil freighter

Figure 21.22 The major areas of energy resources

Like minerals, fossil fuels and uranium are found in many places. The largest users of these fuels are also the industrial nations. Fossil fuels and uranium must be moved, like the minerals, from wherever they are found to wherever they are used. Petroleum is moved the farthest and in great amounts. Coal is moved to where it is used, but this is usually only a relatively short distance. Most of the uranium that is used in nuclear power plants must be moved long distances to the industrial countries. Figure 21.22 shows where most of the large fossil fuel and uranium resources are located in the world. Compare this map with Figure 21.23. You can see that the places where the fuels are used are often not the same as the places where they are found.

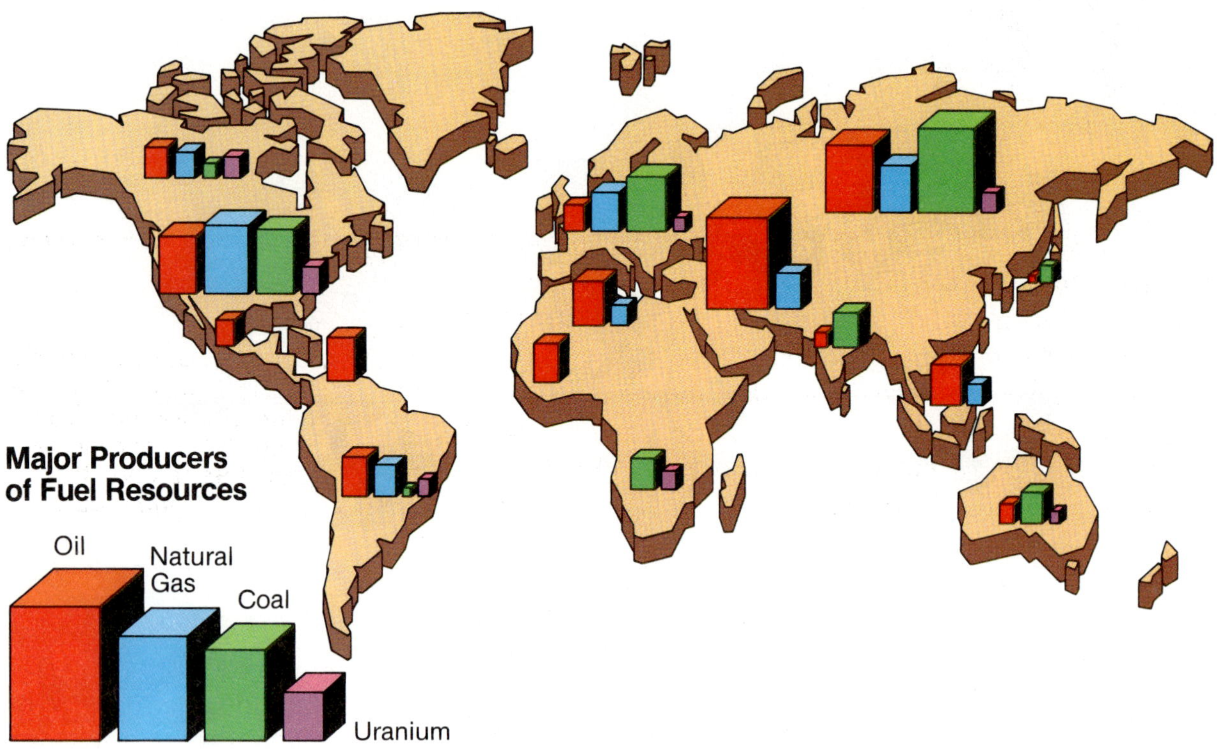

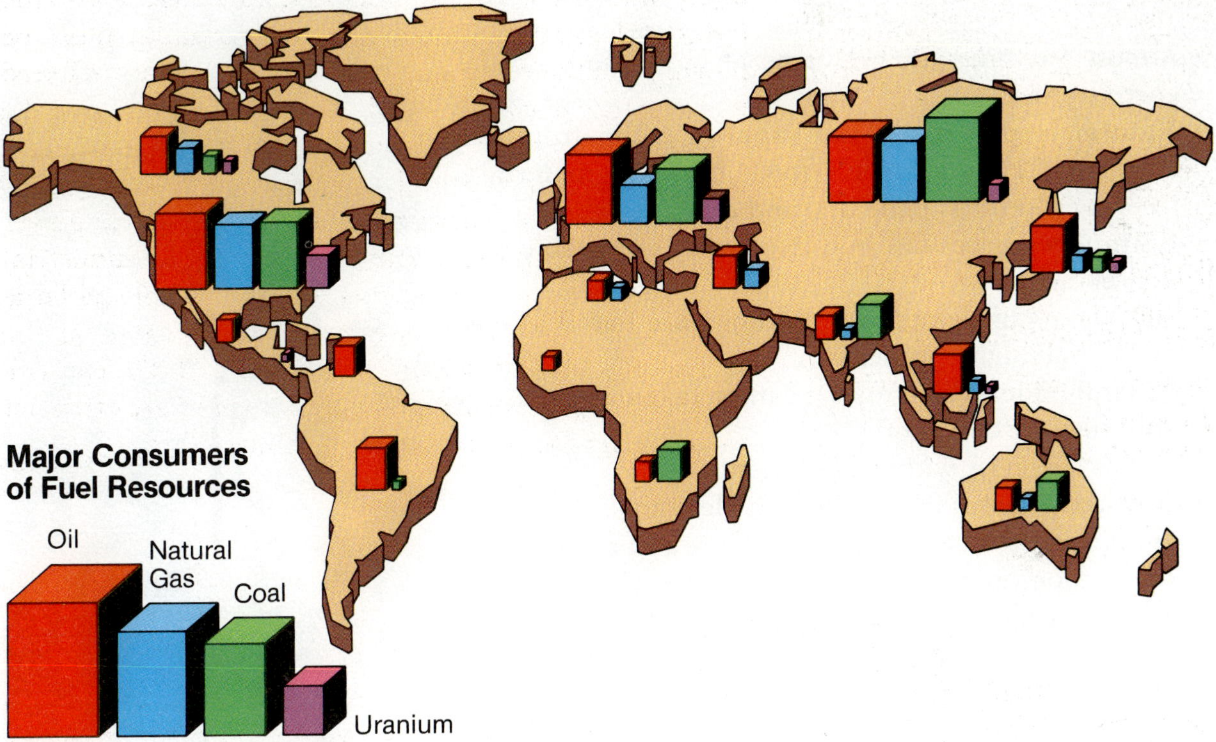

Figure 21.23 The major areas of energy use

How will the expanding needs for mineral resources be met? One way will be to use machines that require less material in their construction and which require less energy to run. Another way is to recycle materials that are scarce and costly. Industry and people worldwide will also learn to conserve materials and reduce wastes. All of these efforts will greatly help to maintain adequate supplies of mineral resources in the future.

Do You Know?
The concentration of some metals in waste dumps is higher than the concentration of these metals in the earth's crust. Someday it may become profitable to mine the waste piles being made today.

Study Questions for 21.4

1. Each continent is not equally rich in minerals. Which continent seems to have the most minerals? Which seems to have the least?
2. Which continent seems to have the most petroleum?
3. List four ways to meet the future needs for the earth's mineral resources.

What Did You Learn?

- Almost anything you see came at one time from the earth's crust.
- Natural resources are the supplies taken from the earth and include water, food, building materials, fiber for clothing, minerals, and energy.
- Natural resources fall into two types: renewable and nonrenewable.
- The metals used for making things are found as elements making up minerals.
- Minerals from which metals can be taken at a reasonable cost are called ores.
- Iron is the most widely used metal.
- Most copper ores contain very little copper.
- Aluminum is used because it is light and strong.
- The major nonmetallic resources are (1) the fossil fuels—petroleum, natural gas, and coal; (2) minerals for fertilizers; (3) minerals for the chemical industry—sulfur and salt; (4) abrasive minerals; and (5) building materials.
- The most valuable of the nonmetallic resources is the fossil fuels.
- Building materials are a valuable nonmetallic resource.
- In every cubic kilometer there are 5×10^{12} metric tons of dissolved elements.
- The most valuable thing taken from seawater is fresh water.
- Mineral-rich nodules are found on the ocean floor.
- The ocean is an excellent source of food.
- Aquaculture is a way of farming the ocean.
- The earth's natural resources are being used up faster than they are being produced in the earth.
- Many mineral resources have to be moved from where they are found to where they are used.
- Future needs for minerals will be met by (1) using less material in the construction of machines, (2) using more efficient machines, (3) recycling, (4) conserving, and (5) reducing waste.

Key Terms

natural resource
renewable resource
nonrenewable resource
ore
nodule
protein
aquaculture

Biography

Floyd Gray (1951–   )

On September 15, 1951, Floyd Gray, a research geologist with the Western Mineral Resource Branch of the United States Geological Survey (USGS), was born in New Orleans, Louisiana. He moved at an early age to Baton Rouge, Louisiana, where he attended public school. His early interest in science was encouraged by his father, who took him through the laboratories in the refinery where he worked. In the fourth grade Floyd's interest in science was sparked by a wonderful science teacher, and in the ninth grade he won a statewide science competition.

In college he started studying journalism, but a summer USGS geology job changed his life. The job took him to Hawaii, where he saw a volcanic eruption. In fact, he had the opportunity to gather data and make observations of that eruption.

Upon returning to the mainland, Gray enrolled in the University of California at Santa Cruz. His two bachelor's degrees were in earth science and anthropology. He earned a master's degree at the University of Massachusetts in Amherst. For his thesis he studied the magmatic ore deposits in southern Oregon.

Currently Gray is working for the USGS, evaluating the mineral potential of rocks of the ocean floor that have been forced onto the continental crust in northern California and southern Oregon. He does field mapping and lab petrology work and makes geochemical analyses of these rocks. He is looking for geochemical tracers to ore deposits.

Why does he like his work? He says, "I like outside work that is challenging. I'm interpreting nature and I like that. I enjoy traveling and the chance to do something rewarding."

TO THINK ABOUT AND DO

Vocabulary

Number your paper from 1 to 5. Unscramble the underlined word or words. Write the word or words next to the appropriate number.

1. nitepor is a substance a body uses for the growth and repair of tissues.
2. uuuaatcqler is a way of farming the ocean.
3. uurrrnlaaeessoct (two words) are supplies taken from the earth.
4. bsseeeeecrrranluwo (two words) add a new supply each growing season or will become available again sometime in the future.
5. sreo are minerals from which metals can be removed.

What Do You Remember?

1. List five things in your classroom that are made of renewable resources. List the resource of each.
2. List five things in your classroom that are made of nonrenewable resources. List the resource of each.
3. Which metal is most widely used? What are its major ore minerals?
4. What are the steps to remove a metal such as iron from its ore?
5. Which of the nonmetallic resources are most valuable?
6. What elements are the big three in fertilizers?
7. Sulfur is widely used in ________ and ________ killers.
8. What mineral is used in making grinding wheels?
9. What materials are mixed together to make concrete?
10. What is the difference between cement and concrete?
11. Name the substances that make up building materials.
12. Why wouldn't you be rich if you owned a cubic kilometer of seawater?
13. Of all the substances that can be taken from seawater, which one is the most valuable to farmers?

Applying What You Have Learned

1. Describe how to calculate the amount of money spent per person per year for a resource.
2. If 1 km^3 of seawater contains \$150,000 worth of gold at \$14.00 per gram, how many grams of gold are in 1 km^3 of seawater?
3. Make a list of nonmetallic resources used in constructing your school.
4. Explain why this statement is not true: the earth's resources will someday be used up.

Research and Investigation

1. Investigate the composition of seawater and make a report on the percentage of concentration for the major elements.
2. Find out how table salt is recovered from seawater. Locate the places where this is done in the United States. If you can, find out which companies obtain salt this way.
3. Give an oral report on the major fishing grounds found in the ocean. Why are they located where they are?
4. Prepare a written report on the uses of fish meal and the future uses of fish-protein concentrate.
5. Interview a local manufacturing industry executive to find the sources of the materials the industry used in its plants.
6. Prepare an illustrated report on how iron is removed from its ore minerals.
7. Prepare an oral report on the mining of copper and how copper is removed from its ore minerals.
8. What will be the future world demand for fertilizer? Will there be enough? What can be substituted for chemical fertilizers?
9. Develop an illustrated talk showing how the mineral-supply picture of the United States has changed in the last 30 to 40 years.

New uses are constantly being found for lightweight, rust-resistant aluminum.

7001
L'eggs
1
Mini Marathon
1027

22 Nonrenewable Energy Resources

Do you have a lot of energy? When you say you feel full of energy, what do you mean? You probably mean that you feel ready and able to do something like play in a big ball game or go for a long bicycle ride. Whatever you do, you will probably make something move. When you pedal a bicycle, for example, you are using your energy to make the bicycle move.

Nothing can move or happen without energy. A bicycle will not move without energy. No light can be turned on without energy. No car can move, no plane can fly, no bell can ring, no breeze can blow without energy.

Where do moving things get their energy? You know that you need food. You know that an electric fan must be plugged into an electrical outlet. You know that cars need gasoline. You, fans, and cars need a source of energy. Food is your energy source. Gasoline is the energy source for a car. Electricity is the energy source for fans, lights, and toasters.

This chapter is about some sources of energy. You will learn about oil, coal, and nuclear sources of energy. You will learn what they are, where they came from, and how they are used. You will also learn how long the oil, coal, and nuclear fuel supplies will last.

Figure 22.1 *opposite:* It takes a lot of energy to run a marathon.

22.1 Petroleum

You will find out

- what petroleum is;
- what the source of petroleum is;
- how petroleum is used;
- how much petroleum the United States has.

Gasoline is but one of many products made from a thick, smelly black liquid. This black liquid is called **petroleum** [puh-TROH-lee-uhm]. *Petroleum* literally means "oil that comes from rocks." For years people called petroleum rock oil. In fact, the first oil well was drilled by the Pennsylvania Rock Oil Company in 1859.

Petroleum is mostly **hydrocarbons** [HY-droh-kahr-buhnz]. Hydrocarbons are made of just two elements, hydrogen and carbon. Both hydrogen and carbon can burn and give off energy. They are the energy sources in petroleum. The hydrogen and carbon combine in hundreds of different ways, making many different hydrocarbons. Petroleum is a mixture of dozens of these different hydrocarbons.

Petroleum is believed to have formed over a period of millions of years. It formed from tiny plants and animals that had been buried. Over long periods of time they decomposed. Petroleum is called a **fossil fuel**, since it formed from these plants and animals of such a long time ago.

Figure 22.2 How fossil fuels are formed

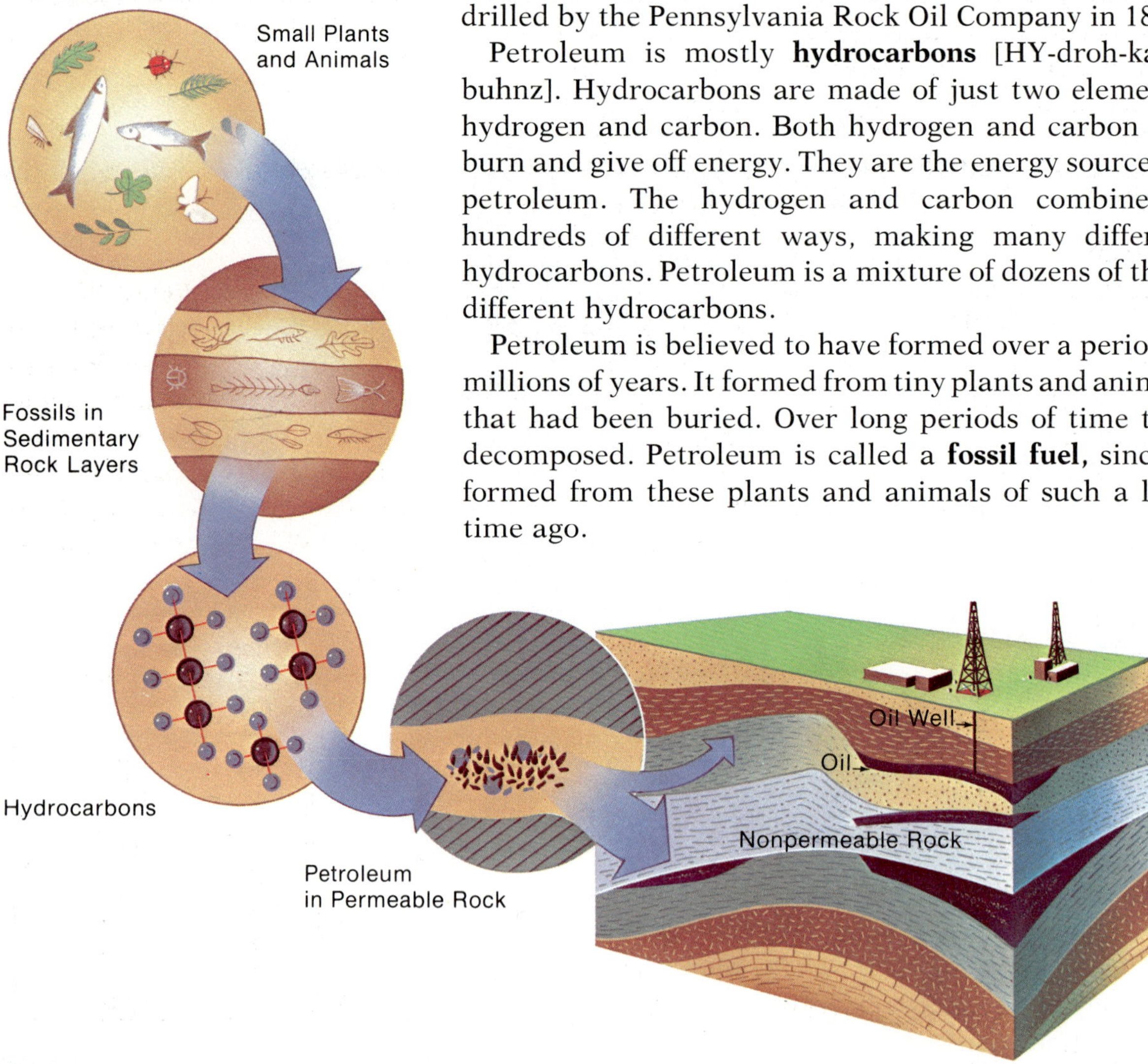

After the petroleum formed, it was forced through porous rock. It moved through the rock until it reached a rock type or rock structure it could not pass through. Here, in the rocks, it collected and formed a large oil pool called an **oil field.**

Petroleum is pumped from wells drilled into oil fields. The petroleum pumped from the wells is called **crude oil.** Oil can be dark or light, depending on the oil field. It also may contain various amounts of sulfur and other substances.

Many oil wells also produce petroleum gas in addition to liquid petroleum. Petroleum gas is a form of naturally occurring hydrocarbons of petroleum. It burns cleanly and can be transported easily by pipelines. Natural gas is widely used as a fuel for heating.

Figure 22.3 Shale may be a source of energy in the future.

ACTIVITY

How Much Oil Will a Pump Remove?

Materials

graduated cylinder
plastic bottle with spray pump
50-cm length of clear plastic hose
clean pebbles
100 mL of any motor oil
cold water
hot water
liquid detergent

Procedure

1. Assemble a model oil well as shown in the diagram. Pour 100 mL of oil into the well.

2. Pump out as much oil as possible. Record the amount recovered.

3. Add 50 mL of cold water to the well and again pump out all you can. Record the additional oil recovered.

4. Repeat step 3 but this time use 50 mL of hot water.

5. Repeat step 3 but this time add 50 mL of hot water with 15 drops of detergent.

Questions

1. How much oil was recovered by pumping?

2. How much oil was recovered by adding water, hot water, and detergent to the well?

3. Do you think oil wells are able to pump all the oil from an oil field? What evidence do you have for your answer?

4. Can you increase the amount of oil recovered from an oil field by adding water or detergents?

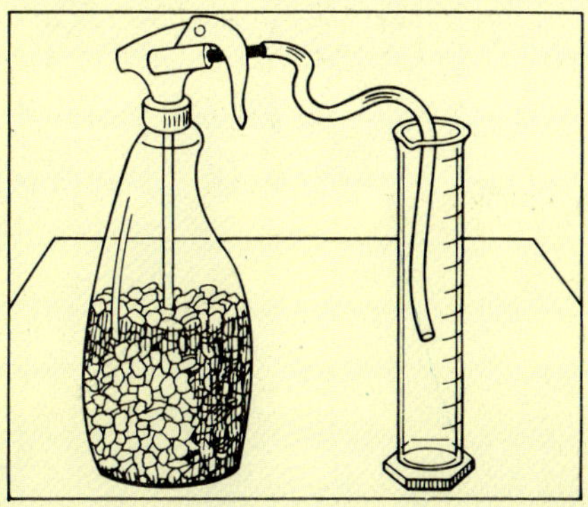

Crude oil itself has little use. It must be separated into the different groups of hydrocarbons which make up gasoline, kerosene, and other products. Crude oil is separated into its many useful products at a **refinery** [ree-FYN-uhr-ee].

The first step in the refining process is to heat the crude oil. Crude oil vaporizes when heated. Since different hydrocarbons condense from the vapors at different temperatures, the hydrocarbons can be separated. Crude oil vapors are piped into a tall, slim tower like the one shown in Figure 22.5. As the vapors rise through the tower, different hydrocarbons condense at different levels and are drawn off. Heavy fuel oils collect in pans low in the tower. Above this, pans collect the kerosene, which has condensed. Still higher, gasoline condenses into pans. Pipes drain the pans. Each group of hydrocarbons, such as gasoline, is now processed further or sold as it is.

Figure 22.4 *above:* An oil refinery like this one turns crude oil into many useful hydrocarbons.

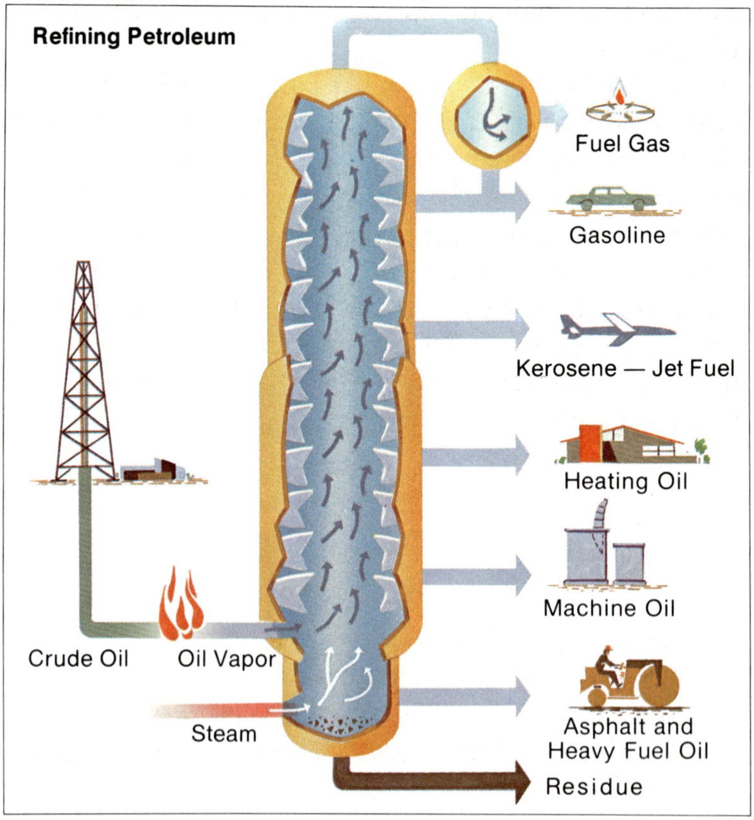

Figure 22.5 *right:* Through special heating and condensation processes, crude oil is separated into its different parts.

How is petroleum used? Petroleum oil and gas presently provide more than two thirds of the total United States energy supply. Petroleum moves almost all of our transportation—cars, trucks, planes, boats, and trains. Petroleum is also used for tractor fuel and to make fertilizers to grow the food you eat as well as to process it. In addition, petroleum is used for heating homes and for running some electric power plants. It is also used to make plastics and synthetic fibers such as nylon, some medicines, and detergents.

How much petroleum is there? Many of the oil and gas fields in the United States are being pumped almost dry. They are not really dry, since some of the oil remains in the rocks. However, present methods of pumping cannot move all the oil out. The part that can be pumped out may be largely gone by the early 1990's. Some way other than pumping will have to be used to remove the remaining petroleum. In addition, other sources of petroleum will need to be developed. Tar sands and oil shale, for example, can be heated to produce petroleum.

Figure 22.6 Will this be a common sign in the future?

It took millions of years for petroleum to form, but it is being used much more rapidly. Once it is used, there will not be any more for millions of years. For this reason petroleum is called a nonrenewable resource.

With conservation, the United States supply of petroleum could last until about the year 2050. All the petroleum could be gone by the year 2000 if it is used the way it has been used in the past. When the petroleum is gone, how will you run the cars, light the lights, and grow the food?

Do You Know?
Even if the earth were a hollow ball full of petroleum, the United States could consume it all in less than 350 years.

Study Questions for 22.1

1. What is petroleum?
2. Where is petroleum found?
3. Make a list of ways that petroleum is used.
4. How long will the United States petroleum supply last?

22.2 Coal

You will find out

- what coal is and how it forms;
- what the kinds of coal are;
- how much coal the United States has.

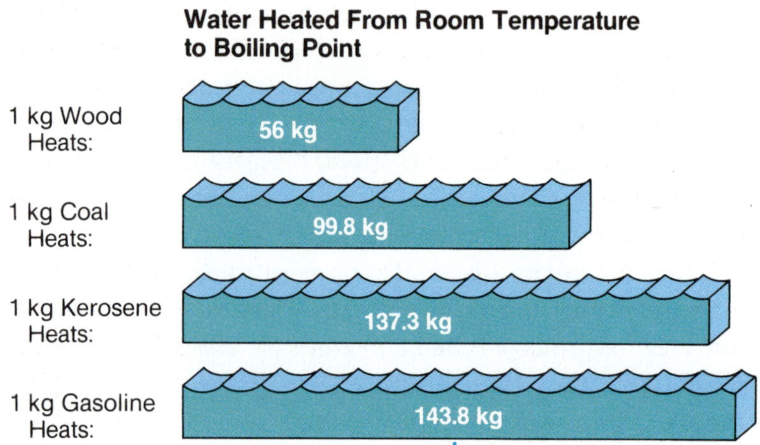

Figure 22.7 *right:* Compare the energy production of wood, coal, and kerosene.

Figure 22.8 *below:* The coal stove, once popular and then forgotten, may be commonly used again in the future.

Wood has been an energy source since the discovery of fire. Wood fires have been used to provide heat and cook food. Wood was one of the first energy sources used in the development of the United States. In fact, wood supplied almost all the energy until the late 1800's. Then people started using **coal,** the black rock that burns.

Burning coal releases about twice the amount of energy that burning the same amount of wood does. So people changed to coal. By 1910, coal supplied about three fourths of the United States energy needs. But then people began to discover the convenience of petroleum. Not only does petroleum burn cleaner than coal, but also it is a liquid fuel. Liquid fuels are easier to transport and to use. The switch was on again—this time to petroleum.

Today coal supplies only about 20 percent of the United States energy needs. However, as the petroleum supply shrinks, people are again looking to coal. Some oil-burning power plants have switched to coal. More ways of replacing petroleum with coal are being sought. Coal may again become a major energy source.

Coal is also called a fossil fuel. Coal formed from plants that lived 250 to 400 million years ago. The plants had to grow and collect under special conditions to form coal. Normally, when plants die, they are eaten by bugs or rot away. The coal-forming plants lived and died in great swamps. When the plants died, they fell into the water and sank. Since they were protected from bugs and rotting away, great thick layers of the dead plants collected in these swamps. In time these thick layers rotted a little and formed a brown material called **peat.** The peat, when dried, can be used as a fuel. This peat should not be confused with the peat moss sold in garden shops.

All coal was at one time peat. Over long periods of time the peat was buried by sediments. As the sediments piled up layer upon layer, they pushed down on the peat. The peat was pressed until it was only one eighth as thick as it once was. This pressure squeezed water from the peat and caused changes in it. The peat was compressed into a brown coal called **lignite** [LIHG-nyt]. More pressure and heating brought about more changes. The brown coal became a blacker, soft coal called **bituminous** [bi-TOO-mih-nuhs] **coal.** Still more pressure and heating made still more changes, and the soft coal became a hard coal called **anthracite** [AN-thruh-syt].

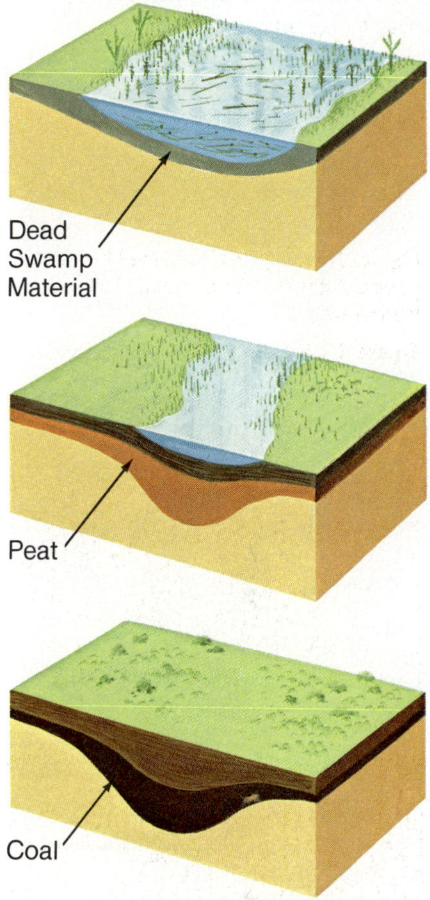

Figure 22.9 *above:* The making of coal

Figure 22.10 Peat bricks *(left)* change in time to anthracite coal *(right)*.

All coal is made of varying amounts of carbon, hydrocarbons, water, sulfur, and clay and sand that collected with the dead plants. It is the carbon and hydrocarbons that burn away as coal is burned. The remaining clay and sand are called ash. The quality of coal is measured by its heating value, how much ash it will produce, and how much sulfur is in it.

Figure 22.11 *above:* Fly ash enters the atmosphere from the smokestacks of a coal-burning power plant.

Figure 22.12 *right:* Acid rain is formed when gases such as sulfur dioxide react chemically with water vapor in the atmosphere.

About two thirds of the coal mined each year is burned to generate electricity. Modern electric power plants grind the coal to a flourlike powder before it is burned. This enables the power plants to get more energy from the coal, but it could also pollute the air with **fly ash.** Fly ash is the fine ash that flies up the chimney into the air when the ground coal is burned.

When coal is burned, it may also release sulfur dioxide. The sulfur dioxide from burning coal can combine with water in the atmosphere to produce **acid rain.** Some people are concerned that acid rain may kill the fish in lakes and streams, stunt plant growth, and even damage buildings.

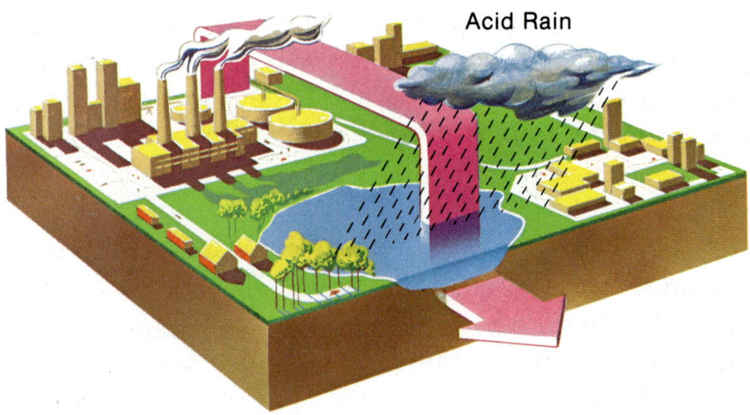

Sulfur dioxide and fly ash are the major air-pollution problems related to burning coal. Industries that burn coal must use the best available devices to remove sulfur dioxide and fly ash from plant exhausts. From 20 to 40 percent of the cost of a coal power plant goes toward pollution-control equipment. This makes electric bills that much higher, but everyone benefits from these efforts to keep the air clean.

Do You Know?
People felt so strongly about pollution from coal in England during the Middle Ages that the punishment for burning coal was death.

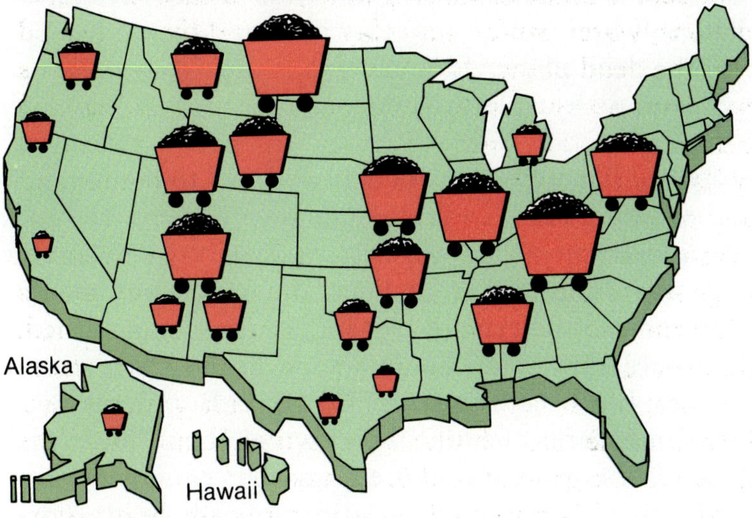

Figure 22.13 *left:* Major known coal resources in the United States

Figure 22.14 *below:* Inside a coal mine

How much coal does the United States have? Coal, like petroleum, is a nonrenewable resource. If coal is used at the present rate, it is estimated that there is more than a 200-year supply. However, some people are talking about increasing the use of coal in place of petroleum and even using coal to make gasoline. If coal is used at an increasing rate to replace shrinking petroleum supplies, the coal will be consumed much earlier. If just six percent more coal is used each year, it could be gone by the year 2050.

Hydrocarbons from coal are also used as a material to make dyes, synthetic fibers, plastics, and chemical products. If the petroleum and coal supply are both consumed by the year 2050, how will people run the cars and light the lights? Where will the hydrocarbons come from to make things? Are there other energy sources that can be used to replace petroleum and coal?

Study Questions for 22.2

1. What is in coal?
2. How is coal formed?
3. What are the major kinds of coal?
4. How long will the coal supply last?

22.3 Nuclear Fuel

You will find out
- how uranium provides energy;
- how a nuclear plant works.

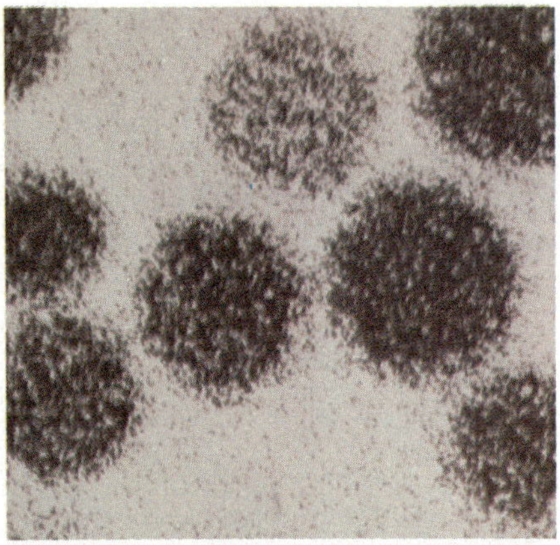

Figure 22.15 Radiation from uranium pellets can be seen on film.

It looked like any other rock. It was cool to the touch. It didn't glow. Yet it could take its own picture. The rock was placed on a sheet of film in a dark room. When the film was developed a day later, the rock could be seen. You can see the type of picture the rock made of itself in Figure 22.15. How could a rock do this?

Nearly 100 years ago, Antoine Henri Becquerel found that some rocks can take their own picture. These rocks give off a form of radiant energy that cannot be seen. Rocks that give off such radiation are called **radioactive** [ray-dee-oh-AK-tihv] rocks.

At Becquerel's suggestion, Pierre and Marie Curie studied the radioactive rocks. They found that the rocks contained the element uranium. They also discovered new elements that were radioactive.

ACTIVITY

What Is Fission?

Materials
graduated cylinder
water glass
125 mL of rubbing alcohol
1 teaspoon cooking oil
50 mL of water
teaspoon
knife

Procedure

1. Pour 125 mL of rubbing alcohol into a water glass. Add 50 mL of water. Stir the solution.

2. Carefully lower one teaspoon of cooking oil to the surface of the alcohol-water mixture. Gently tip the spoon over all at once. A single blob of oil should float in the middle of the mixture as shown in the drawing.

3. Use the knife to carefully push on the drop. See if you can make the drop split. Record your observations.

Questions

1. Describe your observations of the floating oil drop.

2. If the oil-drop model is similar to an atom, how do you know an atom resists fission?

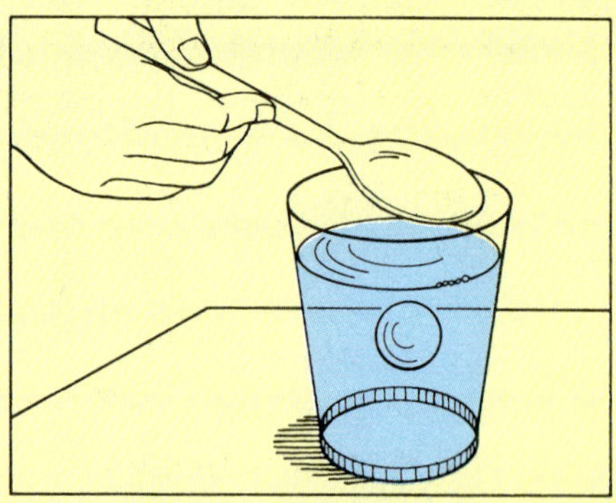

Other people continued to study and work with the radioactive elements. In 1939, Otto Hahn and Fritz Strassmann found that they could split uranium atoms into new and lighter elements. As a result of the splitting, there was a loss of matter. Part of the matter had been changed to energy. This splitting of atoms with the release of energy is called **fission** [FISH-uhn]. Figure 22.16 shows how the uranium atoms fissioned.

Not all atoms can be made to fission. Earlier, you learned that atoms of an element occur in different forms called **isotopes** [EYE-suh-tohps]. Uranium 235 (U 235) is an isotope that can be made to fission. As shown in Figure 22.17, the splitting of the atom not only produces less massive atoms but also releases energy and three neutrons. A neutron is one of the elementary particles of an atom. These neutrons can be used to split *other* U 235 atoms. This will release even more neutrons, which can move out to split still more U 235 atoms. This can keep going on and on by itself and releasing energy. This continuous action is called a **chain reaction.**

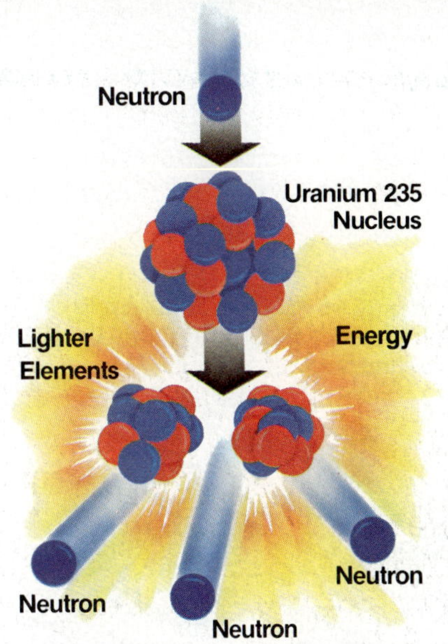

Figure 22.16 *above:* Fission happens when certain atoms split apart, releasing large amounts of energy.

Figure 22.17 *left:* A chain reaction occurs when the splitting of one atom produces neutrons that then cause other atoms to split in the same way.

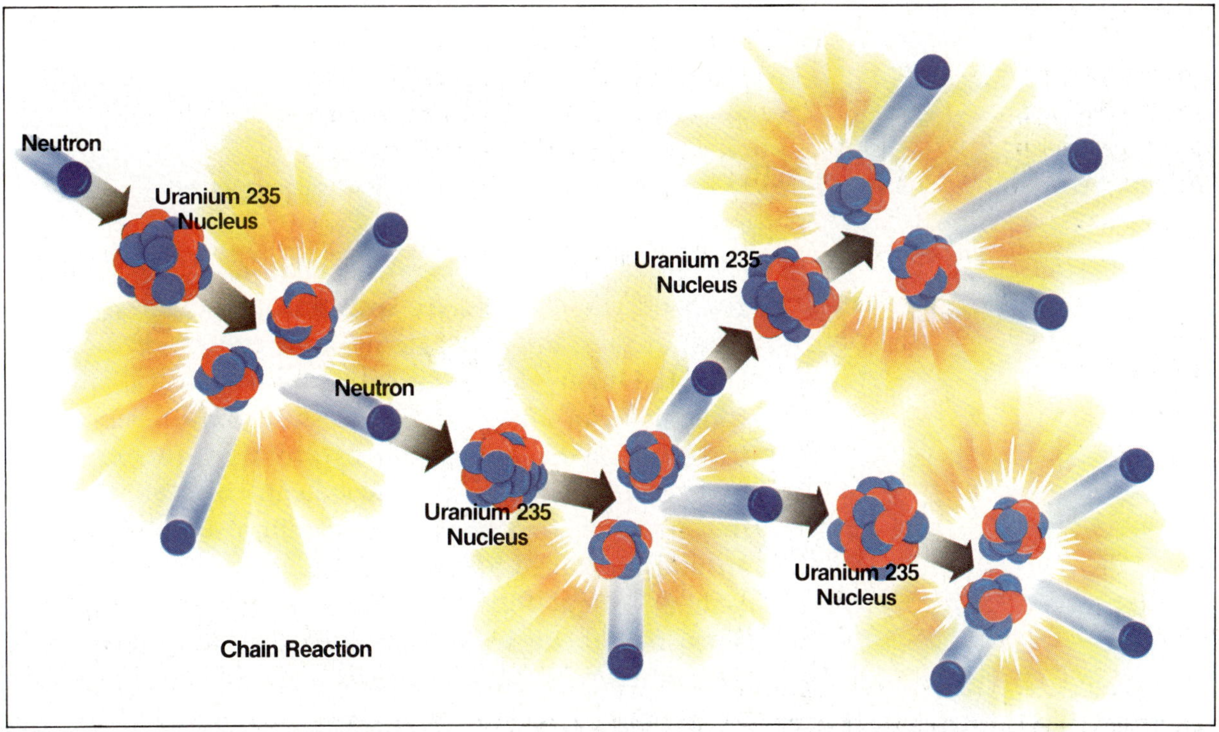

Uranium, as it is mined from the earth, is a mixture of isotopes. It contains both U 238 and U 235 isotopes. It contains about 99.3 percent U 238 and 0.7 percent U 235. U 235 can be fissioned, but U 238 cannot. Before the uranium can be used as an energy source, it must be refined to increase the amount of U 235 isotopes it contains to at least 3 percent. This will enable the fission to keep going by a chain reaction.

Figure 22.18 Compare the energy resources of coal, petroleum, and uranium.

A great amount of energy, called nuclear energy, is given off when atoms are split during a chain reaction. Compare the different amounts of energy that can be obtained from wood, coal, oil, gasoline, natural gas, and splitting uranium. Figure 22.18 shows how far a car can travel, using one kilogram of each of these fuels. Are you surprised to learn that the wood can only get the car to the end of the driveway? The uranium will send the car around the world 560 times!

In a typical fossil-fuel power plant, coal or oil is burned to heat water, producing steam. The steam runs a turbine which drives a generator, producing electricity.

A nuclear electric power plant like the fossil-fuel power plant also requires steam to run a turbine and generator. But a nuclear plant is different in how it produces the steam. To produce the steam, energy is released by fissioning uranium 235 in a special vessel called a **nuclear reactor.** A nuclear reactor is a 20-cm-thick steel vessel that stands about 10 m tall.

The uranium is in tubes in the lower part of the reactor. The top part contains the **control rods** that move up and down. These rods are made of a material that absorbs neutrons. As the rods are lowered, they absorb more neutrons and slow the chain reaction. When the rods are at the bottom, they completely stop the chain reaction. The reactor is inside a large, heavy, concrete **containment building.**

Figure 22.19 *above:* One uranium pellet can supply energy equal to three barrels of oil or one ton of coal.

Figure 22.20 *left:* A nuclear reactor is carefully encased.

Do You Know?
It is totally impossible for a nuclear power plant to explode like an atom bomb.

Water under high pressure is pumped through the reactor. It becomes heated by the energy given off by the fission process. This hot water is used to heat other water outside the reactor and change it to steam. The steam now turns the turbines and generator, producing electricity.

In nuclear reactions various particles and rays are given off from the nucleus. These particles and rays are called nuclear radiation.

Radiation can be dangerous. Too much radiation can harm people, just as too much of some chemicals or even too much sunlight is harmful. Nuclear power plant builders know that accidents are possible. Great care is taken to build in as many features to prevent accidents as possible. Nuclear power plants are built to control and contain radiation.

Figure 22.21 This diagram shows how energy from a fission reaction is used to produce electricity.

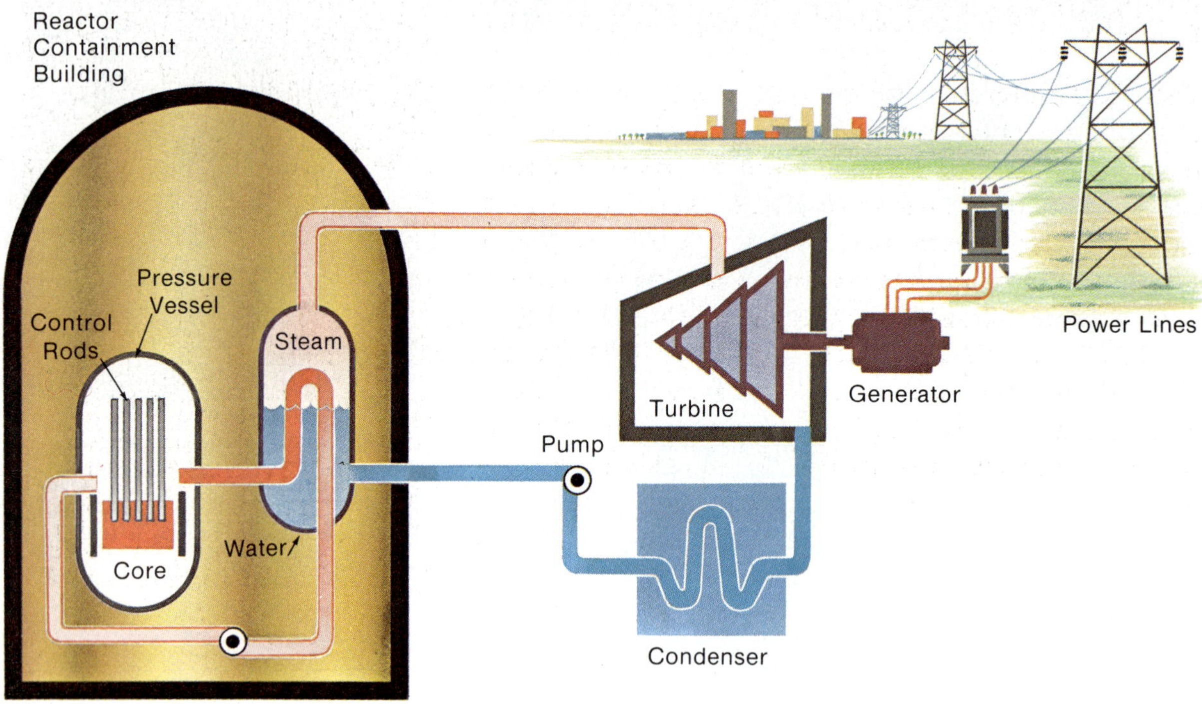

Uranium, like oil and coal, is a nonrenewable resource when used in a fission power plant. The known available uranium in the United States will last only until about the year 2020.

Do You Know?
All the recycled high-level nuclear waste from all United States power plants by the year 2000 would make a pile on one football field about one meter high.

Figure 22.22 A nuclear power plant

Figure 22.23 Energy for city lights coming from the center of atoms

There is a different kind of reactor that makes more fuel than it uses and can extend the uranium supply by thousands of years. It is called a **breeder reactor.** Some people feel there are, however, unanswered questions about health and safety matters in connection with the breeder reactor. Scientists are studying these problems.

Once a nonrenewable energy source is used, it is gone. Estimates indicate that much of the petroleum, coal, and uranium will be gone sometime in the next century. Conservation and new discoveries may extend that time some, but other energy sources will obviously be needed soon.

Study Questions for 22.3

1. What is a fission?
2. What is a chain reaction?
3. How is nuclear fuel used to provide energy?
4. How does a nuclear power plant generate electricity?

What Did You Learn?

- Petroleum is oil from oil-bearing rock.
- Petroleum is a mixture of different hydrocarbons.
- Petroleum formed over millions of years from buried plants and animals. It is thus called a fossil fuel.
- Petroleum is forced through porous rock until it reaches a place it cannot move through, forming an oil field.
- Petroleum pumped from oil fields is called crude oil.
- Crude oil is refined by heating it and then condensing the hydrocarbon vapors.
- Petroleum is used as an energy source for transportation, heating, and generating electricity.
- Petroleum is a nonrenewable energy source. The United States supply may be used up between the years 2000 and 2050.
- Coal is a black rock that burns. It formed from plants that lived 250 to 400 million years ago.
- Coal contains varying amounts of water, hydrocarbons, carbon, sulfur, and clay and sand.
- Clay and sand form ash that is left after the coal burns. Ash and sulfur present environmental problems when coal is burned.
- Coal is a nonrenewable energy source. The United States supply may be used up between the years 2050 and 2200.
- Some elements are radioactive and give off radiant energy.
- Some elements occur in different forms called isotopes. Some isotopes, such as uranium 235, will fission when struck with a neutron.
- When uranium 235 fissions, it produces less massive atoms and more neutrons. These neutrons can cause other U 235 atoms to fission in a continuing reaction called a chain reaction.
- A nuclear reactor is a container in which a controlled chain reaction releases energy. This energy can be used to produce electricity.

Key Terms

petroleum
hydrocarbon
fossil fuel
oil field
crude oil
refinery
coal
peat
lignite
bituminous coal
anthracite
fly ash
acid rain
radioactive
fission
isotope
neutron
chain reaction
nuclear reactor
containment building
control rod
breeder reactor

Career

Petroleum Chemist

Petroleum is a mixture of hydrocarbons and other substances. Crude oil from different oil fields contains various amounts of different hydrocarbons, other substances, and water. Before an oil company can refine the crude oil into gasoline and other products, it must know what substances and how much water are in a shipment of crude oil. A petroleum chemist is a scientist who analyzes crude oil to find out what is in it.

Sonia Hoffner is a petroleum chemist. She works for the Marathon Oil Company in Denver, Colorado. In addition to analyzing crude oil, Sonia analyzes metals. Expensive metals such as platinum are used in the refining process. These metals help refine the gasoline that runs automobiles. By analyzing the metals, Sonia Hoffner helps to verify that her company gets what it pays for. She also analyzes groundwater from near oil wells. This information tells engineers what the rock formations are like.

When Sonia Hoffner was in school, she didn't know that one day she would be a scientist. She was very good in math and enjoyed studying it. A science course that she took increased her curiosity about the world. She learned to ask questions about things she didn't understand in books. With her ability in math and an active curiosity, she went to college and studied to become a chemist.

TO THINK ABOUT AND DO

Vocabulary

Some of the words in the following sentences have scrambled letters. Unscramble the words. Number your paper from 1 to 8 and write the unscrambled word beside the appropriate number.

1. Coal and petroleum are *LOSFIS LESUF*.
2. Hydrocarbons are condensed from vaporized crude oil at a *FERNYIER*.
3. Petroleum, coal, and uranium are *WEEBENRONNAL*.
4. *AOCL* is formed from former plant life.
5. *ISOFINS* occurs when a large atom splits into smaller ones.
6. When one fissioning atom causes other atoms to fission, there is a (an) *ACINH ACINTOER*.
7. A container in which a controlled chain reaction releases energy is a (an) *CAERLUN CAERROT*.
8. The *ROTCLON DOSR* can stop fissioning in a nuclear reactor.

What Do You Remember?

Number your paper from 1 to 10. Write the letter of the correct answer beside the appropriate number.

1. Petroleum pumped from wells is called (a) vegetable oil, (b) mineral oil, (c) crude oil, (d) motor oil.
2. How much of the United States energy supply comes from petroleum? (a) all of it (b) more than half (c) more than two thirds (d) hardly any
3. Without conservation, the United States petroleum supply could be gone by the year (a) 2000, (b) 2050, (c) 2500, (d) wait . . . there is enough petroleum to last forever.
4. The greatest amount of sulfur dioxide and ash wastes are produced by using (a) petroleum, (b) coal, (c) uranium, (d) any; they all are equal.
5. If used at the present rate, the United States has a 200-year supply of (a) petroleum, (b) coal, (c) uranium, (d) wood.

6. Which of the following energy sources releases the most energy per kilogram? (a) petroleum (b) coal (c) uranium (d) wood
7. The ash that remains after coal has burned is from what in the coal? (a) hydrocarbons (b) carbon (c) sulfur (d) clay and sand
8. Which of the following can stop a fission reaction? (a) water (b) chain reaction (c) control rod (d) U 235
9. The difference in coal and nuclear power plants is (a) the way electricity is produced, (b) the energy source, (c) the type of electricity produced, (d) electricity from a nuclear power plants is radioactive.
10. More fuel is made than is used in a (a) fission reactor, (b) chain reaction, (c) breeder reactor, (d) nuclear power plant.

Applying What You Have Learned

1. Make a list of the many ways you use petroleum or petroleum products. Remember that almost all plastics and many synthetic fibers are made from petroleum. Assume petroleum is now used up. Make another list showing how you would replace the petroleum products with other items.
2. What causes acid rain? Why is most of the concern about acid rain focused in New England and eastern Canada?
3. What are the problems with petroleum, coal, and nuclear fuel supplies? How can these problems be resolved?

Research and Investigation

1. Some people feel that synthetic petroleum from coal will replace the dwindling petroleum supply. Research and prepare a report on how coal can be made into gas and liquid fuels. Is this being done now? What are the problems?
2. Investigate how the nuclear fuel supply can be increased. Prepare a report on how the breeder reactor works. Investigate how spent fuel rods can be recycled. Where are these things being done today?

23 Renewable Energy Sources

Imagine living and working on a huge satellite, a space station. Outside, space would always be black, spotted with clearly shining stars. You wouldn't have to wait for sunset or worry about clouds. You could see the stars any time you wished.

Energy would be needed to run the space station. Where do you suppose the space station would get its energy? If the station were dependent on fuel, rockets would have to continually bring new supplies. This would not be practical. What would be needed is a source of energy that would not run out. Otherwise, the station could operate only for a short time.

Now think of yourself on another satellite, Earth. Earth is a satellite of the sun. People on Earth need energy to run their cars and trucks, heat and cool their homes, and light their lights. Most of this energy now comes from oil and coal. But these sources will be used up in time. Like the space-station satellite, people on Earth also need an energy source that will not run out. How many sources of energy do you know of that are endless? In this chapter you will explore some sources that may answer this question.

Figure 23.1 *opposite:* Whether in space or on Earth, people need a dependable source of energy.

23.1 Solar Energy

You will find out

- what a renewable energy source is;
- how a solar collector provides hot water;
- how to heat a house with solar energy;
- how heat from solar energy can be used to generate electricity;
- what a solar cell is;
- why more solar energy isn't used to generate electricity today.

As you pass through the 1990's and into the next century, you will be passing from one energy age to a new one. The old age was based on cheap and once-plentiful fossil fuels. The new age will be based on **renewable energy sources.** Renewable energy sources do not run out because they are constantly replaced.

You know how the sun warms you. Sunlight is an energy source called **solar energy.** Since the sun continues to send more solar energy each day, solar energy is called renewable. All you need to use solar energy is some way to collect it and put it to use.

Figure 23.2 *above:* The sun is the most abundant renewable energy source.

Figure 23.3 *right:* Sunlight provides energy for living things even under the ocean's surface.

One way to collect solar energy is to use a **solar collector.** One type of solar collector is like a small greenhouse. It is a box with a glass or plastic top. Inside the box are tubes that are painted black. Black is used because it absorbs more sunlight than colors do. Water moving through the tubes in the solar collector is heated by sunlight. In some solar collectors a pump moves the water from a storage tank, through the collector tubes, and back to the tank. The tank provides hot water for bathing, washing, or even heating a swimming pool.

New homes and apartment buildings are now designed and located to use solar water heaters. Roof lines are planned to provide the best angle for the solar collectors. The buildings can also be located so they face the direction that gets the most sunlight. Building sites are also carefully checked to be sure there are no high natural obstructions to cast shadows on the collectors.

Do You Know?
Of all the energy used in the United States in the early 1980's, 25 percent was consumed in warming or cooling buildings. The use of passive solar energy could reduce this to 10 percent.

Figure 23.4 *above:* A solar collector absorbs sunlight, which produces heated water.

Figure 23.5 *left:* Solar collectors are being used in more and more places to provide heat and hot water for people.

Solar energy can also be used to heat a house. A house can be built so the house itself collects solar energy. Sunlight enters the house through south-facing windows and is absorbed by materials in the house. The warmed materials then heat the living space during the day and night.

When a house is built to use solar energy and does not use electric pumps, fans, or heaters of any type, the house is called a **passive solar** house. You may think that a passive solar house would just have large windows to let sunlight in. This kind of design will not work because the house will become cold after sunset. A passive solar house has something to collect and store energy during the day. Sunlight warms some dense material such as concrete, brick, or stone. This dense material, or **storage mass,** then gives off the stored energy during the day and night. It warms people by giving off radiant energy. It also warms the nearby air. This sets up convection currents that can carry the warm air to the rest of the house.

Figure 23.6 During the daytime, sunlight coming through the windows heats the air inside and is absorbed by the storage mass. At night the storage mass radiates the stored heat into the house.

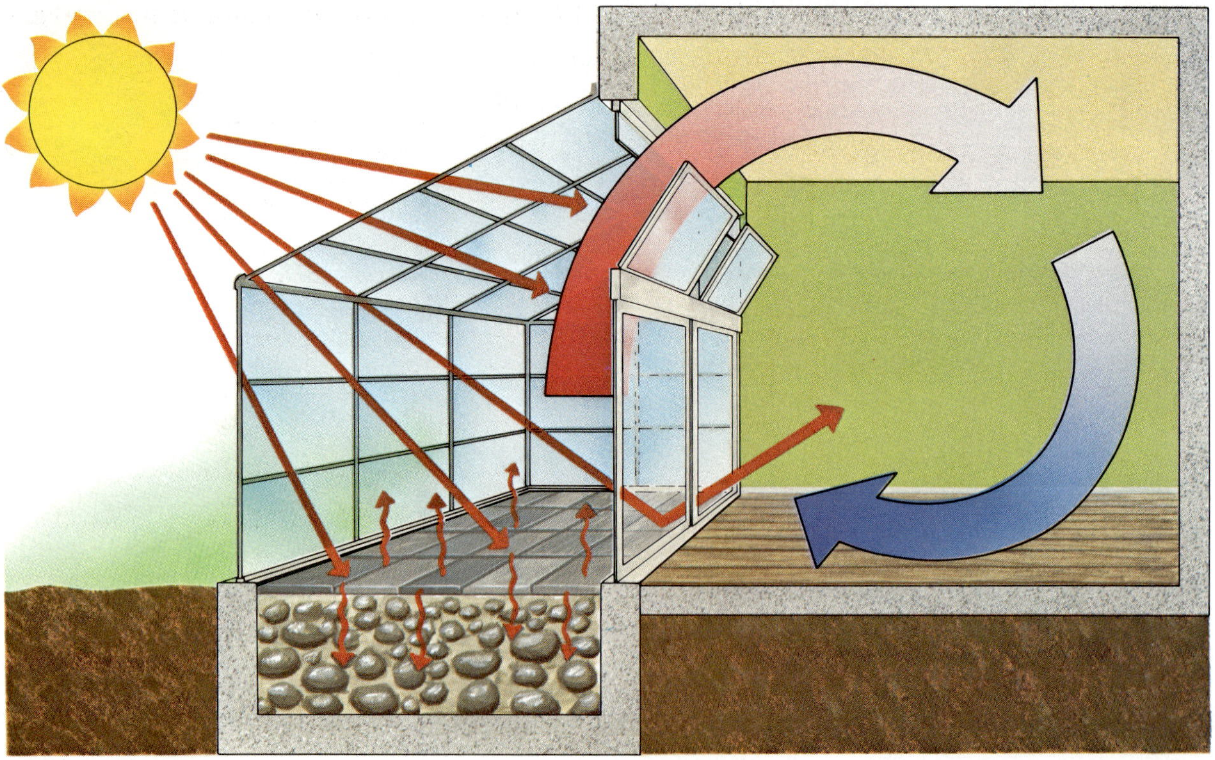

ACTIVITY

What Does a Storage Mass Do?

Materials

8 tin cans painted black
water
2 large cardboard boxes
2 thermometers
plastic wrap
tape
string
rubber bands

Procedure

1. Use two cardboard boxes to build two solar greenhouses like the one shown.

2. Fill each tin can with water and cover with plastic wrap and a rubber band.

3. Place all the cans in the back of one of the greenhouses as shown.

4. Hang a thermometer inside each greenhouse. Cover each greenhouse with plastic wrap and seal it tightly with tape.

5. Place both greenhouses side by side in sunlight so that the sunlight shines into each.

6. Make a record sheet as shown. Record the temperature every five minutes in each greenhouse for a half hour.

7. Now place the two greenhouses in the shade. Record the temperature every five minutes for another half hour.

8. Make a graph of the temperature changes in the two greenhouses.

Questions

1. What did you see happen to the temperature in the two greenhouses?

2. What was the purpose of the water-filled cans? How do your observations show this?

3. Which of the greenhouses is more comparable to a passive solar house?

Do You Know?
Solar power towers are expected to start replacing oil- and gas-fired electric power plants by 1990.

Figure 23.7 *above:* Heat coming from sunlight can be intensified by focusing the beams through a lens.

Figure 23.8 *right:* These mirrors act like hundreds of magnifying lenses and concentrate sunlight on the power tower.

Most electricity in the United States is generated by using steam. Fossil fuels or uranium is used to make the steam in a power plant. Why isn't solar energy used instead of fossil fuels or uranium in a power plant? Consider trying to boil a pan of water by placing it in sunlight. The water does not boil because not enough solar energy arrives at the pan to boil the water. The sunlight must be concentrated, or brought together, before it will boil water. You can concentrate sunlight, for example, by using a magnifying glass. It would take an enormous magnifying glass, however, to provide the energy needed by a power plant to make steam.

Mirrors, however, can be used to concentrate enough sunlight to produce steam for a power plant. Figure 23.8 shows one way this has been done. Mirrors have been designed that follow the sun and reflect sunlight to the top of a tall structure called a **power tower.** Sunlight from 2.5 square kilometers can thus be concentrated on an area of several square meters. Tubes in the top of this power tower are heated to over 500°C by the concentrated sunlight. Steam is produced from the heating in the power tower. This steam is used to turn a turbine to generate electricity just like in other power plants.

Do You Know?
The first major use of solar cells was to provide energy for the *Vanguard* satellite in 1958.

Figure 23.9 *left:* Solar cells.

Figure 23.10 *below:* Solar cells provide electricity in unusual situations.

Sunlight can also be converted directly into electricity in **solar cells.** A solar cell is made of silicon and tiny amounts of arsenic and boron and produces an electric current when sunlight falls on it. When rays of sunlight strike the top part of a solar cell, electrons are knocked loose. These electrons move to the bottom part. The top part now has an excess of positive charges, while the bottom layer is loaded with negative charges. The movement of charged particles produces an electric current.

You may wonder why solar energy isn't used more to generate electricity. The problem is the cost. Solar energy is free, but the devices necessary to convert it to your use are very expensive. A power tower, with its mirrors, for example, costs about one third more than a comparable nuclear power plant. You could use solar cells to produce all of your electric power. However, the cost of the solar cells would make electricity several times more expensive than it is today. Many people think the cost of solar cells will be less in the future.

Study Questions for 23.1

1. What are the characteristics of a renewable energy source?
2. Describe how a solar collector can be used to heat water.
3. Describe how a house can be heated with solar energy.
4. How does a power tower generate electricity?
5. Describe a solar cell.
6. Why isn't solar energy used more to generate electricity?

23.2 Water, Wind, and Plants

You will find out
- how energy is obtained from moving water;
- how the wind provides energy;
- why plants can be considered a source of energy.

Have you ever tried to swim or paddle a canoe in a fast-moving stream? Did you ever try to walk against a strong wind? Have you ever had to move back from a roaring campfire? Moving water, wind, and wood all contain energy. How are they used as energy sources?

Moving water provides energy by turning a waterwheel or turbine. Waterwheels have been used for many years to move a millstone to grind grains into flour and to run machines in factories. There was a problem, however. The waterwheels had to be located where there were waterfalls, and some streams where the waterfalls were located ran dry during part of the year. The building of dams and the development of electricity solved this problem.

Figure 23.11 *above:* The old-fashioned waterwheel is a reminder of a time when energy demands were small.

Figure 23.12 *right:* Dams control the flow of river water so that electric power can be generated on a regular basis.

Today moving water is used mostly to generate electricity. Water stored behind huge dams falls through large pipes in the dam to a powerhouse. Here the falling water turns turbines, which turn generators that make electric energy. Since electric energy moves through wires, energy can now be sent to homes and factories located elsewhere.

Do You Know?
The first waterwheel that generated electricity in the United States was used to provide electricity for a mine near Ouray, Colorado, in 1890. The equipment was built by George Westinghouse.

Wind power has been used for centuries to move ships, grind grain into flour, and pump water for crops and cattle. Wind was turning generators and producing electricity as early as 1910. The wind provides energy by moving blades of a wind machine or windmill. The moving blades turn a shaft that can run a water pump or do other work.

The wind seems stronger in some places than in others. Big wind machines for making electricity are usually located on mountains and near the seashore, which seem to have stronger and steadier winds.

The parts of plants or things made from plants can be used for energy. This source of energy is called **biomass** [BY-oh-mas]. Biomass includes wood, leaves, stems, and other plant parts as well as wood chips, trash, and even animal wastes.

Biomass can be burned for energy. It can be burned to heat houses, cook with, make hot water, and even generate steam. It can be used for just about anything that oil or coal can. It just isn't so convenient, and it does not give off as much energy for the same amount of material.

In addition to being burned directly, biomass can be converted into fuels. When biomass is rotted in a closed container without air, methane gas is made. This gas can be used for heating, cooking, or running engines.

Biomass that has sugar or starch in it can be changed to alcohol. When this kind of biomass is mixed with yeast and water, the yeast makes alcohol from the sugar. This alcohol can be separated and used as a fuel. When mixed with gasoline, the fuel is called gasohol. People use too much energy today to depend just on biomass. However, biomass is an important energy source that can be used to supplement other energy sources.

Figure 23.13 Wind is an old source of energy that is being used in new ways.

Do You Know?
Biomass provides about 2 percent of all United States energy needs.

Figure 23.14 Chemical changes occur in biomass-producing substances that can be burned for energy.

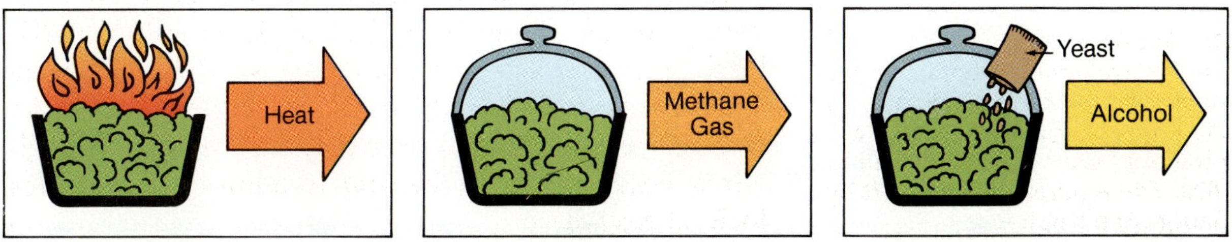

Where did moving water, wind, and plants get their energy? They got their energy from the sun as you will see. Using energy from moving water, wind, or wood is really an indirect use of solar energy. How is this so?

As part of the water cycle, the sun warms the ocean, evaporating water. The water vapor rises and moves in over the land and forms clouds. Then the water falls as precipitation from the clouds. The water that falls in higher elevations rushes downhill across the land and back to the sea, completing the cycle. Energy can be

ACTIVITY

Which Nut Has the Most Energy?

Materials

safety goggles
gloves
shelled peanut, walnut, and pecan
knife
wire coat hanger
test tube
aluminum foil
clay
straight pins
graduated cylinder
balance
thermometer
matches
water

Procedure

1. Cut a wire coat hanger and form it as shown. The bottom should be 10 cm across. The loop at the top should hold a test tube.

2. Cover the wire with aluminum foil as shown. Carefully place the test tube in the top of the loop.

3. Trim a walnut and a pecan until they have the same mass as the peanut.

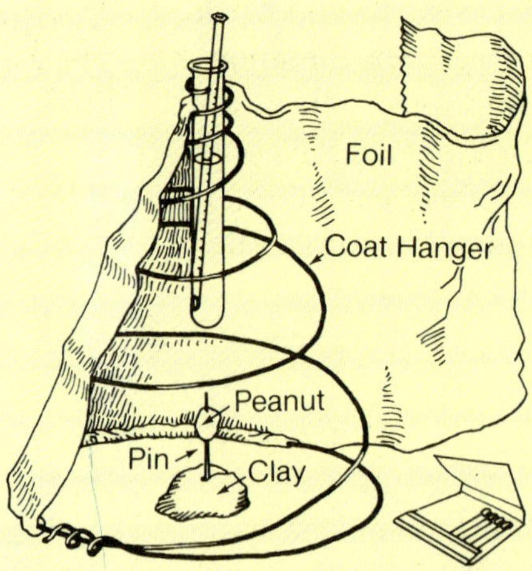

4. Push a pin into the peanut. Position the pin in a piece of clay as shown.

5. Pour 10 mL of water into the test tube. Measure and record the temperature of the water.

6. Put on safety goggles and light the peanut with a match. Place it under the test tube when it starts to burn.

7. When the peanut is through burning, measure and record the temperature of the water again.

8. Repeat steps 5–7 with the walnut and the pecan.

Questions

1. How much did the temperature increase from burning each nut?

2. What do your observations tell you about the energy in each nut?

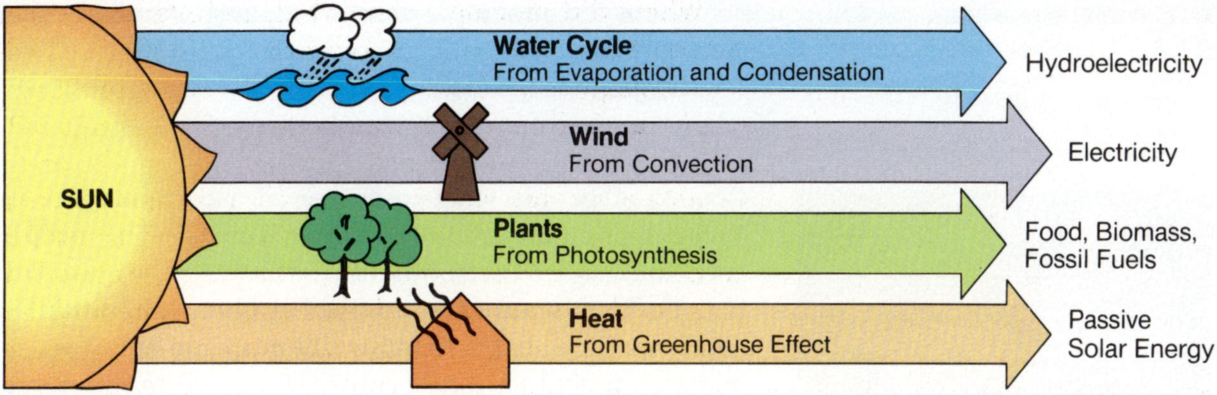

obtained from the moving water. Since the supply of moving water is continuously replaced, it is a renewable energy source. However, it is sunlight that is the original energy source that begins the water cycle. The energy in the moving water is indirect energy from sunlight.

The wind is another form of indirect solar energy. The wind blows because of energy the earth receives from the sun. Air is warmed because of energy from sunlight. The warm air rises and starts a convection current. This movement of the air produces wind. Energy can be obtained from the wind. Moving air, like moving water, is indirect energy from sunlight.

Now think about this a minute. All of your food, your clothes, and even the fossil fuels have been made possible by the sun. They can all be traced back to plants, and green plants use light from the sun. Green plants use sunlight to make the stems, leaves, roots, wood, and fruit of the plant. The sun's energy is stored in the plant. It is given off when the plant is burned. A plant, then, can be thought of as a living solar collector that stores the sun's energy. Using plants is another indirect use of solar energy.

Figure 23.15 *above:* In one way or another the sun supplies most of the energy used on Earth.

Figure 23.16 *below:* An orange is really a package of the sun's energy that has been stored as food by the tree.

Study Questions for 23.2

1. How is the energy from moving water used to generate electricity?
2. How does the wind provide energy?
3. Describe how energy from moving water, wind, and plants is indirectly solar energy.

23.3 Geothermal Energy

You will find out
- what geothermal energy is;
- how geothermal energy can be used.

Groups of people start to gather at a certain place in Yellowstone National Park in Wyoming. The people keep looking at their watches. What are they waiting for? They are waiting to see Old Faithful. Old Faithful is a **geyser** [GY-zuhr], a fountainlike eruption of hot water and steam. Old Faithful erupts for several minutes and is then still. Since the geyser erupts regularly about every hour, it was named Old Faithful.

Geysers and **hot springs** are found near old volcanoes. Surface water moves down through rocks to a place where the rocks are very hot. Convection then pushes the now hot water back to the surface, forming hot springs. In some rock formations steam is generated beneath the water. As the steam pressure builds up, it

Figure 23.17 *above:* The energy that heats water into steam and produces Old Faithful comes from within the earth.

Figure 23.18 *right:* The water in hot springs is heated by rocks below the surface of the earth.

periodically is released by forcing the water and steam to erupt through a surface vent, forming a geyser. All of this takes place near the surface. The rocks that heat this water are fewer than several kilometers below the surface.

Hot rocks can heat water and produce steam. How much energy can you get from this kind of source? Scientists think that it may have 30 times more energy than the United States uses each year. As you will see, however, not all of this energy can be used. Since this energy is from the land or earth's crust, it is called **geothermal** [jee-oh-THUHR-mel] **energy.**

Do You Know?
The worldwide use of geothermal energy for generating electricity is expected to be 17 percent greater each year until the year 2000.

Figure 23.19 Forms of geothermal energy shown here are being developed as new energy sources for people.

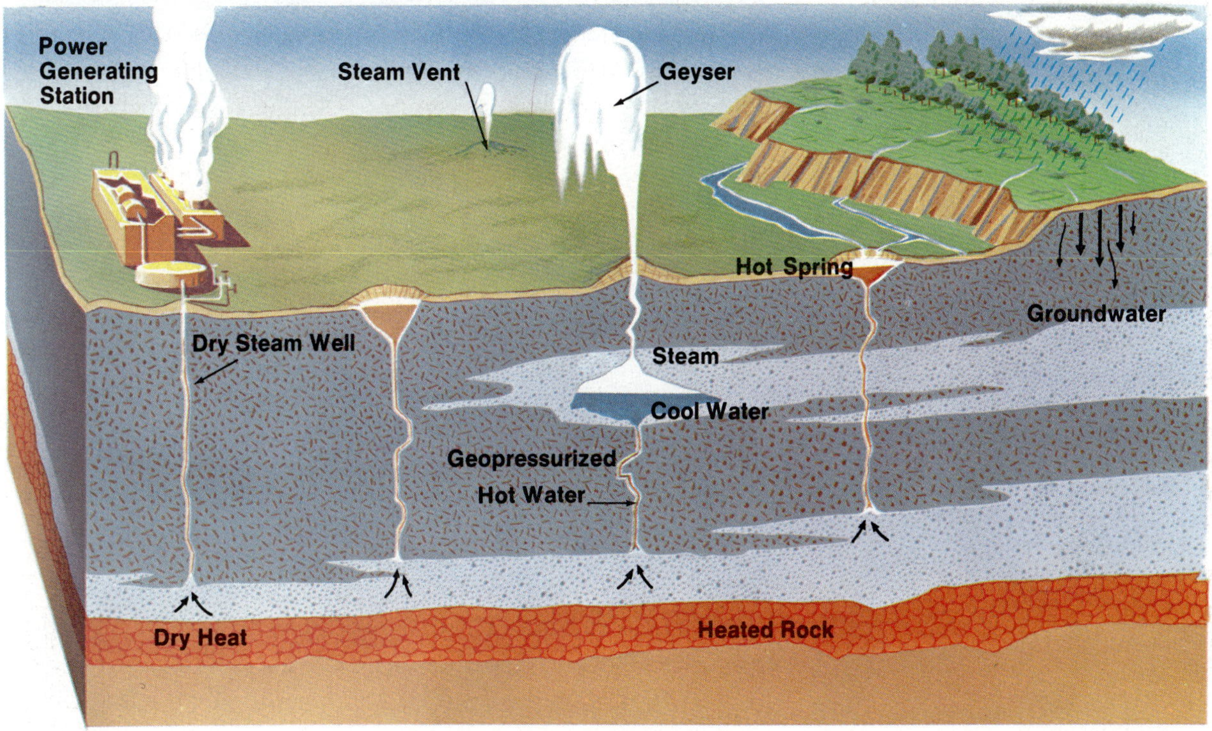

Each place where geothermal energy is found is different. There are differences in the rocks and in the size and temperature of the area. There are also differences in the amount of water present. These differences result in different forms of geothermal energy. Geothermal energy is generally found in one of four forms: (1) dry steam, (2) hot water, (3) geopressurized hot water, and (4) hot, dry rock.

Dry steam makes up only 0.5 percent of all geothermal energy in the United States. Steam is piped from wells to turbines. The turbines turn generators, producing electricity. Dry steam runs an electric power plant north of San Francisco, called The Geysers. More than half of San Francisco's electricity comes from this plant. Two more places are known to have dry steam. Both of these places are in national parks in California and Wyoming. Since they are located in parks, this steam cannot be used for power.

Figure 23.20 *right:* The Geysers dry-steam power plant

Figure 23.21 *below:* A geothermal hot water source in Iceland.

Hot water is heated by the hot, underground rocks. Many states have this geothermal resource. It is being used in California, Utah, New Mexico, Oregon, Idaho, and other states. Geothermal hot water, like geothermal dry steam, can be used to generate electricity. Geothermal hot water, however, requires a different type of power plant. The hot water is used to heat fluid with a low boiling point. Freon is often used, since it boils at about 30°C. The boiling Freon then turns a turbine which turns a generator, generating electricity.

Geothermal hot water can also be used any place heating is required. It is used for heating in Iceland, New Zealand, and Italy. It is also used in Klamath Falls, Oregon, and Boise, Idaho, to heat buildings.

Figure 23.22 *left:* Today's scientific research may lead to greater use of dry-rock geothermal energy in the future.

Figure 23.23 *below:* An electrically powered car

The two other forms of geothermal energy have yet to be developed. *Geopressurized hot water* occurs in deep reservoirs along the coast of Louisiana and Texas. This hot water is sealed off from the surface and is under pressure from the overlying rock. This water contains much dissolved natural gas. *Hot, dry rock* is the most common form of geothermal energy. There is much hot, dry rock in the western United States. Someday energy from hot, dry rock may be used by pumping water underground. After the water is heated underground, it can be pumped out and used as hot water.

Which of these renewable energy sources will you be likely to use in the future? The most likely answer is probably all of them. As fossil fuels are used up, other energy sources will be developed. It does not appear that any one of the new sources will be as abundant or as cheap as petroleum once was. It is likely that many different energy sources will be used instead. Costs, availability, and adaptability will largely determine which energy source is used. Maybe your future work will be in finding new energy sources and how to use them.

Study Questions for 23.3

1. What is geothermal energy? Describe four forms of it.
2. Describe two ways geothermal energy can be used to generate electricity.
3. Other than electric power, how is geothermal energy used?

What Did You Learn?

- A renewable energy source is one that never runs out because it is constantly replaced.
- Solar energy is energy from sunlight.
- One type of solar collector is like a small greenhouse in which tubes of water are heated by the sun.
- A passive solar house uses solar energy for heating and has no electric pumps, fans, or heaters.
- A storage mass is a dense material that absorbs solar energy and then gives off energy during the day and night.
- A power tower is a tall tower where sunlight is concentrated by mirrors.
- A solar cell is a device that converts sunlight directly into electricity.
- Moving water provides energy by turning a waterwheel.
- Water from behind a dam falls through pipes to turn a turbine.
- Wind provides energy by turning the blades of a wind machine.
- Biomass is material from any living thing that can be used as an energy source.
- Biomass can be burned for energy.
- Biomass can be converted into fuels.
- Geothermal energy is energy from below the surface of the earth.
- There are four forms of geothermal energy: (1) dry steam, (2) hot water, (3) geopressurized hot water, and (4) hot, dry rock.
- Dry steam is rare but can be used by a turbine.
- Geothermal hot water can be used to generate electricity or used for heating buildings.
- Geopressurized hot water is water under pressure. It contains dissolved natural gas.
- Hot, dry rock is the most common form of geothermal energy.

Key Terms

renewable energy source
solar energy
solar collector
passive solar
storage mass
power tower
solar cell
biomass
geyser
hot spring
geothermal energy

Career

Solar-Cell Technician

Did you ever wonder how solar cells are made or what is the best way to make them? Most solar cells are made of silicon disks that are coated with certain chemicals that are baked and processed further with other chemicals. A solar-cell technician prepares the disks for each process. A solar-cell technician needs at least a high school diploma and courses in technology processes, mathematics, and electronics.

Erma Egurrola is a solar-cell techcian working for a company that makes solar cells in Phoenix, Arizona. She is trying to find methods of preparing solar cells that will reduce the cost of the cells. Presently she prepares solar disks by coating them with a chemical, spinning them, and baking them at 80°C for 15 minutes. The cells are then inspected, baked again but at a higher temperature, and etched. Afterward they are cleaned several times and rinsed and dried. Since Erma Egurrola works in research and development of solar cells, she also explores different methods of preparing the cells.

For a career as a solar-cell technician Erma Egurrola says that knowledge of math and electronics is helpful, but experience with careful, exact procedures is necessary. She also says that her job is interesting and enjoyable, since she has an opportunity to experiment in many different ways to perfect the process of making solar cells. She knows that her job is important. If she perfects the procedures, low-cost solar cells will reduce the shaky dependence on costly fossil fuels.

TO THINK ABOUT AND DO

Vocabulary

Write the numbers 1 through 10 on a sheet of paper. Write the word or words that best fit the description beside the appropriate number.

passive solar house	renewable energy
geyser	power tower
storage mass	solar cell
solar collector	biomass
solar energy	geothermal energy

1. a small greenhouselike box that heats water
2. plants and materials from plants
3. dense material that absorbs solar energy
4. that which has no moving parts and makes electricity from sunlight
5. tall structure that mirrors reflect sunlight to
6. energy from within the earth
7. uses solar energy, no pumps, no fans, and no heaters
8. a fountainlike blast of geothermal hot water
9. energy from the sun
10. any form of energy that does not run out

What Do You Remember?

Number your paper from 1 to 10. Write the letter of the correct answer beside the appropriate number.

1. A house that is warmed with solar energy must have a (a) mirror, (b) solar cell, (c) power tower, (d) storage mass.
2. A device that makes electricity directly from sunlight is a (a) mirror, (b) solar cell, (c) power tower, (d) storage mass.
3. If you want to heat water with solar energy, you need a (a) solar collector, (b) solar cell, (c) solar mass, (d) all of the above.
4. A power tower has what in common with coal- or oil-fired electric power plants? It produces (a) pollution, (b) ash, (c) steam, (d) all of the above.
5. Water power (a) has been used for years, (b) can be

used to make flour, (c) is used to make electricity, (d) all of the above.

6. Wind power can be used as a source of energy if you (a) buy a wind machine, (b) do not use other energy sources, (c) live where the wind blows, (d) live near a river.
7. More solar energy isn't used today because (a) no one owns the sun and can't sell it, (b) it is impossible to use, (c) it's too expensive, (d) it doesn't exist yet.
8. Biomass is an energy source when you (a) burn it, (b) convert it to methane, (c) convert it to alcohol, (d) any of the above.
9. Geothermal energy comes from (a) the earth's core, (b) within several kilometers of the surface, (c) about halfway between the core and the surface, (d) above the surface.
10. The most common form of geothermal energy is (a) dry steam, (b) geothermal hot water, (c) geopressurized hot water, (d) hot, dry rock.

Applying What You Learned

1. A passive solar house cannot have carpet or rugs on the floor. What is the reason for this?
2. Suppose you have solar cells at your house. What problem would you have if you wanted to produce all of your own electricity?
3. What problems might develop if everyone started burning wood to heat homes?

Research and Investigation

1. Talk to people who have passive solar houses or solar water heaters. Prepare a report on how people have to change their living habits when they use solar energy.
2. Interview someone from your local electric power company. Prepare a report on plans to use solar energy to generate electricity.
3. Discuss passive solar houses and solar water heaters with a city planner. Find out if your city has building codes governing the placement of solar collectors. Report your findings to your class.

Your Environment, Earth 24

Astronauts looking back at Earth from their spacecraft often exclaim, "Oh, what a beautiful sight!" Among all the planets, Earth is the blue one because of its atmosphere and vast oceans. Earth, indeed, is a beautiful place to live.

Is it possible to make the earth more beautiful? Can people improve on nature? Look at the photograph on the left. Is the scene more beautiful because of the changes made by people? What would this scene look like without the changes? People do improve on nature as you can see.

People can also do things to make the earth more unsightly. Causing erosion of cropland and sometimes covering the land with buildings and pavement are ways people make the earth ugly. Failing to dispose of solid wastes properly and polluting the air, water, and land are some other ways.

In this chapter you will study how people and the things they do interact with the earth. You will learn that people have a choice. They can make the earth a more beautiful and healthy place to live if they want to do it.

Figure 24.1 *opposite:* With careful planning, people can improve the beauty of the earth.

24.1 Land

You will find out

- how land is used;
- what soil is and how it is made;
- what makes good growing soil;
- how farmers protect soil.

Figure 24.2 Land may be used in many ways, depending on the needs of the people who use it.

The surveyors in Figure 24.2 are working on a project that will make use of some land. What do you think they are helping to build? Perhaps a new housing development or a new university is planned. Maybe the project is a highway or a park. There are many possibilities, since land can be used in many different ways.

How do people use the land? Building is one way. People build homes, cities, factories, and schools. Land is also occupied by hospitals, offices, shopping centers, and athletic fields. Can you think of another use of land that you depend on every day?

Look at Figure 24.3. This is a farmer's field somewhere in Nebraska. There are no crops growing in the field now, but the farmer is getting ready to plant. The field will be planted, and the crop grown and harvested. Wheat is grown in fields like this. Wheat is used to make the bread you eat. In fact, most of the food you see on your table is grown on farms and then prepared for you to eat. Farms also produce dairy products and raise hogs, cattle, sheep, and poultry.

Figure 24.3 A field stands ready for planting.

Probably the most important use of the land all over the world is for crop production. You may eat the crop directly as in the case of potatoes. You may also eat the crop indirectly if you eat meat. Some crops are used to feed the farm animals that eventually may become your food.

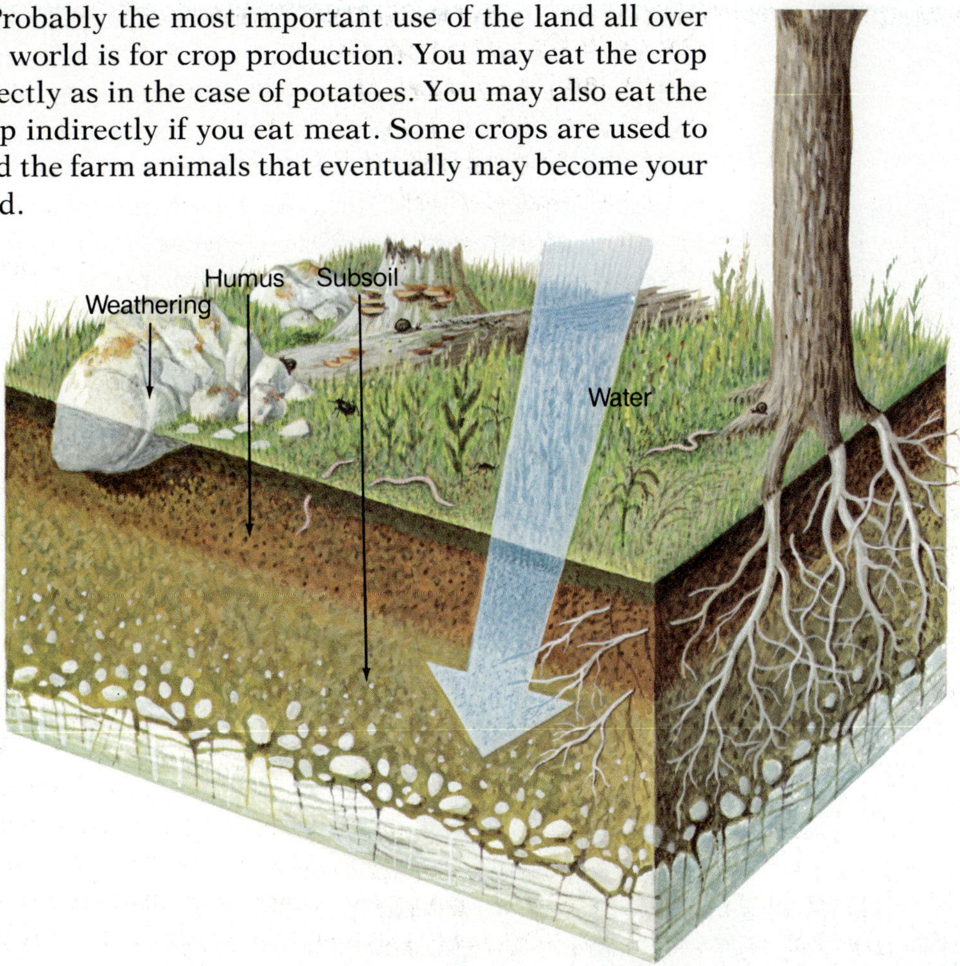

Figure 24.4 Processes that produce soil from the earth's crust occur constantly.

The farmer's crops grow in **soil.** Soil is the upper layer of the earth's crust in which plants grow. Three processes are important in making soils. Figure 24.4 shows how these processes work. First, weathering breaks up and changes the rock. Second, plants and animals growing and living in the weathered rock add **humus** [HYOO-muhs]. Humus is partially or completely decayed plant and animal matter. It adds the black color to soil. The third process is related to the movement of water through the soil. The water carries minerals and decayed plant material into the **subsoil,** the region beneath the layer containing humus. The soil you see on the next page, in Figure 24.5, was made that way.

Do You Know?
One person out of ten in the United States lives on a farm. The other nine out of ten people depend on the farmer for their food.

Figure 24.5 Thickness and quality of the layers in a soil profile vary from one place to another.

Look at Figure 24.5 carefully. Notice that the soil is layered. Geologists call this arrangement of soil layers a **soil profile.** The top layer containing humus is black. The middle layer is lighter in color. This layer includes minerals and decayed plant material that has been moved downward by water. The top of the lower layer often shows different larger-sized pieces of solid rock.

Soil profiles are different in different places. This is because the conditions that help produce soil can vary. The kind of rock from which soil is made is important. Climate, time, plant and animal influences, and the slope of the land all contribute to making different soils.

Good growing soils must be **fertile** [FUHR-tuhl]. A fertile soil is one that is capable of growing crops because of the materials it contains. It must contain nitrogen, potassium, and phosphorus. These three elements must be present in soluble form, which means dissolved in water, in order for plants to use them. To be a good growing soil, it must also have the right **texture** and **structure.** Soil texture is how the particles making up the soil are arranged. They must let water and air enter and leave the soil. The structure allows water to move through the soil layers.

Figure 24.6 *left:* This soil has the right texture for growing.

Figure 24.7 *right:* This soil has poor texture.

Why do farmers want to maintain good soil? Soil is a factory for farmers where they produce something that can be sold. Farmers can grow plants on the same land for hundreds or thousands of years if soil quality is maintained. In Egypt, some fields have grown a crop every year for more than 7,000 years.

Figure 24.8 Wheat is a major crop in the United States and must be grown in well-maintained soil.

ACTIVITY

Which Soil Holds More Water?

Materials

2 small containers the same size (Tuna-fish cans work fine.)
2 pieces of cloth
string or 2 rubber bands
pencil or pen
2 types of soil
water and container
graduated cylinder
2 pie tins

Procedure

1. Completely fill the two containers, each with a different type of soil.

2. Secure the cloth over the top of the containers.

3. Pour water into a graduated cylinder and write down the volume. Add the water to container 1 until it is filled. Determine how much water you used.

4. Follow the same procedure for container 2.

5. Arrange each container upside down in a separate pie tin. Wedge a pencil or pen under one edge of the can. Measure the water that drains into each tin.

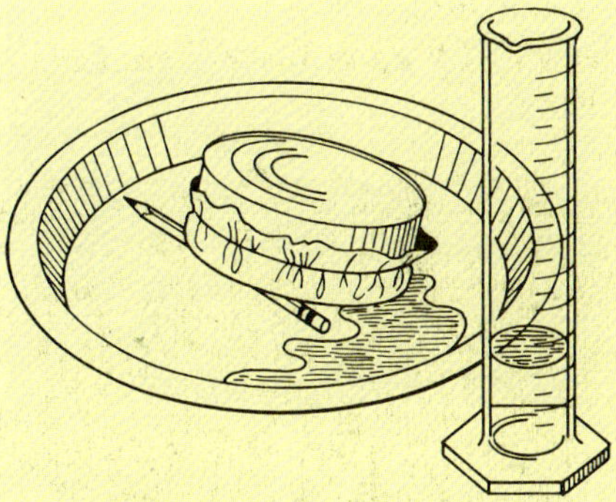

Questions

1. How much water did you add to each can?

2. Calculate how much water remained in each can of soil by subtracting the amount of water that drained out of the can from the total amount of water that was added to it.

Figure 24.9 *right:* Severe erosion from flooding

Figure 24.10 *below:* Overgrazing by animals removes plant cover. Soil may then be easily eroded.

What factors must farmers think about when planning how to maintain good soil quality? Erosion is the major threat to good soil. Figure 24.9 shows how water eroded a field. Wind also removes the soil. At present the United States is losing about four billion tons of soil annually by erosion. About 25 percent of the loss is caused by wind. The other losses are by water and by movements of the soil due to gravity. What is even more important about the loss of this soil is that three billion tons of it are lost from cropland.

Land that has a natural cover of trees or grass usually does not erode. All the roots of the plants help hold the soil together. When the forests are removed or when the land is overgrazed or left bare between crops, this natural protection does not exist. Farmers must carefully plan how they use the land. Animals are moved about to avoid overgrazing. Cover crops or new crops are planted to avoid leaving the land bare. Forest land is selectively harvested rather than cleared. Strip planting or contour planting is also used to prevent water erosion. If the rows went up and down the slopes, rain would create small streams and carry away the soil.

Irrigation can also cause a loss of soil quality. The water can change the soil texture and structure, making it harder for plants to grow. Water in the soil can provide a means for salts in the subsoil to move upward. The salts come to the surface and are deposited when the water evaporates. Irrigation water can also remove salts from the land. If such water is carried to other places where it is used again, the water may become too salty. Sometimes irrigation water can concentrate harmful chemicals from pesticides and insecticides in the soil. Farmers try to use good quality water and follow good drainage practices.

Do You Know?
As much as 90 percent of the fertilizer put on fields in some areas is not used by plants but carried away with surface runoff and underground water movement.

Figure 24.11 Rows of contour planting cut across hillsides and make water erosion less likely.

Farms are the major source of food for the world's people and a major use of land. It is important to you and your family that good soil is available for growing food. Without this precious resource people could not exist.

Study Questions for 24.1

1. Name two major ways that land is used.
2. What is soil?
3. Describe the processes that produce soil.
4. What makes good growing soil?
5. Describe what farmers do to protect the soil.

24.2 Air

You will find out

- how air is used;
- what pollution is;
- what causes air pollution;
- how air can be kept clean.

Figure 24.12 Green plants use carbon dioxide during photosynthesis. They use oxygen for other life processes.

Earth, like several of the other planets in the solar system, has an atmosphere. Scientists often say that our atmosphere is special. Why would a mixture of invisible gases be special?

Think about the gases found in the air. You already know that most living things require oxygen to live, including humans. Can you think of another atmospheric gas that some organisms use? Carbon dioxide is important to green plants. They use carbon dioxide during **photosynthesis** [foh-toh-SIN-thuh-sis]. Photosynthesis is the process green plants follow to manufacture food, using sunlight, carbon dioxide, and water. While manufacturing food, the plants produce oxygen. Both plants and animals use air for life processes.

Another important function of the atmosphere is protection. Air forms a shield against damaging ultraviolet rays from the sun. Ultraviolet light can destroy simple life forms such as bacteria and harm other organisms too. **Ozone** [OH-zohn] is another gas present in the atmosphere that screens out harmful ultraviolet rays. You have already learned a third function of air. The atmosphere moves, and the shifting winds contribute to the climate and weather patterns on Earth.

Figure 24.13 *right:* Ozone gas, which is made from oxygen, stops most ultraviolet light from reaching the earth's surface.

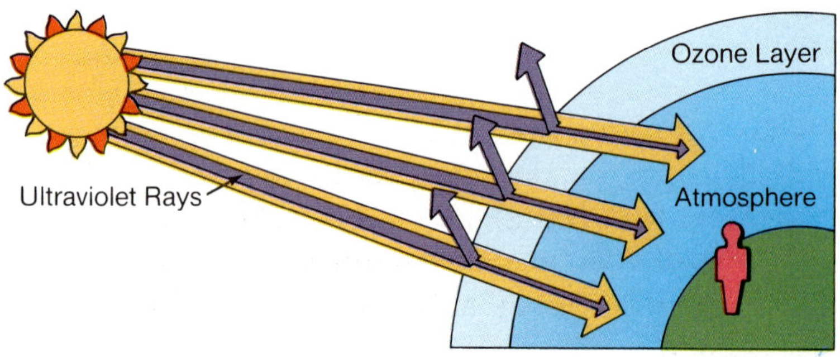

Do You Know?
Some of the most dangerous air pollutants, such as sulfur dioxide and carbon monoxide, are invisible.

It is obvious that air is an important resource. Is it possible to damage the air? Unfortunately, the answer is yes. The air can be damaged by **pollution** [puh-LOO-shuhn]. Pollution occurs when irritating or dirty substances accumulate in air. These irritants, or **pollutants,** might be chemical gases, dust and waste particles, and smoke. Figures 24.14 and 24.15 show some examples of air pollution. Another kind of pollution comes from aerosol cans. When sprayed, they release chemicals that destroy ozone. How might that affect life on Earth? Air pollutants also occur naturally as plant pollens or volcanic gases and dust.

How do you recognize pollution? Sometimes people can easily detect pollutants in their surroundings through their senses. For example, your eyes may sting and become watery after you have been in polluted air. Other times scientific instruments indicate if pollutants are accumulating. Pollution occurs when natural processes do not act fast enough to remove the irritants.

What can be done to reduce air pollution? People need automobiles and trucks. There have to be power plants, industries, and cities. If a way can be found to decrease pollutants, the air can become cleaner and healthier. One way is to pass laws regulating pollution. These laws limit the kinds of fuels which can be burned, where they can be burned, and when. Cars, trucks, and factories can also be made to use cleaner fuels. Manufacturers are changing processes and products to limit the release of harmful gases into the air. Of course, nothing can be done about erupting volcanoes. However, in places where people make a difference, air can be made cleaner.

Figure 24.14 *top:* Air pollution may come from people and their machines.

Figure 24.15 *bottom:* Air pollution may also come from nature.

Study Questions for 24.2

1. Name three uses of air on Earth.
2. What is pollution and how is it recognized?
3. What are two major pollutants of the air?
4. Describe some steps being taken to reduce air pollution.

24.3 Water

You will find out
- how water is used;
- what causes water pollution;
- how water can be protected.

Figure 24.16 Water is the earth's most precious resource. Life could not exist without water.

Earth is sometimes called the water planet. Why do you think this is so? About 70 percent of the earth's surface is covered by water. Life on the earth depends on water. About 60 percent of the earth's oxygen is produced through photosynthesis by microscopic green plants that live in the water. People also obtain food by fishing from the earth's waters and harvesting crops of sea plants. Water is a very important resource.

If you could measure the amount of water you use in a year, how much would it be? An average of 6,500 L of water per day is used by each person. That becomes more than two million liters in a year! Consumption of water for personal needs such as drinking, cooking, and bathing is only one way water is used. These activities account for about 650 L daily per person, which is certainly a large amount of water. But much more water is consumed by agriculture and industry. Water is used to irrigate fields of crops and maintain animals. Water is also used for cooling power plants, in factory machines, and in manufacturing processes.

Rivers, lakes, streams, and underground water are the sources from which fresh water comes. These sources are constantly replenished in the water cycle by rainfall. It is important to protect these sources and the ocean in order to preserve our water supply.

Water can become polluted in ways similar to the pollution of air, through an accumulation of irritants. What are some ways water can be polluted? In Figure 24.17, you can see an example of the pollution of water. Major pollutants of water are (1) human wastes, (2) chemicals from cities and industry, (3) hot water from industry and power plants, (4) fertilizers, insecticides, pesticides, herbicides, and animal wastes from agriculture.

Figure 24.17 *left:* Pollution is spoiling many freshwater sources around the world.

Figure 24.18 *below:* Parts of the chiseled surface of Cleopatra's Needle in New York City have been eroded away by acid rain.

Another kind of water pollution is acid rain. Acid rain occurs when rain falls through air which has become polluted with certain industrial and combustion chemicals. The acid rain falls into lakes and streams, making it difficult for plants and water animals to live. It also erodes buildings, statues, and paint.

What can be done about water pollution? Careful treatment of waste water is one answer to the problem. Industries must also make their chemical wastes harmless before they are released. Fewer and less dangerous chemicals can be used in agriculture. Lessening air pollution will decrease acid rain. These steps are not easy. Protecting water from pollution is expensive and inconvenient. But clean water, like clean air, depends on people and their willingness to make important choices. A little effort by people can prevent a lot of water pollution.

Study Questions for 24.3

1. Name three major ways that water is used.
2. What are some of the causes of water pollution?
3. Describe how water pollution can be decreased.

24.4 Making Choices

You will find out

- what solid waste is;
- how solid waste is disposed of;
- why recycling is important;
- what must be considered when finding ways to protect the environment.

Figure 24.19 Is it junk or is it a treasure?

You may have heard the remark that "One person's junk is another person's treasure." It is a very old saying. An old car that will not run is someone's junk. But to an auto hobbyist its parts and pieces are valuable. When you think of **solid waste**, that idea is true. Solid waste is anything that is considered to be useless, unused, unwanted, or discarded. It can be such things as paper, bottles, cans, rubber, plastics, and metals.

ACTIVITY

How Much Waste Does Your Earth Science Class Produce in a Week?

Materials

large box to collect waste
bathroom scale

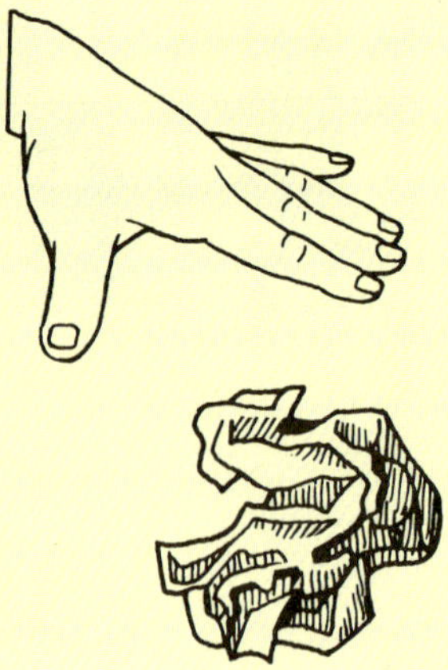

Procedure

1. On Monday morning, remove all old trash from the classroom.
2. During the week, place all trash produced by the class in a large box.
3. On Friday, separate the trash into categories of (1) paper or paper products, (2) metals, (3) glass, (4) plant materials, (5) plastics, and (6) miscellaneous. Weigh each kind of waste.

Questions

1. How much waste of each kind was produced?
2. What was the total amount of waste?
3. Calculate how much waste was produced per student per day.
4. Calculate how much waste per student would be produced per year.
5. Estimate the total amount of waste for your school.

In the United States enough solid waste is produced each day to cover almost 300 football fields to a depth of 3 m! Why is there so much? One reason is the way things are packaged. Packages make things easy to ship. They protect whatever is inside, but the wrappings are eventually thrown away as waste. Another reason is that people are accustomed to disposable consumer goods. It is easier and cheaper to replace things than to reuse or repair them. Of course, there are also more people now than there were years ago. More people produce more solid wastes.

What happens to the solid waste people produce? Most of it will end up in one of the following places: (1) a sanitary landfill, (2) incinerators, (3) a recycling plant, or (4) sometimes even in the ocean. Which method do you think is best? Sometimes people discard waste as trash onto streets or open land where it may remain for a long time.

Figure 24.20 *above:* Even the ocean suffers from the burden of solid waste pollution.

Figure 24.21 *left:* The accumulation of solid waste is a problem people must find new ways to solve.

Do You Know?
The average American produces nearly a ton of solid wastes each year.

Many believe the current sites for waste disposal have reached a limit. Are there other ways to handle solid waste that might even be useful? One old method being tried in a new way is **incineration** [ihn-sihn-uh-RAY-shuhn]. Incineration is the burning of waste at high temperatures in furnaces. People are now using the heat that incinerators produce to warm buildings or produce electric power.

Recycling is another way to handle solid waste and make it useful. Recycling does not actually dispose of the waste. It provides a way to use the discarded materials again. For example, bottles are collected to be cleaned and refilled. Or they, along with cans, can be reprocessed, and the glass and metal used in other ways. Waste paper is recycled by many industries. Scrap metals can be melted down and cast into new, usable forms. Recycling may prove to be the best way people have to prevent themselves from being covered up by piles of solid waste. Recycling also conserves valuable resources. Do you know why? When paper, metal, and other wastes are recycled, there is less danger of using up our wood and mineral resources. Also, with a lower need for new materials, less energy and money are used for refining processes. With recycling, a waste is no longer a waste.

Figure 24.22 *above:* An incinerating plant burns solid wastes. It may also use the energy that is produced to supply other needs.

Figure 24.23 *right:* Wastes become useful when they are recycled. What things can you recycle?

Protecting the land, air, and water is a complicated challenge. How will everyone share resources while preserving them for the future? This is not an easy question to answer. The solution to the problem requires people to look at many facts and then make choices. In some cases the choices will be difficult.

Some of the factors to think about are cost, convenience, consequences, future needs, and fairness to people with different interests. For example, the amount of farmland needed to grow food for the world's growing population must double in the next 50 years. But the best farmland is also attractive to developers because it is often flat and well drained. Each year over 400,000 hectares of good farmland become new sites for building. People need to find ways to allow both the farmers and the developers to prosper.

People must consider how pollution affects them as well as how they can lessen those effects. Polluted land, air, and water are unpleasant and ugly. Pollutants endanger the natural balance of life on Earth because they may be poisonous or cause disease. Also, careless use of natural resources will waste them. People must remember that protecting natural resources may sometimes slow the growth of industries and cities. Recycling and regulation of fuels and waste treatment is costly. Will people accept the inconvenience of becoming better recyclers?

Not everyone agrees to the solutions of the pollution problems. Educating people about pollution will help them make decisions. The decisions they make for the future must benefit everyone and protect the earth too.

Figure 24.24 Careful choices for the future must be made about the best ways to use land.

Study Questions for 24.4

1. Define solid waste and give some examples of different types.
2. Describe ways in which solid waste can be disposed.
3. Why is recycling an important process?
4. What factors must people consider when making choices about environmental problems?

What Did You Learn?

- The most important use of the land all over the world is for crop production.
- Soil is the upper layer of the earth's crust in which plants grow.
- The three processes that make soil are (1) weathering, (2) plants and minerals living in weathered rock, and (3) the movement of water through the soil.
- The five factors related to the making of soil are (1) climate, (2) the kind of rock from which the soil is made, (3) time, (4) plant and animal influences, and (5) the slope of the land.
- To be good for growing, soil must contain nitrogen, potassium, and phosphorus as well as be the right texture and structure.
- The major factors farmers think about for maintaining good soil quality are (1) erosion, and (2) irrigation.
- To reduce erosion, the farmer protects the soil with (1) cover crops, (2) strip planting, and (3) contour planting.
- Plants and animals use air for their life processes.
- Pollution is the accumulation of one or more irritants in the environment.
- The pollutants of the air include chemicals, gases, dust and waste particles, and smoke.
- Pollution occurs when natural processes do not act fast enough to remove the irritants.
- One way to reduce pollution is to pass laws regulating pollution.
- Water is a source of oxygen and food.
- The pollutants of the water are (1) human wastes, (2) chemicals from cities and from industry, (3) hot water from industry and power plants, (4) fertilizers, insecticides, pesticides, herbicides, and animal wastes from agriculture.
- Solid waste is anything that is considered to be useless, unused, unwanted, or discarded.
- Solid waste usually ends up in a sanitary landfill, the ocean, incinerators, or a recycling plant.
- Solid waste can be used by incinerating it to make energy and by recycling.

Key Terms

soil
humus
subsoil
soil profile
fertile
soil texture
soil structure
photosynthesis
ozone
pollution
pollutant
solid waste
incineration

Biography

Dr. Bahe Billy

Dr. Bahe Billy was born on the Navaho Indian Reservation in Leupp, Arizona, in 1937. As a child he worked with his father helping with farm chores and with irrigation tasks. It was during this time that Dr. Billy became interested in soil and what happens when water enters and leaves the soil. At this early age he noticed the accumulation of salts in irrigated soil and wondered why this happened. Later he studied this very problem for his Master's thesis.

Dr. Billy studied agronomy at Utah State University in Logan, Utah. His Master's and Ph.D. degrees were earned at the University of Arizona at Tucson, Arizona. He is married and has seven children.

Currently Dr. Billy is teaching in Navaho Community College, Shiprock, New Mexico. During the summer months he serves as a consultant on soil problems. His work contributes to the solution of soil problems and irrigation problems in his part of the United States. For example, one project involved collecting soil samples to study the fertility of soils. Another project was determining the irrigation time for certain fields.

Dr. Billy recently received the Tanner Award, an Indian Alumni Award, from the University of Arizona in recognition of his soils work and teaching efforts on the Navaho Indian Reservation.

TO THINK ABOUT AND DO

Vocabulary

Find the missing word. For each of the descriptions below, choose a word that best matches it. Write each word in the appropriate box. Then you will know the missing word.

1. that which is found in land, air, and water
2. that which is made up of layers of soil
3. that which is found below the humus layer
4. that in which plants grow
5. the burning of waste at high temperature

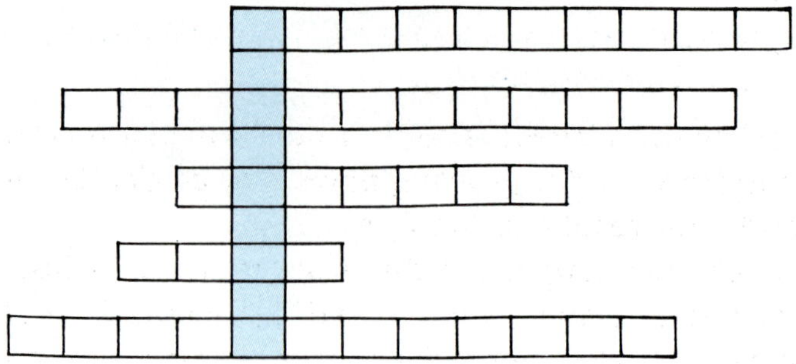

What Do You Remember?

1. Why is a farmer's field important?
2. What is humus?
3. How do plants and animals help in making soil?
4. How does soil texture and soil structure help in making a good growing soil?
5. Complete the sentence "Farmers can grow plants on the same land for a long time if . . ."
6. Why do farmers want to stop erosion? How can they do it?
7. How can irrigation cause a loss of soil quality?
8. List eight things that can be called irritants in the environment.
9. Give an example of a natural pollutant.
10. What happens in recycling?

Applying What You Have Learned

1. Describe the steps taken for food to get from a farmer's field to your dinner table.
2. What has to be done to soil to keep it producing a good crop for a long time?
3. What is meant by "overgrazing by animals" and what effect does this have on the soil?
4. Explain how water can become too salty to be used in irrigation.
5. What pollutants are found in both air and water pollution?
6. Describe a situation in which you recognized pollution.

Research and Investigation

1. Interview local growers to find out how soil in your area is prepared for growing flowers or garden crops. Report your results to the class.
2. Find out what happens to the trash from your school. What kind of disposal site is used? How much does it cost per year to haul off the trash from your school?
3. Contact a TV weather station that broadcasts the weather in your area. Find out how the air in your city is checked for pollution. What time of year is your city most polluted? Has your city had a pollution alert? What causes most of the air pollution where you live?

25 Earth Science Coming Attractions

Have you ever made a prediction? Perhaps you have predicted who will win a big game or that you will get a new bicycle for your birthday. Many people try to predict the weather. Geologists predict the likelihood of finding petroleum by examining rock formations. These kinds of predictions are possible because they are based on some evidence, or data. Predicting what scientific discoveries will be made in the future is not as likely. Instead, scientists prefer to think about the future in terms of things they would like to know. As you study this chapter, think about the things you would like to learn about the universe, about the earth, and about the place where you live.

Will meteorologists be able to make long-range weather forecasts? Will they be able to tell what kind of weather you will have on your birthday or six months from now? Will geologists discover new sources of hydrothermal energy under your school? Will people build cities in outer space or in the ocean? Will you be a scientist that answers some of the questions people ask about the future?

Figure 25.1 *opposite:* What new discoveries and inventions will there be in earth science in the future?

25.1 The Future in Space

You will find out

- what possibilities exist for exploration in space;
- what space observatories and platforms are;
- what new questions astronomers are asking about the universe.

Figure 25.2 Traveling in space is one way people can explore space's mysteries further.

Would you like to visit the moon or live on Mars? People have dreamed about such adventures for centuries. Now, more than ever before, such dreams may be closer to coming true. People have traveled in space, walked on the moon, and photographed planets by using space probes. What possibilities are there for future space activities?

There are two ways that people on Earth can extend their world into space. One way is to send space probes like the Mariner and Voyager spacecrafts out into the solar system and beyond. These probes will radio back even more information than has already been learned. The information may unlock secrets about the birth of the solar system and the nature of other planets. By studying these things, scientists may also learn more about life on Earth. In addition, space probes may help locate mineral resources on nearby planets or on the Asteroids. Someday these resources may be brought to Earth.

The second way people can reach further into space is through continuing space travel. Someday, this may lead to the establishment of a colony on the moon. Can you think of a reason why people would want to be on the moon? Space observations are easier to make outside Earth's atmosphere. The moon may be a perfect location for future astronomers, geologists, and researchers.

Figure 25.3 Space probes bring information back to Earth from places too far away for people to visit.

Do You Know?
The space telescope is capable of detecting a 100-watt light bulb from eight million kilometers away.

Some scientists believe that Mars may someday be a good place for an earth colony. Could we make Mars habitable? Why would we want to do that? Perhaps engineers will want to live on Mars to mine its mineral resources. Others may want to go there for adventure. It is possible that Mars may become a new home for people who can adapt to living there. Do you want to be a Martian?

Figure 25.4 *top:* This is an artist's idea of what another inhabitable planet may be like. Would you want to visit here?

Figure 25.5 *bottom:* A type of space platform designed for orbiting Earth

How can scientists answer the questions they ask about space? Astonomers believe they will learn more about the universe in the next ten years than has been learned throughout history. One way they will do this is by using a space telescope positioned above the atmosphere, in Earth orbit. The space telescope will be able to search the skies 24 hours a day because outer space has a night sky at all times. Actually, the space telescope will be part of a complete observatory. Astronomers hope space observatories will be like giant eyes on the universe, used to gather information and solve new mysteries.

The space telescope will be placed in orbit by the space shuttle. The shuttle is really a fancy truck that carries materials into space. Someday the shuttle will carry materials into orbit so they can be assembled into a space platform. A space platform may look like the drawing in Figure 25.5. It will contain everything that is needed for people to live and work in space.

Do You Know?
Some materials that cannot be mixed under surface gravity conditions can be mixed in space. This suggests that stronger, lighter, and more temperature-resistant metal alloys may be produced in space.

Figure 25.6 What questions about our galaxy and others will be answered in the future?

How would you use a space platform? Of course, a space platform can be an observatory. However, scientists also believe that space platforms will be useful as manufacturing sites. A space platform enjoys almost zero gravity. Over 150 possible products have already been identified that could be more readily manufactured in zero gravity. They range from perfectly round steel ball bearings to ultrapure chemicals for treating disease. Manufacturing in space will be reserved for the kind of work that cannot be done in Earth's gravitational field. Many new products will be created on space platforms in the future. Someday you will be buying products or using medicines that were made partly in space.

What questions will astronomers of the future be asking? You have already learned about some of the questions by studying this book. For example, what more can be learned about black holes? What happens to time in a black hole? Some astronomers think time slows down or stops in a black hole. Are some black holes actually giant galaxies that are disappearing? Also, the old question of whether or not life exists elsewhere remains to be answered.

The nature and shape of the universe are popular topics among astronomers. Their ideas about the nature of the universe vary. Some describe the universe as being closed. This means it will stop expanding someday and collapse into itself. Others describe the universe as being open. They believe it will continue to expand forever. Many astronomers argue that the universe has a definite limit to its size, but some believe it does not. There are even some astronomers who suggest that there may be other universes besides the one that can be seen.

Sometimes people refer to space as a frontier like the wild west used to be. This is because they are excited by the adventures and new experiences that space study promises. Can answers be found to some of the questions raised about space? No one knows for sure, but curious people will always look for answers and ask new questions.

ACTIVITY

The Magical Mobius Strip

Materials
paper
tape
scissors
wide-tipped felt marker

Procedure
1. Cut two strips of paper each about 27 cm long and 2.5 cm wide.
2. Tape the ends of one strip together to make a loop.
3. With the other strip, make a twist and then tape the ends together without untwisting it. You have made a Mobius strip.
4. Using the marker, follow along the edge of the paper loop until you return to the starting point. *Do not lift the marker from the paper's edge.*
5. Repeat step 4 on the Mobius strip.

Questions
1. How are the resulting marks you made on the two loops different?
2. Why are the results different?

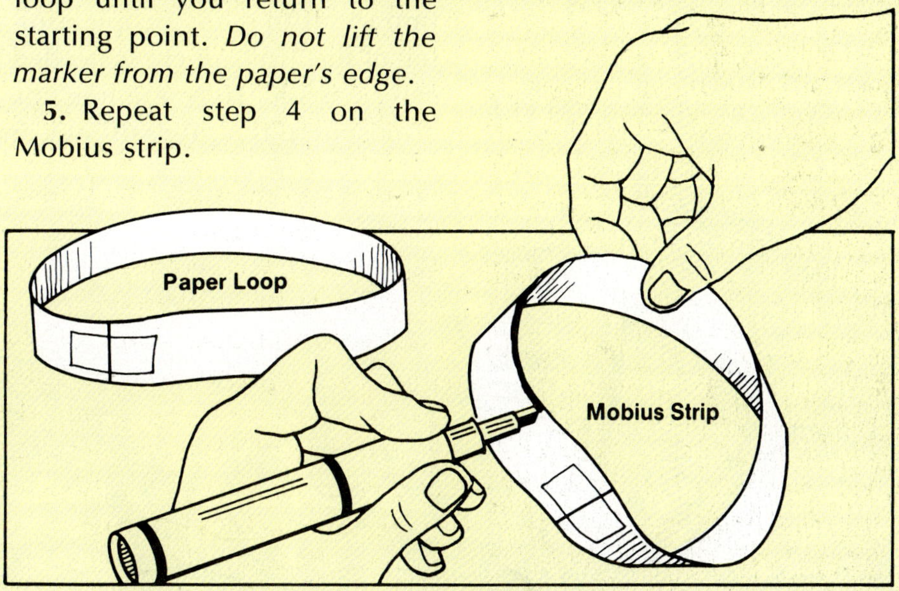

Study Questions for 25.1

1. What are two ways the solar system can be explored in the future?
2. How may space observatories and platforms be useful?
3. Name three questions that have not been solved in astronomy.

25.2 Questions About the Weather

You will find out

- what the four major questions about the atmosphere are;
- how new satellites may answer some of these questions.

Do You Know?
If the weather is predicted accurately only five days in advance, about 15 billion dollars a year could be saved by farmers, construction workers, and industry.

Suppose in the future people knew exactly what the weather was going to be. How would that change their lives? You obviously would not plan to have a picnic on a rainy day. Builders would know exactly when to pour new concrete so it could not be ruined by rain. Farmers would plant, irrigate, and harvest crops based on weather forecasts.

Having accurate long-range weather predictions is important to many people. But a complete understanding of the atmosphere is needed before such predictions can be made. Here are four major questions about the atmosphere that are still not understood:

Figure 25.7 Flowers cannot change their growing patterns no matter what the weather brings.

1) *What are the weather triggers?* What makes a stable weather condition become unstable?

A storm cloud brings rain and wind to an area, then it breaks up. Another very similar cloud nearby brings rain, wind, and a tornado. What triggered the tornado?

A bubble of warm air rises over the mid-Atlantic, then cools and goes away. Another bubble rises and becomes a hurricane. What was the hurricane trigger?

Figure 25.8 How does the ocean affect the development of storm clouds and the weather they bring?

2) *Exactly how do the atmosphere and ocean interact?* The interaction between the air and water of Earth may influence the weather. For example, every few years a large pool of warm water appears off Ecuador and Peru when the easterly wind ceases. Severe winters seem to occur in the eastern part of the United States these same years. Could measuring the temperature of the ocean off the coast of Peru be a way of predicting weather in the United States? More needs to be learned about how the temperature of ocean water affects the atmosphere and weather changes.

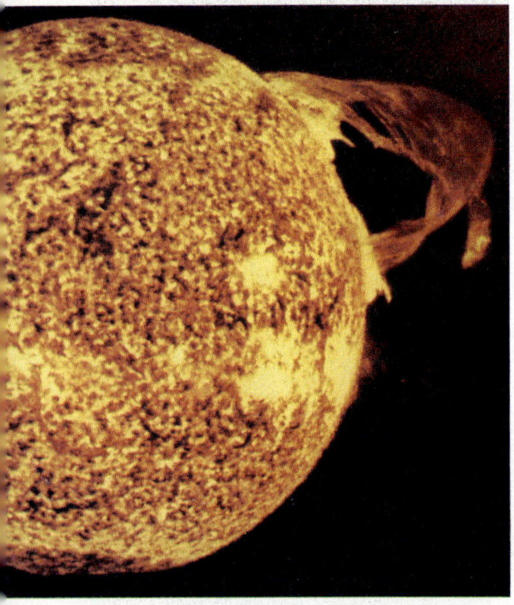

3) *Do changes in the sun's activity change the weather?* Sunlight is the source of energy that moves the atmosphere. Changes in the sun, such as sunspots, seem to indicate changes in the weather. Why? Solar flares and sunspot activity change the electromagnetic energy reaching Earth. Some scientists believe there is a relationship between the changes in this energy and changes in the atmosphere.

4) *How do very small-scale changes in the atmosphere affect its large-scale behavior?* According to some meteorologists, small-scale weather activity may put a limit on accurate, long-range forecasting. They believe that local occurrences, such as small gusts of wind, may have an impact on large-scale atmospheric motions. Since it is now impossible to observe all the small changes over a large region, uncertainties in weather information exist.

Figure 25.9 *above:* What key do sunspots and solar flares hold to understanding the earth's weather?

Figure 25.10 *right:* Can the flutter of a butterfly's wings be enough of a disturbance to affect local weather patterns? Would the flutter of a thousand butterflies' wings be enough?

Some of the questions about the atmosphere, however, may be answered by a new satellite instrument. NASA launched *Seasat* to test a radar instrument that can measure wind speed and direction. The radar measures wind speed to within two meters per second by watching waves on the surface of the ocean. When regularly placed in orbit, Seasat's instruments will help meteorologists learn about small-scale changes in the atmosphere. This will also speed up the understanding of how the atmosphere and oceans interact. Then maybe long-range weather predictions will become a reality.

Figure 25.11 *top:* A painting of NASA's *Seasat; bottom:* Information from *Seasat* was used to make this relief map of the ocean floor.

Study Questions for 25.2

1. What are some of the questions that remain to be answered about the atmosphere?
2. How will new weather satellites answer some of these questions?

25.3 Answers from the Ocean

Figure 25.12 More people need more food, more energy, and more space. Can the ocean help solve such problems?

You will find out

- what problems population growth may bring in the future;
- how the ocean can be used to increase food supplies;
- what the potential of the ocean is for supplying future energy needs.

Suppose the world population keeps growing. What will the future be like? Making predictions about the future is difficult, but recognizing possible problems is not. A growing population means more and more people each year who will need food, energy, and space to live. Even today there are many people who do not have enough food. Some cities are severely overcrowded, and there is concern about dwindling fuel resources.

Since the ocean covers 70 percent of the earth's surface, a good place to look for solutions to these problems may be the ocean. Can the ocean provide more food, new energy sources, and even places to live? The world's fishing boats already harvest millions of tons of food from the oceans. Yet, the potential source of food in the ocean is billions and billions of tons. There are several ideas about how to increase the amount of food from the ocean. One of these ideas is to grow sea animals and plants in ocean farms. These techniques are being used today on a limited basis. In the future, much more food can be obtained from the ocean by greater use of these methods.

Figure 25.13 *right:* Food from the sea can increase food supplies worldwide.

A new kind of food from the ocean may require that you change your eating habits. Many parts of the ocean contain small animals and plants that float near the surface. They are so small that you need a microscope to see them. These tiny plants and animals are called **plankton** [PLANK-tuhn]. Plankton contains protein, carbohydrates, and vitamins. It is a good food source and is already used as an ingredient in some foods.

Figure 25.14 Tiny plankton floating in seawater, provide food for larger organisms.

Do You Know?
There is a potential supply of fuel in the ocean sufficient to release 100 times the national energy use for 10 billion years.

The ocean may provide an answer to the energy problem. Seawater contains a form of hydrogen that can be used as a fuel. Hydrogen atoms combine together, or fuse, to form helium atoms. This **fusion** [FYOO-shuhn] **reaction** produces large amounts of energy.

Each liter of seawater has a few milliliters of the hydrogen fuel. However, even this tiny amount will release about the same energy as 300 liters of gasoline.

The problem is that fusion research is very expensive. Controlling fusion may, in fact, be one of the most expensive tasks ever undertaken. Things that haven't been invented yet will be needed. If the research succeeds, it will solve the energy problem for a long, long time. The ocean will make this possible.

Figure 25.15 Hydrogen from seawater can fuel a fusion chamber like this and actually provide many more times the energy than people need.

Figure 25.16 Will people someday use the open spaces on the ocean floor to build uncrowded cities?

Food and energy are just two benefits people may obtain from using the ocean wisely. What other possibilities does the future hold? One idea is to build cities underwater or on the surface of the ocean. New cities will provide more living space for people. Another possible use of the ocean is as a source of fresh water. **Desalinization** [dee-SAL-uh-nuh-zay-shun] plants that take salt out of seawater already exist. Yet, they operate in small numbers and are expensive. In the future, they may be used more effectively. The ocean can also be a rich source of minerals if mined properly. Imagine what other answers to problems the ocean may provide. There is still much that is not known about the ocean. New discoveries will lead to new questions and provide challenges for future oceanographers. Will you be one of them?

Study Questions for 25.3

1. Name three problems that continued population growth may bring.
2. Describe ways the ocean can be used as a food source.
3. How can the ocean supply needed energy?
4. Name two other possible uses of the ocean in the future.

25.4 Geological Discoveries in the Future

You will find out

- why it is important to scientists to measure continental drift;
- how geologists expect to measure continental drift;
- what some new methods to prospect for minerals are;
- how computers are helping geologists.

Figure 25.17 A laser telescope.

Geologists using the plate tectonic theory can explain why mountains, volcanoes, and earthquakes occur where they do. The theory is also used in searching for minerals and oil. Yet, the idea remains a scientific theory and not a confirmed scientific fact. If only someone would make a direct measurement of continental drift! A direct measurement would go a long way toward establishing plate tectonics as a scientific fact. Would you like to be that person? You could become famous.

If continental drift could be measured, geologists would be able to answer some important questions. They would know (1) how fast the plates are moving, (2) in what direction the motion is, (3) whether or not the plate motions are constant, and (4) whether or not an entire plate moves as a single unit.

How is it possible to measure a movement of as little as 5 cm/year between points thousands of kilometers apart? One way is to use a special kind of mobile telescope and computer. Wherever it is placed on Earth, the telescope can tell its precise location by transmitting a laser beam to the moon. A reflector on the moon will send the laser beam back to Earth where the exact location of the telescope is calculated by the computer. The telescope will be moved to different places on Earth where the procedure will be repeated. Ten years later the telescope will return to each of these spots and new measurements made. The computer will be able to tell if any of the spots move.

Do You Know?
Taking pictures of the earth by radar records differences in surface roughness. Geologists use these pictures to find faults, anticlines, folds, domes, drainage patterns, and stratification.

It is expected that a movement of as little as 5 to 10 cm per year between the United States and Africa will be detected. Measurements of plate movement that have been confirmed many times over will be strong evidence for accepting plate tectonics as a scientific fact.

Geologists will be working on many other projects as well. Some will be interested in prospecting for minerals and oil. There are two methods they will use in the search. One process is called **remote sensing.** An example of this process is the identification or measurement of something distant through photographs. In looking for mineral deposits, geologists will use special pictures of Earth taken from space platforms or satellites. Remote sensing will help geologists discover new metallic mineral deposits and oil fields. Hot spots of geothermal energy, gold deposits, and even fishing grounds can also be discovered through remote sensing. Someday, you may have a fish for dinner that was caught by using this technique.

Do You Know?
Special photographs taken from satellites can be designed to identify different types of rocks and soils associated with hidden mineral and petroleum deposits.

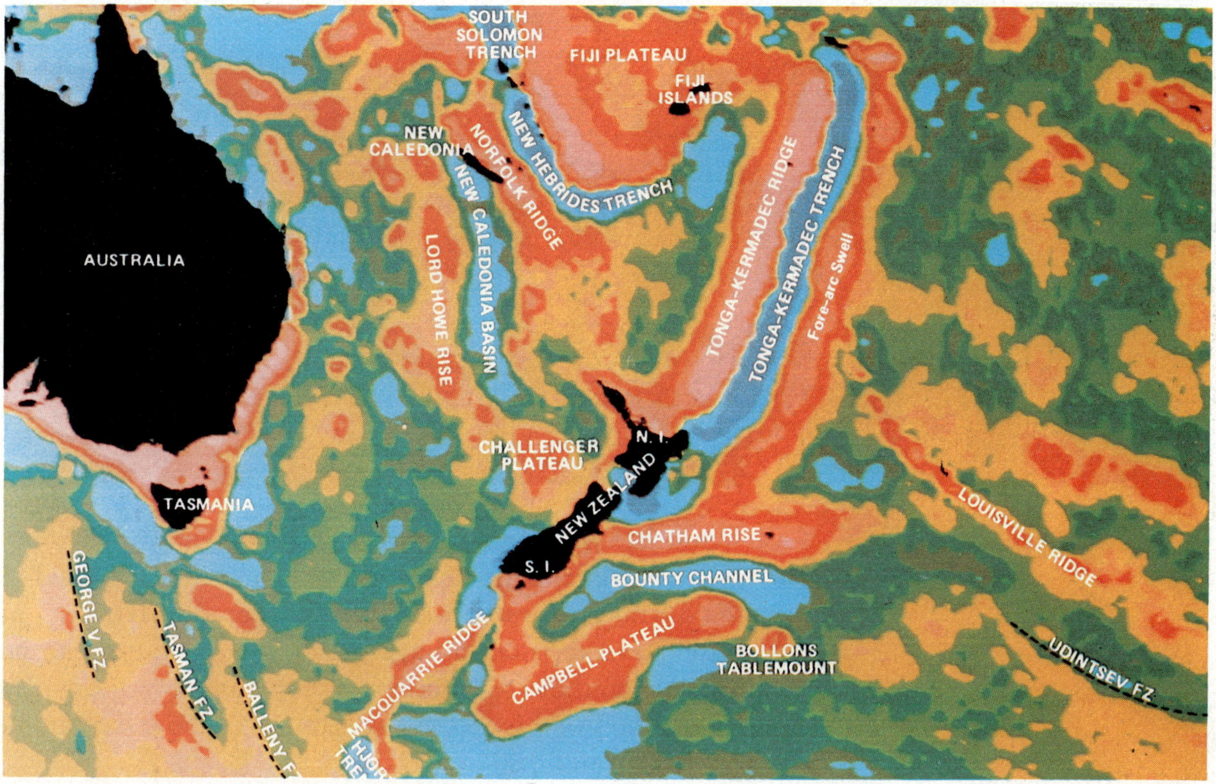

Figure 25.18 Scientists use remote sensing photos like this one to study the earth's features.

A second way of locating geologic features on Earth's surface is **image enhancing.** Image enhancing is a technique that uses computers to strengthen certain portions of photographs. These photographs are taken from space. They are sent to Earth by radio signals and put into computer language. Then they are printed out as colored photographs by the computer. The computer adjusts the contrast between colors in the photograph the same way that you adjust your television. Image enhancing makes it possible to see geologic features such as rock formations and potential mineral deposits.

Figure 25.19 Image enhancing uncovers startling features on the earth.

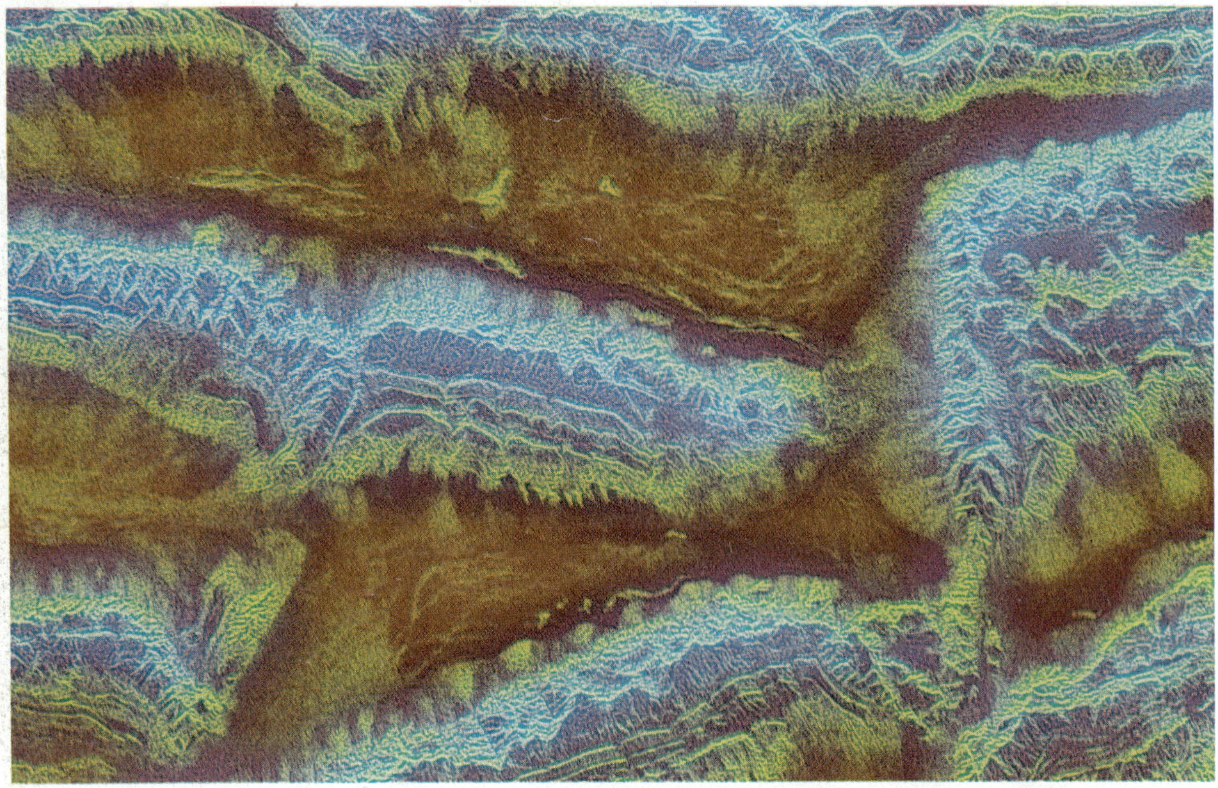

Another use for computers is to make model Earth systems. For example, computers can make a model of the drainage basin of a stream. If a large rain occurs at some point in the drainage basin, the computer can be used to predict (1) whether or not a flood is likely, (2) where the flood is likely to occur, or (3) how damaging the flood might be.

Geologists are finding that computers are essential in the search for oil. To find oil, waves from small explosives or mechanical strikers are sent through the earth. The wave energy travels into the earth and strikes the rock layers. Some of the wave energy is reflected back to the surface where it is recorded on a seismograph. The record from the seismograph is then fed into a computer. The computer printout of the seismic data can show faults and potential oil traps deep underground. In the future, the fuel used in your car will likely come from very deep oil traps found by using a computer.

Figure 25.20 *below left:* This is a computer readout of a seismograph record of the ocean's floor.

Figure 25.21 *below:* New ideas from people will open the doors of discovery to the future.

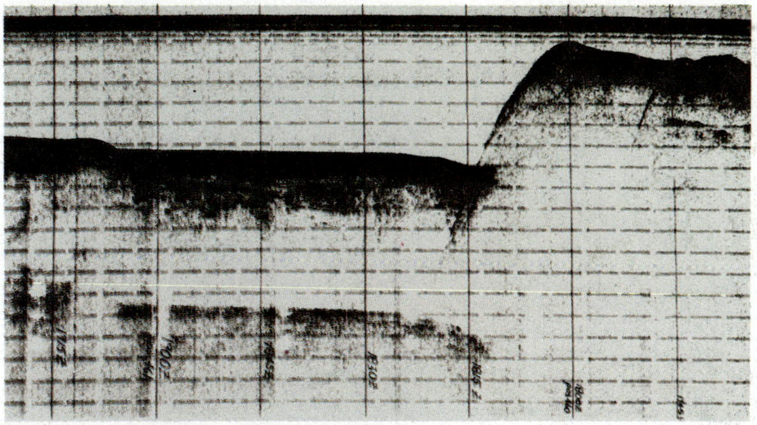

What will be tomorrow's discoveries in the earth sciences? No one really knows, but you can be sure that there will be many surprises. Some of these surprising discoveries will occur soon. Many new scientific achievements that will change your lives and the world in which you live will come later. The reason for this is that the students of today will be the ones who will find the science discoveries of tomorrow.

Study Questions for 25.4

1. Describe how geologists expect to measure continental drift.
2. What are two new methods of prospecting for minerals that will be used in the future?
3. What are some uses that geologists make of computers?

What Did You Learn?

- Space probes may help locate mineral resources on nearby planets or on the Asteroids.
- Space travel may lead to colonies on the Moon or Mars.
- Astronomers believe that by using the space telescope they will learn more about the universe in the next ten years than has been learned throughout history.
- Scientists believe that space platforms will be useful as manufacturing sites.
- Complete understanding of the atmosphere is needed before accurate, long-range forecasts can be made.
- Weather triggers, small-scale weather changes, the sun's influence, and interactions between the atmosphere and the ocean must still be understood.
- Some of the unanswered questions about the atmosphere may be answered by a new satellite, *Seasat.*
- The ocean may provide additional food sources of sea animals, sea plants, and plankton.
- A fusion reaction produces large amounts of energy.
- Geologists would like to make a direct measurement of the movement of crustal plates.
- Scientists are working on a special kind of telescope with a computer that will be able to measure the direction and amount of plate motion if it exists.
- Remote sensing photographs can be made more useful by computer image enhancing.
- Computers are used by geologists to enhance photographs, make model Earth systems, and to search for oil.

Key Terms

plankton
fusion reaction
desalinization
remote sensing
image enhancing

Career

Remote Sensing Scientist

Are you concerned about pollution? Would you like to study what an increasing population does to land and water resources? Would you like to identify natural hazards at the same time you find new mineral and fuel sources? Would you like to provide information that would help farmers increase food production? These questions are about problems that affect everyone's existence on Earth. The answers come from information provided by a remote sensing scientist.

Remote sensing involves the mapping and analysis of Earth's land and oceans using airplane and satellite images. These images include ordinary photographs as well as computer analyzed temperature and radar images from satellites.

Aulis Lind is a remote sensing scientist. She does research on remote sensing of Earth's environment. This includes studying conventional photographs taken from high-flying airplanes and satellite images. Sometimes this means just finding out what the images mean about the environment. Aulis Lind also teaches others interested in Earth's environment how to use the imaging tools of the space age.

Remote sensing is very exciting to Aulis Lind because there are constantly new challenges. There are new data from satellites that need analysis. There are new sensors being developed that provide new kinds of information. There are new resource problems that require solutions. And there are new opportunities to help others understand remote sensing. Aulis Lind says that remote sensing of the environment allows her to become involved with places all around the world—places where much remains to be discovered. Remote sensing is a new and fascinating career.

TO THINK ABOUT AND DO

Vocabulary

Some of the words in the following sequences have scrambled letters. Unscramble each word and write the correct word on a sheet of paper.

1. A new TILTASEEL will help meteorologists understand the weather.
2. NOANTKLP may be a common food source of the future.
3. The joining together of light elements occurs in a NOIFUS reaction.
4. TREEOM GNNISES is the process of identifying or measuring something from a distance.
5. GEIMA HNGENACNI is a technique for strengthening certain portions of a photograph to show detailed features.

What Do You Remember?

Number your page from 1 to 9. Write the correct answer beside the appropriate number.

1. Why will space telescopes be able to work continuously?
2. What are two future tasks for the space shuttle?
3. What kind of manufacturing tasks will be reserved for space platforms?
4. What new information will be provided by a satellite of the future? (a) cloud over (b) rainfall amounts (c) wind patterns (d) location of hurricanes
5. Which is not an unanswered question about the weather? (a) What makes stable weather unstable? (b) How do the ocean and atmosphere interact? (c) What makes the wind blow? (d) How do changes in the sun change the weather?
6. What is being considered as a new food source from the ocean? (a) fish (b) shellfish (c) seaweed (d) plankton
7. A form of hydrogen from ocean water can be used as a fuel in what type of reaction? (a) burning (b) fusion (c) fission (d) electrical

8. How will the moon be used to indicate where a mobile telescope is placed on Earth?
9. How are computers used by geologists?

Applying What You Have Learned

1. Explain the statement: The moon may be a perfect location for future astronomers, geologists, and researchers.
2. Describe how new satellite instruments may change weather predictions in the future.
3. Describe two ways that the ocean may provide new food sources in the future.
4. Describe how the ocean may provide energy in the future.
5. Describe how scientists expect to use a mobile space telescope in measuring crustal plate movements.

Research and Investigation

1. Make a collection of pictures on one of these topics: (a) different kinds of space telescopes or (b) different kinds of space platforms.
2. Find out what plankton tastes like. Read *Kon Tiki,* a book in which an explorer ate plankton.
3. Investigate how the first cloud pictures from satellites changed meteorology.
4. Assemble a collection of space photographs of Earth that have been enhanced. Write a description of each, explaining how enhancement has made it more useful.

APPENDICES

GUIDE TO THE METRIC SYSTEM

The International System of Measurement (SI) is used by scientists throughout the world. Each measurement unit in this system is based on 10 and may be multiplied by 10 to obtain the next larger unit or divided by 10 to obtain the next smaller unit. The basic unit of length is the meter; a kilogram is the basic unit of mass; and the basic unit of volume is the liter.

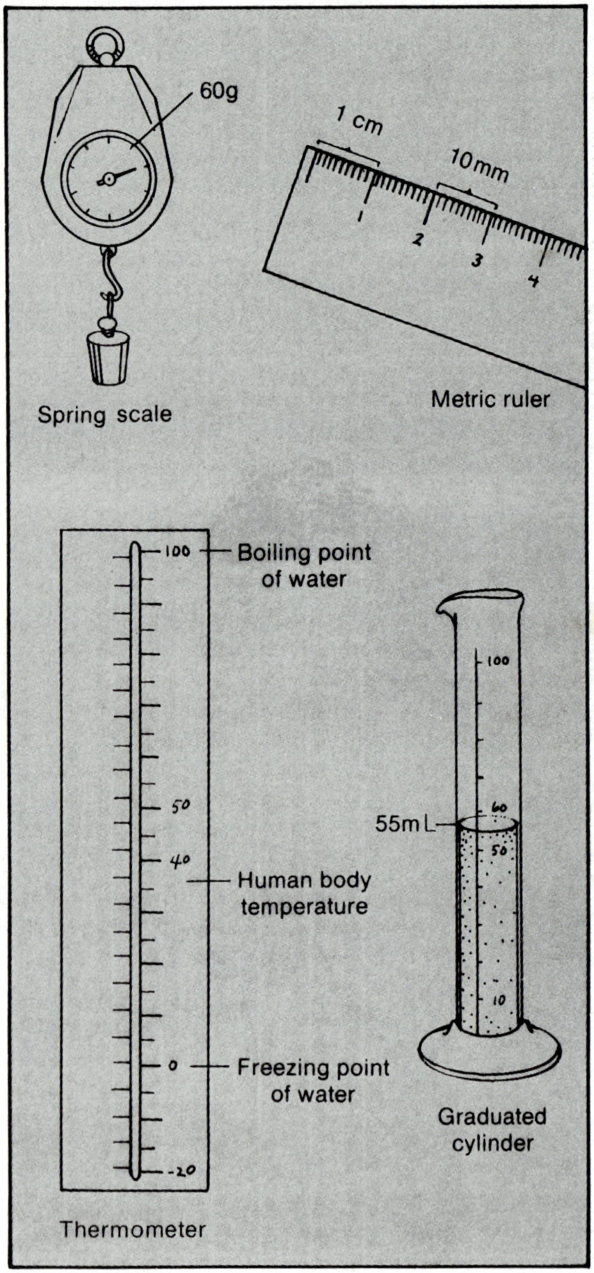

Metric System Prefixes

Greater than 1:
- kilo (k) = 1,000
- hecto (h) = 100
- deka (da) = 10

Less than 1:
- deci (d) = 0.1
- centi (c) = 0.01
- milli (m) = 0.001
- micro (μ) = 0.0000001

Commonly Used Metric Units

Length: The distance between two points

1 millimeter (mm) = 1,000 micrometers (μm)
1 centimeter (cm) = 10 millimeters (mm)
1 meter (m) = 100 centimeters (cm)
1 kilometer (km) = 1,000 meters (m)
1 light-year = 9,460,000,000,000 kilometers (km)

Area: A measure of the size of a two-dimensional surface

1 square centimeter (cm^2) = 100 square millimeters (mm^2)
1 square meter (m^2) = 10,000 square centimeters (cm^2)
1 square kilometer (km^2) = 1,000,000 square meters (m^2)
1 hectare (ha) = 10,000 square meters (m^2)

Volume: The amount of space an object takes up

1 milliliter (mL) = 1 cubic centimeter (cm^3)
1 liter (L) = 1,000 milliliters (mL)

Mass: The quantity of matter in a body

1 gram (g) = 1,000 milligrams (mg)
1 kilogram (kg) = 1,000 grams (g)
1 metric ton = 1,000 kilograms (kg)

Temperature: The relative measurement of hotness or coldness

0° Celsius (°C) = freezing point of water
37° Celsius (°C) = human body temperature
100° Celsius (°C) = boiling point of water

RELATIVE HUMIDITY (%)

Dry-bulb temperature (°C)	Difference between wet-bulb and dry-bulb temperatures (°C)																			
	1	2	3	4	5	6	7	8	9	10	11	12	13	14	15	16	17	18	19	20
0	81	64	46	29	13															
1	83	66	49	33	17															
2	84	68	52	37	22	7														
3	84	70	55	40	26	12														
4	85	71	57	43	29	16														
5	86	72	58	45	33	20	7													
6	86	73	60	48	35	24	11													
7	87	74	62	50	38	26	15													
8	87	75	63	51	40	29	19	8												
9	88	76	64	53	42	32	22	12												
10	88	77	66	55	44	34	24	15	6											
11	89	78	67	56	46	36	27	18	9											
12	89	78	68	58	48	39	29	21	12											
13	89	79	69	59	50	41	32	23	15	7										
14	90	79	70	60	51	42	34	26	18	10										
15	90	80	71	61	53	44	36	27	20	13	6									
16	90	81	71	63	54	46	38	30	23	15	8									
17	90	81	72	64	55	47	40	32	25	18	11									
18	91	82	73	65	57	49	41	34	27	20	14	7								
19	91	82	74	65	58	50	43	36	29	22	16	10								
20	91	83	74	66	59	51	44	37	31	24	18	12	6							
21	91	83	75	67	60	53	46	39	32	26	20	14	9							
22	92	83	76	68	61	54	47	40	34	28	22	17	11	6						
23	92	84	76	69	62	55	48	42	36	30	24	19	13	8						
24	92	84	77	69	62	56	49	43	37	31	26	20	15	10	5					
25	92	84	77	70	63	57	50	44	39	33	28	22	17	12	8					
26	92	85	78	71	64	58	51	46	40	34	29	24	19	14	10	5				
27	92	85	78	71	65	58	52	47	41	36	31	26	21	16	12	7				
28	93	85	78	72	65	59	53	48	42	37	32	27	22	18	13	9	5			
29	93	86	79	72	66	60	54	49	43	38	33	28	24	19	15	11	7			
30	93	86	79	73	67	61	55	50	44	39	35	30	25	21	17	13	9	5		
31	93	86	80	73	67	61	56	51	45	40	36	31	27	22	18	14	11	7		
32	93	86	80	74	68	62	57	51	46	41	37	32	28	24	20	16	12	9	5	
33	93	87	80	74	68	63	57	52	47	42	38	33	29	25	21	17	14	10	7	
34	93	87	81	75	69	63	58	53	48	43	39	35	30	28	23	19	15	12	8	5
35	94	87	81	75	69	64	59	54	49	44	40	36	32	28	24	20	17	13	10	7

PROPERTIES OF THE PLANETS

CHARACTERISTICS	Mercury	Venus	Earth	Mars	Jupiter	Saturn	Uranus	Neptune	Pluto
Radius (Earth = 1)	0.382	0.949	1.0	0.533	11.2	9.45	4.1	3.88	0.24
Volume (Earth = 1)	0.056	0.855	1.0	0.151	1,404.9	843.91	68.92	58.41	0.014
Mass (Earth = 1)	0.055	0.815	1.0	0.107	318	95.2	14.6	17.2	0.002
Surface Gravity (Earth = 1)	0.377	0.905	1.0	0.377	2.54	1.07	0.87	1.14	0.03
Orbital Period	88^d	225^d	365^d	1.88^y	11.9^y	29.5^y	84.0^y	165^y	248^y
Spin Period	58.7^d	243^d	23.9^d	24.6^h	9.92^h	10.7^h	23.9^h	18^h44^m	6.39^d
Mean Distance from Sun (Astronomical Units)	0.39	0.72	1.0	1.52	5.2	9.54	19.2	30.1	39.4
Mean Density (gm/cm³)	5.4	5.3	5.5	3.9	1.2	0.6	1.2	1.6	0.8
Number of Satellites	0	0	1	2	16	20*	5	2	1
Maximum Visual Magnitude	−1.9	−4.4	——	−2.8	−2.5	−0.4	+5.6	+7.9	+14.9
Mean Orbital Speed (km/s)	47.8	35.0	29.8	24.2	13.1	9.7	6.8	5.4	4.7

^{d} = Earth Days, ^{y} = Earth Years, ^{h} = Hours, ^{m} = Minutes. *Latest studies indicate number to be higher than 20.

Mineral Chart

Material Name	Composition	Hardness	Specific Gravity	Color	Cleavage	Luster	Streak	Uses
Albite	Al, Na, O, Si	6	2.62	Grayish White	2/93°, 87°	Glossy	White	Ceramics
Apatite	Ca, Cl, F, H, O, P	5	3.2	Green, Brown	Poor	Glossy	White	Fertilizer, Jewelry
Augite	Al, Ca, Fe, Mg, O, Si	5–6	3.3	Green	None	Glossy	Greenish Gray	Rock-Forming Mineral
Barite	Ba, O, S	3.0–3.5	4.5	Clear, White, Reddish	1 Good 2 Poor	Glossy to Earthy	White	Drilling Mud
Bauxite	Mixture of Clays	—	2.0–2.5	Gray, Brown	None	Dull to Earthy	Colorless	Ore of Aluminum
Biotite	Al, Fe, H, K, Mg, O, S	2.5–3.0	2.8–3.2	Black, Green, Brown	1 Very Good	Glossy, Pearly	Colorless	Rock-Forming Mineral
Bornite	Cu, Fe, S	3	5.0	Bronze, Purple, Blue, Black	None	Metallic	Grayish Black	Ore of Copper
Calcite	Ca, C, O	3	2.72	White, Gray, Yellow, Colorless	3/not 90°	Glossy	Colorless	Cement, Fertilizer
Copper	Cu	2.5–3.0	8.9	Copper Red	None	Metallic	Metallic	Wire, Jewelry
Fluorite	Ca, F	4	3.2	Lt. Green, Blue, Yellow, Purple	4 Good	Glossy	Colorless	Flux in Steelmaking
Galena	Pb, S	2.5	7.5	Lead Gray	3/90°	Metallic	Lead Gray	Ore of Lead
Garnet	Variable	7	3.5–4.3	Red, Brown, Green, Black	None	Glossy to Resinous	Colorless	Abrasives, Jewelry
Graphite	C	1–2	2.3	Black	1 Very Good	Metallic to Earthy	Black	Lubricant, Pencils, Electrodes, Rubber

Gypsum	Ca, H, O, S	2	2.3	Colorless, White, Gray, Yellow	1 Good	Glossy to Pearly, Silky	Colorless	Plaster, Wallboard
Halite	Cl, Na	2.5	2.2	White, Colorless	3/90°	Glossy to Dull	Colorless	Nutrient, Chemical Industry
Hematite	Fe, O	5.5–6.5	5.26	Reddish Brown to Black	None	Metallic	Reddish Brown	Ore of Iron
Hornblende	Al, Ca, Fe, H, Na, Mg, O, Si	5–6	3.25	Dark Green to Black	2/56°, 124°	Glossy to Silky	Colorless	Rock-Forming Mineral
Limonite	Fe, H, O	5–5.5	3.3–4.7	Yellow Brown to Brown	1 Very Good in Crystals	Dull to Vitreous	Yellow to Brown	Pigment
Magnetite	Fe, O	6	5.2	Black	None	Metallic	Black	Ore of Iron
Muscovite	Al, H, K, O, Si	2–2.5	3.0	Colorless, Shades of Green, Gray, or Brown	1 Very Good	Glossy, Silky, Pearly	White, Colorless	Electric Insulation
Olivine	Fe, Mg, O, Si	6.5	3.2–4.4	Olive Green, Grayish Green	None	Glossy	White, Gray	Gemstone
Potassium Feldspar	Al, K, O, Si	6	2.6	White, Gray, Pink	2/90°	Glossy, Pearly	White	Ceramics, Porcelain
Pyrite	Fe, S	5	4.3	Brassy Yellow	None	Metallic	Black	Source of Sulfur
Quartz	O, Si	7	2.7	Colorless, White, Varied	None	Glossy to Dull	White	Glass, Gemstones
Serpentine	Fe, H, Mg, O, Si	2–5	2.2	Green, Yellow	None	Greasy, Waxy, Silky	White	Source of Asbestos
Talc	Fe, H, Mg, O, Si	1	2.7–2.8	Pale Green, White, Gray	1 Very Good	Pearly, Greasy, Dull	White	Talcum Powder, Lubricant
Topaz	Al, F, H, O, Si	8	3.5	Clear, Yellow, Blue, Pink	1 Very Good	Glossy	Colorless	Gemstone

Key

Al = Aluminum	C = Carbon	Cu = Copper	H = Hydrogen	Na = Sodium	Pb = Lead
Ba = Barium	Ca = Calcium	F = Fluorine	K = Potassium	O = Oxygen	S = Sulfur
	Cl = Chlorine	Fe = Iron	Mg = Magnesium	P = Phosphorus	Si = Silicon

GLOSSARY

PRONUNCIATION KEY

The glossary contains all the Key Terms in *Heath Earth Science* and their definitions. Some of these Key Terms are followed by their phonetic spelling, which shows you how to pronounce them. The words are divided into syllables and respelled according to the way each syllable sounds. The syllable that has the emphasis when the word is spoken appears in capital letters. The guide below is based on the Key to Pronunciation in *The World Book Encyclopedia.*

Sound	Respelling of Sound	Example	
hat, map	a	animal	AN-uh-muhl
age, face	ay	space	spays
care, air	ai	hair	hair
father, far	ah	charge	chahrj
child, much	ch	chocolate	CHAWK-luht
let, best	eh	energy	EHN-uhr-jee
equal, see,	ee	leaf	leef
machine, city		dream	dreem
term, learn,	ur	earthworms	URTH-wurmz
sir, work		turkey	TUR-kee
it, pin	ih	system	SIHS-tuhm
ice, five	eye, y	iodine	EYE-uh-dyn
coat, hook	k	cloud	klowd
hot, rock	ah	cotton	KAHT-n
open, go,	oh	program	PROH-gram
grow		row	roh
order, all	aw	corn	kawrn
		fall	fawl
oil, voice	oy	poison	POY-zn
house, out	ow	fountain	FOWN-tuhn
say, nice	s	soil	soyl
she	sh	motion	MOH-shuhn
cup, butter,	uh	bulb	buhlb
flood		blood	bluhd
full, put,	u	pulley	PUL-ee
wood		wool	wul
rule, move,	oo	dew	doo
food		shoe	shoo
pleasure	zh	treasure	TREHZH-uhr
		vision	VIHZH-uhn
about	uh	sofa	SOH-fuh
taken	uh	paper	PAY-puhr
pencil	uh	fossil	FAHS-uhl
lemon	uh	carbon	KAHR-buhn
circus	uh	syrup	SIHR-uhp
curtain	uh	mountain	MOWN-tuhn
section	uh	action	AK-shuhn

A

abrasion [un-BRAY-zhuhn]: process of erosion during which particles carried by the wind polish or break up things hit by the particles

acid rain: rain produced when sulfur dioxide from burning coal combines with water in the atmosphere

air mass: body of air over thousands of square kilometers, having uniform temperature and moisture

air-mass weather: weather that is about the same from day to day, with slow, gradual changes

alluvial [uh-LOO-vee-ahl] **fan:** large, fan-shaped deposit left by a stream

altostratus [al-toh-STRAT-uhs]: stratus clouds that occur between 2 km and 7 km above the ground

amber [AM-buhr]: transparent yellow-orange fossil formed from hardened tree sap

analyze [AN-uh-lyz]: to separate things into parts for study

angular unconformity [uhn-kuhn-FOHR-mih-tee]: break in the rock record in which the older layers slope at a different angle from those of the younger layers

anthracite [AN-thrah-syt]: hard coal formed from bituminous coal by pressure and heat

anticlines [AN-tih-klyns]: layers of rock that have been bent upward into an arch

aquaculture [AH-kwa-KUL-chur]: process of farming the ocean

aquifers [AH-kwih-fuhrz]: permeable rocks with groundwater in them

arid: dry

artesian [ahr-TEE-zhuhn] **wells:** wells from which water flows without pumping

astrogram [AS-troh-gram]: message sent into space by a radio telescope

astronomers [uh-STRAHN-uh-muhrs]: scientists who study objects in space

atmosphere: invisible sea of air around the earth

atmospheric pressure: pressure made by moving air molecules hitting a surface

atom: building block of all matter

B

barometer [buh-RAHM-uh-tuhr]: device that measures atmospheric pressure

barrier bars: long sandbars usually parallel to a coast

barrier islands: islands formed from large barrier bars, usually parallel to a coast

batholith [BATH-uh-lith]: huge underground mass of rock formed by the cooling of a magma pocket

biological weathering: process by which rocks are broken up by plant material and organisms

biomass [BY-oh-mas]: parts of plants or things made from plants that can be used for energy

bituminous [bi-TOO-mih-nuhs] **coal:** soft coal formed from lignite by pressure and heat

black holes: objects with intense gravitational fields, thought to exist in space

boundary: zone where plates come together

brachiopods [BRAY-key-oh-pods]: ancient marine shellfish

breakers: wave tops that bend forward and break

breeder reactor: nuclear reactor that makes more fuel than it uses

bright-line spectrum: pattern made when light passing through a prism shows narrow, bright lines

C

cast: fossil formed when a mold is later filled with mud or mineral matter

celestial sphere [suh-LES-chuhl SFEER]: imaginary surface of the sky to which stars seem to be attached

cementation [see-mehn-TAY-shuhn]: crystallization of mineral matter that holds sedimentary layers together

Cenozoic [see-nuh-ZOH-ihk] **Era:** geologic era of recent life

cephalopods [SEF-uh-luh-PODS]: ancient, coil-shelled animals

Cepheid [SEE-fee-uhd] **variable stars:** stars that regularly change in brightness

chain reaction: continuous action that keeps going on by itself and releasing energy

chemical weathering: process by which substances in the atmosphere combine with materials on the earth and slowly change

chromosphere [KROH-muh-sfeer]: region of the sun that extends above the photosphere

cinder cone: volcano built of rock and ash

cirrostratus [STRAT-uhs]: stratus clouds that occur more than 7 km above the ground and are made of ice crystals

cirrus [SIHR-uhs]: basic cloud shape with hairlike appearance and made of ice crystals

cleavage [KLEEV-uhj]: ability of some minerals to break along smooth parallel planes

climate: average weather over a long period of time

coal: black rock that burns

cold front: leading edge of a moving cold air mass

color: property of minerals

comets [KAHM-its]: small, icy, dusty bodies that orbit the sun

compaction [kuhm-PACK-shuhn]: pressing together of materials by the weight of layers above

composite volcano, or **stratovolcano:** volcano built up from combination of lava, solid rock, and ash

condensation nuclei [NOO-klee-eye]: particles present in the air onto which water droplets have been condensed

condensing [kahn-DENS-ihng]: more molecules returning to the liquid state than leaving as water vapor

constellations [kahn-stuh-LAY-shuns]: named groups of stars

containment building: large, heavy concrete structure that holds a nuclear reactor

continental: describing a climate influenced by air masses from large land areas

continental divide: geologic feature that separates regions that drain into one ocean from regions that drain into another

continental drift: movement of continents across the earth's surface

continental glaciers: glaciers that cover a large area of a continent

continental polar: describing an air mass having a polar land area as its source region

continental shelf: somewhat shallow area that surrounds a continent

continental slope: area beyond a continental shelf where depth increases rapidly

continental tropical: describing an air mass having a tropical land area as its source region

continuous spectrum: pattern made when light passes through a prism and is separated into its colors, as in a rainbow

contour [KAHN-toor] **line:** line that connects points of the same elevation on a map

control rods: rods made of material that absorbs neutrons, used to slow a chain reaction

convection: process of warmer air being pushed upward

convection cell: pattern of air movement created when warm air rises because of convection, and then moves back down onto a cooler surface

converging [kuhn-VUHR-jing] **boundaries:** boundaries between plates that are coming together

core: center part of the earth

Coriolis [kawr-ee-OH-lihs] **effect:** tendency of all

objects moving in the northern hemisphere to curve to the right, and in the southern hemisphere to curve to the left, because of the earth's rotation

corona [kuh-ROH-nuh]: outermost part of the sun's atmosphere

correlation: comparison of the ages of rocks in different locations

crest: highest part of a wave

crude oil: petroleum pumped from wells

crust: outside layer of the earth

crystals [KRIS-tuhls]: solids with repeating patterns of packed ions

cumulonimbus [kyoo-myuh-loh-NIHM-buhs]: piled-up cloud that often produces stormy weather

cumulus [KYOOM-yuh-luhs]: basic cloud shape with piled-up appearance

currents: moving streams of water in a body of water

cycle [SY-kuhl]: series of events that keeps happening over and over

D

dark-line spectrum: pattern made when light passing through a prism shows narrow dark lines

data [DAY-tuh]: information recorded from observations

deflation [dee-FLAY-shuhn]: process of erosion during which wind picks up material and moves it, often leaving shallow bowllike features on the surface

deltas [DEL-tuhs]: deposits at the opening of a stream into a lake or the ocean

density: comparison of how heavy something is to the space it occupies

desalinization [dee-SAL-uh-nuh-zay-shun]: taking salt out of seawater

dew: water vapor that has condensed on a surface as a liquid when the dew point is above 0°C

dew point: temperature at which condensation occurs

diffraction grating [di-FRACK-shuhn GRAY-ting]: piece of glass or plastic with many narrow, parallel slits, which makes a spectrum of light when light passes through it

dikes: magma that has cooled and hardened in vertical cracks

diverging [dy-VUHR-jing] **boundaries:** boundaries between plates that are separating

Doppler [DAHP-luhr] **effect:** change in pitch heard because of the change in distance from a sound source

Doppler shift: shift of spectral lines in the spectrum made by a moving object

drainage patterns: distinctive forms of streams and tributaries as viewed on a map

drought [drout]: period when there is not enough rain

E

earthquakes: sudden vibrations or waves in the earth

echo sounding: way to measure the depth of the ocean floor with a device that sends a sound signal through the water

electromagnetic radiation [ee-lek-troh-mag-NET-ik ray-dee-AY-shuhn]: energy in the form of waves from space

electron [ee-LEK-trahn]: particle with a negative electric charge that flies around the nucleus of an atom

electron microscope [ee-LEK-trahn MY-kruh-skohp]: instrument for magnifying objects too small to be studied with the optical microscope

element: material made of only one kind of atom

ellipse [ee-LIPS]: oval shape

epicenter [EP-uh-sen-tuhr]: place on the surface of the earth directly above the focus of an earthquake

epochs [EP-uhks]: smallest dimensions of geologic time

eras [IHR-ruhs]: largest dimensions of geologic time, named for fossils

erosion [ee-ROH-zhuhn]: process of loosening and carrying away materials on the earth's surface by the action of water and wind

esker [ES-kuhr]: snakelike ridge of layered deposits left by glacial meltwater

evaporating [ee-VAP-uh-rayt-ihng]: more molecules going into the air as water vapor than returning to the liquid state

frontal weather: changes caused by a front moving by

frost: water vapor condensed on a surface as a solid when the temperature is at or below 0°C

full moon: phase of the moon during which its sunlit side faces entirely toward the earth

fusion [FYOO-shuhn] **reaction:** process by which hydrogen atoms combine to form helium atoms

F

fault: break or crack in part of the earth's crust; the plane along which rock breaks and slips

fault-block mountains: landmass noticeably higher than its surroundings and bounded by one or more faults

faulting: breaking of crustal rock into separate masses, as in mountain building

fertile: describing the quality of soil capable of growing crops because of the materials it contains

fetch: distance the wind blows over water

fission [FISH-uhn]: splitting of atoms with the release of energy

fly ash: fine ash that flies up the chimney into the air when ground-up coal is burned

focus [FOH-kuhs]: place along a fault where an earthquake begins

fog: cloud at or near the earth's surface

folded mountains, or **complex mountains:** mountains with zones of folded sedimentary rock alongside metamorphic and igneous rock

folding: bending of layered rocks

forecast: prediction about coming weather

fossil [FAHS-uhl] **fuel:** energy source formed from decomposed plants and animals

fossils [FAHS-uhls]: traces of plants or animals preserved in rock

fracture: breaking of a mineral when it doesn't have cleavage

fresh water: water that is not salty

front: leading edge of a moving air mass

G

galaxies [GAL-uhk-sees]: large assemblies of stars

geologic column: sequence of rock layers placed in order of age

geologic cross section: view of the vertical arrangement of rock layers, as seen in a road cut

geologic cycle: cycle that describes gradual, continual changes in the earth's crust; builds and erodes mountains

geologic map: map that shows geologic features of the land

geologists [jee-AHL-uh-jists]: scientists who study the solid outer and liquid inner parts of the earth

geothermal [jee-oh-THUHR-mel] **energy:** energy in the form of heated water and steam from the land or the earth's crust

geyser [GY-zuhr]: fountainlike eruption of hot water and steam

giants: stars that lie just above the main sequence band on the Hertzsprung-Russell diagram

glacier [GLAY-shuhr]: large mass of ice above the snow line

globular clusters: close groupings of stars

Glossopteris [glahs-SAHP-tur-us]: plant whose fossils are found on five continents

graph: diagram or picture of organized data

gravitation: force of attraction between objects

greenhouse effect: warming of the atmosphere; happens when energy from sunlight is absorbed by the earth's surface and released as infrared radiation

groundwater: water from an underground zone where all pore spaces are filled with water

H

half-life: time taken for half the atoms in a radioactive sample to change into simpler unchanging atoms

hardness: resistance of a mineral to being scratched

Hertzsprung-Russell [HUHRTZ-spruhng-rus-sehl] **diagram:** graph that compares brightness and temperature of stars

histogram [HIS-tuh-gram]: type of graph

horizontal fault: faulting that causes blocks to move sideways past each other

hot springs: water heated within the earth and pushed to the surface by convection

humid: moist

humidity [hyoo-MIHD-uh-tee]: amount of water vapor in the air

humus [HYOO-muhs]: partially or completely decayed plant and animal matter

hurricane: large, violent, circular storm that forms over tropical waters

hydrocarbons [HY-droh-kahr-buhns]: substances made of hydrogen and carbon

I

igneous [IG-nee-uhs] **rocks:** rocks formed by the process of cooling liquid rock

image enhancing: technique that uses computers to strengthen certain portions of photographs

incineration [ihn-sihn-uh-RAY-shuhn]: burning of waste at high temperatures in furnaces

index fossils [FAHS-uhls]: distinctive fossils used to determine the age of rocks

invertebrates [in-VUR-tuh-bruhts]: animals without backbones

ions [EYE-uhns]: electrically charged atoms

isotopes [EYE-suh-tohps]: atoms of an element that occur in different forms

J

Jurassic [juh-RAS-ihk]: geologic period of the Mesozoic era

L

laboratory: place equipped for scientific study and testing

lake: standing body of fresh water so deep that sunlight cannot reach the bottom

landslide: mass movement of sliding rock materials

lava: magma that has reached the earth's surface

law of crosscutting relationships: fact that young features in rock cut across older features

law of superposition: fact that younger layers of sediment are always deposited on top of older layers

lignite [LIG-nyt]: brown coal made of compressed peat

loess [LOH-uhs]: fine dust deposited by winds over a large area

longshore current: current pushed along the shore by waves

lunar eclipse [ee-KLIPS]: blocking of the moon's light when the earth passes between the sun and the moon

luster [LUHS-tuhr]: glow of reflected light coming from a mineral's surface

Lystrosaurus [li-struh-SAWR-us]: reptile whose fossils are found on three continents

magma: liquid rock deep in the earth

magnetic reversal: change in direction of the magnetic field of the earth

main sequence stars: stars of the main sequence band on the Hertzsprung-Russell diagram

major landform regions: large areas that contain similar or related landforms, soils, rock types, and rock structures

mammals: warm-blooded, hairy animals, the females of which nurse their young

mantle: layer of the earth below the crust

marias [MAHR-ee-ahs]: flat, dark, sealike areas of the moon made by ancient lava flows

marine: describing a climate influenced by air masses from ocean areas

maritime polar: describing an air mass having polar waters as its source region

maritime tropical: describing an air mass having tropical waters as its source region

marker layer: layer of rocks having a distinctive characteristic that identifies rocks of a similar age

mass movement: downward movement of weathered earth materials

matter: anything that takes up space and has mass

meanders [mee-AN-duhrs]: bends in level areas of streams that wind back and forth

mechanical weathering: process by which rocks are broken apart by the freezing of water in cracks

Mesozoic [mez-uh-ZOH-ihk] **Era:** geologic era before recent life

metamorphic [meht-uh-MAWR-fihk] **rock:** rock that was once igneous or sedimentary rock but has been changed

meteorologists [mee-tee-uh-RAHL-uh-jists]: scientists who study the atmosphere

microclimates: local patterns of climate

minerals: naturally occurring substances, generally having a crystalline structure, that have never been part of a living organism

Moho [Moh-hoh]: boundary between the crust and the mantle of the earth

moisture index: measure of the moisture gained by precipitation minus the moisture lost by evaporation and plant use

mold: fossil formed when water dissolves shell or bone around which sand or mud has already hardened

molecules [MAHL-uh-kyools]: invisible particles of materials

mountain: any part of the earth's surface that stands much higher than its surroundings

mountain chain: mountain ranges and systems connected together

mountain range: series of mountains

mountain system: series of mountain ranges

mudflow: mass movement of fine materials

N

natural resources: supplies taken from the earth, including water, food, building materials, fiber for clothing, minerals, and energy

nebulae [NEB-yoo-lie]: thin, wispy clouds of glowing gas in outer space

neutron [NOO-tron]: particle with no electric charge; found in the nucleus of an atom

new moon: phase of the moon during which its dark side faces completely toward the earth

nodules [NAHJ-ools]: small lumps of minerals on the ocean's floor

nonrenewable resource: resource that is not available on the earth again once it is taken and used

normal fault: faulting that causes the earth's crust to be extended

nuclear reactor: thick steel vessel in which energy is released by fissioning uranium 235

nucleus [NOO-klee-uhs]: small inner sphere at the center of an atom

O

observation [ahb-zuhr-VAY-shuhn]: inspection of an object or event by using the senses

ocean basin: bottom of the ocean floor

oceanographers [oh-shuh-NAHG-ruh-fers]: scientists who study the ocean

oil field: large oil pool collected in porous rock

open clusters: loose groupings of stars

optical microscope [AHP-ti-kuhl MY-kruh-skohp]: instrument for magnifying objects too small to be seen by the naked eye

ores: minerals from which metals can be taken at a reasonable cost

oxbow lakes: lakes formed when one bend in a stream meets another and cuts off a meander

oxidation [AHKS-uh-DAY-shuhn]: combination of a substance with oxygen

ozone [OH-zohn]: gas in the atmosphere that screens out harmful ultraviolet rays

P

Paleozoic [pay-lee-oh-ZOH-ihk] **Era:** geologic era of ancient life

Pangaea [pan-JEE-ah]: supercontinent that broke apart into separate continents

passive solar: descriptive of solar-energy use that relies on the collection and storage of solar energy

peat: thick layers of brown, rotted plant material

period of revolution: length of time it takes to complete one trip around the sun

periods: intermediate dimensions of geologic time

permeable [PUHR-mee-uh-buhl]: describing rocks that allow water to move through them

petrified [PET-ruh-fyd] **wood:** fossil resulting when silica is substituted for wood cells that have decayed

petroleum [puh-TROH-lee-uhm]: thick, smelly, black-liquid fossil fuel

phases of the moon: changes in appearance of the moon during a month

photosphere [FOH-tuh-sfeer]: visible, granularlike surface of the sun

photosynthesis [foh-toh-SIN-thuh-sis]: process by which green plants use sunlight, carbon dioxide, and water to manufacture food

plankton [PLANK-tuhn]: tiny plants and animals that float in the ocean

plastic: solid material that can flow and change shape over a long period of time

plates: large pieces of the earth's crust

plate tectonics [tehk-TAHN-ihks]: theory and study of massive plates thought to move over Earth's surface

plume: localized hot spot of material rising from the earth's interior

polar: describing a climate near the poles with a cold average temperature

polar easterlies: surface winds generally blowing east to west in polar regions

pollutants: irritating or dirty substances that accumulate in air or water, such as chemical gases, dust and waste particles, and smoke

pollution [puh-LOO-shuhn]: accumulation of irritating or dirty substances in air or water

pond: standing body of fresh water shallow enough for sunlight to reach the bottom

pore space: space between soil particles

porphyry [PAWR-fuhr-ee]: igneous rock with large crystals surrounded by small ones

power tower: tall structure onto which sunlight is reflected by mirrors

Precambrian [pree-KAM-bree-uhn] **Era:** geologic era before ancient life

precipitation [pree-sihp-uh-TAY-shun]: water that falls to the earth's surface as solid or liquid; process that occurs when chemical reactions form a solid that settles out of solution

prevailing westerlies: surface winds generally blowing west to east near latitude 30 degrees north or south of the equator

primary waves: vibrations during an earthquake that cause material to move back and forth

principle of uniformity: idea that the earth's features formed long ago were made by the same processes that operate today in the same way

prism: transparent body that ends light waves passing through it

prominences [PRAHM-ih-nehn-ses]: blasts of flaming gas that leap from the surface of the sun

proteins: substances that the body uses for the growth and repair of tissues

proton [PROH-tahn]: particle with a positive electric charge, in the nucleus of an atom

protoplanet [PROH-toh-plan-iht] **nebular model:** idea that explains the origin of the solar system

pulsars [PUHL-sahrs]: objects in space that emit radio signals in strong pulses

Q

quarter phase: phase of the moon during which half its lighted side is visible

quasars [KWAY-sahrs]: objects in space that emit strong radio signals

R

radioactive [ray-dee-oh-AK-tihv]: giving off radiation

radioactive decay [ray-dee-oh-AK-tihv dee-KAY]: natural, regular process by which certain radioactive elements change into different elements

radiocarbon [ray-dee-oh-KAHR-buhn] **dating:** method for determining the age of material that was once part of a living organism

radiometric [ray-dee-oh-MET-rihk] **time:** method used when dating the age of rock with an atomic clock

refinery [ree-FYN-uhr-ee]: processing plant that separates crude oil into its many useful products

reflecting telescopes [ree-FLEK-ting TEL-uh-skohps]: instruments with lenses and mirrors used to study objects in space

refracting telescopes [ree-FRAK-ting TEL-uh-skohps]: instruments with lenses used to study objects in space

relative humidity [hyoo-MIHD-uh-tee]: amount of water vapor in the air compared to the total that could be in the air at a certain temperature

relative time: dating method that compares two events by asking which event came first

remote sensing: process used to identify or measure something distant through photographs

renewable resources: resources that become available on the earth again after they have been used up

reptiles [REP-tyls]: cold-blooded, air-breathing animals that lay eggs

reservoir: lake formed by building a dam on a stream or river

reverse fault: faulting that causes the earth's crust to be shortened

ridges: underwater mountain ranges

rip current: usually narrow, strong current that heads out to sea from a shore

rock cycle: endless changes of rock from one kind to another

S

sandbar: pile of sand that is underwater near a shore

sand dunes: deposits of sand made by the wind

scientific model: idea or construction that shows the way something is

sea-floor spreading: movement of the ocean floor outward from a diverging boundary

seamount: submarine volcano that does not rise to the surface

secondary waves: vibrations during an earthquake that cause material to move from side to side

sedimentary [sehd-uh-MEHN-tuh-ree] **rocks:** rocks formed when sediments are pressed and cemented together

sedimentary structures: patterns formed in sediment before it becomes solid rock

sediments [SEHD-uh-muhnts]: layers of bits and

pieces of crumbled rock that settle to the bottom of the ocean

seismic [SIZE-mihk] **waves:** vibrations that move outward from the focus of an earthquake

seismograph [SIZE-muh-graf]: instrument that measures seismic waves

semiarid: describing a climate with between 25 cm and 50 cm of precipitation a year

shield volcano: volcano built of lava flows

shore: area near land where waves are active

shoreline: place where land and water meet

sills: magma that has cooled and hardened in horizontal cracks

sink holes: funnel-shaped depressions caused by the collapse of underground caves

smog: fog combined with exhaust gases and smoke

snow line: edge of places where snow does not completely melt in summer

soil: upper layer of the earth's crust in which plants grow

soil creep: slow mass movement of soil

soil profile: arrangement of soil layers

solar cells: devices made of silicon, arsenic, and boron that produce an electric current when placed in sunlight

solar collector: device that uses sunlight to heat water moving within tubes

solar eclipse [ee-KLIPS]: blocking of the sun's light when the moon passes between the sun and the earth

solar energy: sunlight

solar system: all objects, including planets and moons, asteroids, and comets, that are in orbit around the sun

solid waste: anything solid that is considered to be useless, unused, unwanted, or discarded

source region: birthplace of an air mass

spectroscope [SPEK-troh-skohp]: instrument used to study light

spectrum [SPEK-truhm]: pattern made by light that has passed through a prism

spit: sandbar connected to the shoreline

stationary front: nonmoving zone or boundary between air masses

storage mass: dense material that absorbs sunlight and later gives off the energy as infrared radiation

storms: rapid and violent weather changes

stratus [STRAT-uhs]: basic cloud shape with spread-out appearance

streak: colored mark left by a mineral when it is rubbed on a streak plate

structure: quality of soil that allows water to move through soil layers

subduction [suhb-DUHKT-shuhn] **zone:** converging boundary where one plate buckles downward into the earth

subsoil: region of the earth's crust beneath the layer containing humus

sunspots: dark, relatively cool markings on the surface of the sun

supergiants: stars found in the upper-right section on the Hertzsprung-Russell diagram

surf: zone where breakers occur

surface runoff: water that moves across the surface

surface waves: vibrations that cause the earth's surface to move up and down

swell: not very high, long, wide wave that can travel a great distance

synclines [SIN-klyns]: layers of rock that have been bent downward to form a trough

synthesize [SIN-thuh-syz]: to combine things into a whole

T

temperate: describing a climate between polar and tropical regions that is neither very cold nor very hot

temperature: measure of the average energy of motion of molecules

terminal moraine [TUHR-mi-nuhl muh-RAYN]: deposit of rock material at the furthest point reached by a glacier

texture of soil: quality of the arrangement of soil particles

thunderstorm: intense rainstorm with thunder, lightning, and often strong winds and hail

tidal marsh: broad, flat muddy area

tides: changes in water levels caused by changes in gravitation between the earth and the moon

till: rocks deposited in an unsorted way when a glacier melts

tornado: storm with funnel- or rope-shaped cloud and violently swirling winds

trade winds: surface winds generally blowing east to west near latitude 30 degrees north or south of the equator

transform fault: boundary between plates that are sliding past each other

trenches: long, narrow, and deep submarine valleys

trilobites [TRY-luh-byts]: ancient shelled animals

tropical: describing a climate near the equator with a hot average temperature

trough [trawf]: lowest part of a wave

tsunami [tsoo-NAH-mee]: wave made by an undersea earthquake

U

unconformity [uhn-kuhn-FOHR-mih-tee]: break in the rock record

undertow: strong current that heads out to sea

V

valley glacier: glacial ice confined to a valley

volcanic neck: mass of igneous rock that remains after erosion has removed the volcanic cone

volcano: mountain formed from liquid rock

W

warm front: leading edge of a moving warm air mass

watershed: area of land drained by a river

water table: level below which the ground is full of groundwater

water vapor: water in the form of a gas

wave height: vertical distance between the crest and the trough of a wave

wavelength: horizontal distance between two wave crests

weathering: process by which exposed rocks are broken down into smaller bits and pieces

white dwarfs: stars found below the main-sequence band on the Hertzsprung-Russell diagram

wind: horizontal movement of air

INDEX

NOTE: Page numbers in **boldfaced** type indicate a definition.

ACKNOWLEDGMENTS

Illustration

Leo Abbett 14, 55, 110 top, 132 top, 182, 221, 285, 357, 366-367, 418, 442, 482
Robert Anderson 141, 161, 467
Lloyd Birmingham 66, 68, 138, 149, 171 top, 174-175, 191 top, 258 bottom, 284 bottom, 297
Gail Burroughs 9, 12-13, 16, 32 bottom, 45, 48, 71 bottom, 94, 97, 99 bottom, 104, 111, 113, 130, 135, 171 bottom, 177 bottom, 191 bottom, 198, 215, 257 bottom, 260, 269, 280, 295, 301 bottom, 341, 347, 353, 371, 379, 385, 395, 413, 419, 433, 440, 455, 460, 475, 495
Michael Carroll 44, 60, 62, 63, 71, 73, 78, 493
Dave Hannum 339 bottom, 340, 343, 344, 348, 349, 363, 364, 378, 380, 381, 382, 383, 394, 415, 422, 423, 424, 425, 436, 439, 454, 459, 461, 478
Walter Hortens 200, 250, 251, 253, 255, 256, 258 top, 259, 261, 420, 432, 434, 438, 444, 463
Leonard Morgan 212, 213, 214
Kim Poor 116, 136, 137
Patrick Russell 27, 28, 33, 110 bottom, 132 bottom, 134, 148, 150, 152, 156, 157
Jeff Stock 8, 30-31, 32 top, 35, 36, 92, 93, 99 top, 119, 231, 240, 270, 271, 292, 293, 294, 298, 300, 301 top, 305, 339 top, 346
James Teason 131, 151, 169, 172, 177 top, 178, 179, 192-193, 194, 195, 201, 202, 203, 218, 236, 237, 268, 270-271, 272, 274, 275, 276, 278, 279, 281, 282, 283, 284 top, 296, 299, 303, 306-307, 318, 321, 322, 328, 329, 338, 361, 392, 393, 398, 437, 473
Gary Torici 26, 37, 49, 64, 65, 67, 88, 89, 95, 96, 98, 100, 101, 114, 126, 127, 128, 129, 168, 180, 241, 359 right, 362, 441
Lily Yamamoto 358

Photography

After-Image: Norm Clasen 118; William Herbeck 292
Alaska Pictorial Service: Steve McCutcheon 248, 262, 265
J.C. Allen and Son 476
Alpha: L. Balterman 439; Jim Howard © 1980 411; Dick Luria © 1982 424
American Museum of Natural History 384 left
Eric Anderson 10, 90 bottom, 183 bottom
Arecibo Observatory, National Astronomy and Ionosphere Center 54
Peter Arnold, Inc.: Peter Arnold 159 bottom, 462 left; Vic Cox 188, 204; Bob Evans 500 right; W. H. Hodge 383 bottom; Jacques Jangoux 108, 479 bottom; Manfred Kage 4 right; James H. Karales 429; Stephen J. Krasemann 112 bottom, 333; Yoram Lehmann 159 bottom; Harvey Lloyd 401 right; Michael Melford 445 right; Bill O'Connor 279; Horst Schäfes 437 left; John Serafin 196 top; Clyde H. Smith 192 bottom
Atomic Industrial Forum 440
Atoz Images: Adolph F. Rohrer 410; National Weather Service 138
Bechtel Power Corporation 182
Lester V. Bergman and Associates 283 right
The Bettmann Archive 19 right, 103, 364
Black Star: Dennis Brack 423; John Lopinot 190; Ted Spiegel 181 left; Jim Sugar 319 left; Dick Swanson 170 right; Werner Wolff 453 left
Boeing Company 51, 493
Robert T. Bradbury 68
David Breed 3 right insert
James N. Butler 192 top, 197 bottom right
California Institute of Technology and Carnegie

Institution of Washington 24, 38, 39, 48; Robert Paz 263
Carlye Calvin 106, 120
Bruce Coleman Inc.: Jane Burton 384 right; Jack Dermid 176 bottom right; Harry Engels 146 top left; J. Fennell 340; Wedigo Ferchland 228 right, 208; Kenneth Fink 342 top left; Keith Gunnar 94, 117 top; M. Timothy O'Keefe 316 bottom; Jack M. Stephens 326 left; Werner Stoy 197 top right, 297 bottom; Norman Tomalin 483 right
Michael Collier 3 right, 230 right, 253 right, 324 left, 334, 342 top center left, 342 bottom left, 345, 350, 356, 376 top, 376 bottom right, 389, 396 bottom left, 400 top, 400 bottom right; Collier/Condit 296
James M. Cribbs 199 left
Culver Pictures 27 top left
Dallas Chamber of Commerce 152 bottom
E. R. Degginger 115 left, 148 top, 172 top left, 183 top, 196 bottom right, 197 bottom left, 228 left, 233 right, 244, 247, 286, 327 left, 377 right, 438, 462 right, 464 left, 470, 486, 496
Phil Degginger 92, 95, 226, 266, 336 left
Dennis Di Cicco 2 left, 41
Donald Dietz 170 left
DPI: James H. Karales 481 left; Alon Reininger 465 right; Larry Smith 472 left; Steven C. Wilson 187
DRK Photo: Wayne Lankinen 290, 308
Earth Images: Terry Domico 297 top right
William E. Ferguson 123, 155 top, 176 top right, 178 left
Field Newspaper Syndicate 5
Florida Department of Commerce 144, 160, 461
Focus on Sports 105, 430, 446
La Fondation St-Thomas 79
Daniel Forster 84
FPG: Tom Carroll 414 left; W. Choroszewski 437 right; Dave Maenza 87; J. Zimmerman 408, 426, 443
Gamma-Liaison: Alain Mingam 419
General Electric Corporation 91 left
Geological Society of London 19 left
Lee Gordon 14, 217 center, 218, 219 right, 254 left, 259
Susan Griggs Agency: Adam Woolfitt 354, 368
Hale Observatories 28, 29 top row, 42, 56
Harbor Branch Foundation: Chris Chulamanis 164
Robert Harding Picture Library 312
Todd Harrington, Explosives Supply Company 254 right
Grant Heilman: William Felger 321; Grant Heilman 151 top, 155 bottom left, 159 top, 176 top left, 176 bottom left, 198, 323 right, 377 left, 396 bottom center, 399, 406, 475; Thomas Horland 485; Alan Pitcairn 152 top; Runk/Schoenberger 375 right, 474 bottom left, 478
Phil and Loretta Hermann 163
Michal Heron 2 right insert
Dr. Ernest Hildner 31
Historical Pictures Service, Chicago 223
Eric and David Hosking: Douglas Wilson 501
The Houghton Library, Harvard University 27 top right
Harold R. Hungerford 7, 8, 229, 294 top right, 299, 325 top right, 337, 378 top
Hurricane Center, Florida 139 right
Randall Hyman 277 left
The Image Bank: Joe Azzara 484 left; Morton Beebe 480; Michel Casino 414 right; Geoffrey Gove 481 right; Steve Krongard 458 right; Bruce McAllister 453 right
Imagery 133

The Image Works: Mark Antman 436, 456 left, 459, 465 left, 469
Indiana Limestone Institute: Bill McDonald 237
Thomas Ives 137, 146 top center
Jet Propulsion Laboratory 76, 77 right, 499 bottom, 505, 506, 507 left
Paul E. Johnson 16, 457 left
Lou Jones 479 top
Michael Jones 403
Breck P. Kent 146 bottom left, 146 bottom right, 173, 210, 212, 214, 215, 216, 217 top left, 217 bottom left, 219 left, 219 center, 220, 222, 225, 230, 232 top left, 232 bottom left, 232 right, 233 left, 235, 236, 239, 241, 243, 253 left, 283 left, 322, 342 bottom center left, 362, 375 left, 380, 381 top
Ken Launie 27 right
John Lemker 319 right
Lick Observatory 34
Talbot Lovering 6, 15
Alex MacLean 199 right
Tom Magno 89 right, 109, 126, 174, 415, 416 bottom
Mammoth Cave National Park: Bob and Zona Cetera 175
Maximum, Inc. 89 left
McDonald Observatory 504
Dennis Milon 27 bottom, 29 bottom
David Muench 119, 148 center, 155 bottom right, 314, 330
Museum of Comparative Zoology, Harvard University 374, 376 bottom left, 378 center, 378 bottom
National Aeronautics and Space Administration viii, 18, 22, 50 right, 52 top, 53, 69, 74 left, 77 left, 82, 102, 252, 277 right, 353, 405, 492 left, 494, 498 left, 507 left, 511
National Aeronautics and Space Administration, Langley Research Center 75 top
National Oceanographic and Atmospheric Administration 507 left
The National Radio Astronomy Observatory viii, 18, 45
Norfolk Southern Corporation 457 right
Pacific Gas and Electric Company 464 right
Palomar Observatory photograph 44
Tom Pantages 67, 81
Jay Pasachoff photo; rights owned by Pasachoff Educational Trusts 2 left insert, 35, 50 left, 52 bottom
Photophile: Sal Maimone 489
Photo Researchers: Louise K. Broman 382; Stephen Dalton 498 right; Susan Ray Field 317 right; Lowell Georgia 325 bottom, 421 left; Georg Gerster 477; Dan Guravich 416 top; Bill Henry 320; Richard Hutchings 417; George Jones 500; Lou Jones 479 top; Tom McHugh 435; Larry Mulvehill 75 bottom, 497; Virginia P. Weinland 278
Photri 134, 450, 466, 484 right, 490, 492 bottom, 503, 308
Planetary Image Center, Lunar and Planetary Institute 371
Carl Purcell 146 bottom
Rainbow: Dan McCoy 2 right, 33, 50, 86 left, 151 bottom, 502; Hank Morgan 166, 184
Raytheon Corporation: Louis Benoze 191
John Reader © National Geographic Society 372, 386
Rockwell International 456 right
Jeff Rotman 3 left, 452 right
Clive Russ 112 top, 392
William Sacco 117 bottom
Peter Schweitzer 452 left

Scripps Institution of Oceanography 420
John S. Shelton 134, 194, 196 top left, 240, 255, 293 bottom right, 293 bottom left, 294 bottom, 295 left, 298, 300, 302, 317 left, 323 left, 324 right, 326 right, 336 right, 396 bottom right, 400 bottom left
Shostal Associates: 139 left, 143, 311, 390, 402, 412, 421, 449, 458 left; J. Beckner 304; Fischer 201; E. Simensen 139 left
Smithsonian Institution 325, 360
Soil Conservation Service: 474 top left; Joe Larson 474 bottom right
Sovfoto: Tass 74 right
Tom Stack and Associates: Tom Bean 98, 155 center; Bill Borsheim 146 top right; Ron Church 207; Dennis Hogan 433; Jim McNee 434; Gary Milburn 445; Keith Murakami 196 bottom left; Bob Wallace 21
Deidre Delano Stead 30, 90 top, 91 right
Stock, Boston: Daniel Brody 179 right; Owen Franken 124, 140; Antonio Mendoza 148 bottom; Peter Menzel 181 right; John Running 180
Taurus Photos: Gary Cregmile 115 right; Arthur Rochelle 483 left
Will Tenney 86 right
Tersch Enterprises 494, 511
Bill Tillery 41 left
Audrey N. Tomera 318
Union Carbide, Nuclear Division 440
United Feature Syndicate © 1974 59
United Press International 205
United States Department of Agriculture: 397, 472 right; Forest Service 396 top
United States Geological Survey: E. Segerstrom 297 top left
United States Naval Observatory 22
University of Florida, Agricultural Station 113
Vassar College Library 47
Andrew Watt 6 left
West Light: Bill Ross 507
Woodfin Camp, Inc.: Dan Budnick 316 top; Kenneth Garrett 200, 272, 274, 287
Woods Hole Oceanographic Institution: Rod Catanach 3 left insert; John Edmond, M.I.T. 289
Jerome Wyckoff 197 top left, 293 top, 294 top left, 295 right, 327 right, 342 right, 346, 401 left
Alpha: Richard Laird 445 top

1 2 3 4 5 6 7 8 9 0